Spontaneous Order

Spontaneous Order

How Norms, Institutions, and Innovations Emerge from the Bottom Up

H. PEYTON YOUNG
James Meade Professor of Economics Emeritus, University of Oxford
Senior Research Fellow, Nuffield College Oxford

OXFORD
UNIVERSITY PRESS

OXFORD
UNIVERSITY PRESS

Great Clarendon Street, Oxford, OX2 6DP,
United Kingdom

Oxford University Press is a department of the University of Oxford.
It furthers the University's objective of excellence in research, scholarship,
and education by publishing worldwide. Oxford is a registered trade mark of
Oxford University Press in the UK and in certain other countries

Published in the United States of America by Oxford University Press
198 Madison Avenue, New York, NY 10016, United States of America

British Library Cataloguing in Publication Data

Data available

Library of Congress Control Number: 2024936404

ISBN 9780198892908

DOI: 10.1093/oso/9780198892908.001.0001

Printed and bound by
CPI Group (UK) Ltd, Croydon, CR0 4YY

For Patrick and Benjamin

Foreword

This volume brings together my work on evolutionary game theory and its applications to economics, learning behavior, philosophy, computer science, and distributed control. Throughout, the central question that I address is whether the interactions of many dispersed agents, acting with limited information and bounded rationality, coalesce into a coherent form of social order. If so, what types of order are most likely to be realized and how long does the process take? I develop a unified framework for answering this question based on concepts from stochastic dynamical systems theory. The dynamic processes that underpin the analysis are inherently stochastic because individual behaviors are idiosyncratic and constantly being buffeted by random shocks. Nevertheless, their aggregate behavior can be systematically characterized, and it can be shown that some states are much more likely to emerge than others.

The first part of the book applies this theory to the evolution of social norms, bargaining behavior, economic contracts, and more generally to the evolution of economic and social institutions. The second part of the book treats various forms of individual learning behavior and their implications for long-run aggregate behavior. Very simple forms of trial-and-error learning, which require little or no information about others' behavior, lead to system-wide equilibria without any of the agents ever intending it or even knowing what the overall process is. The key features of these adaptive rules and their coalescence into equilibria are borne out in experimental settings. The theory applies not only to socio-economic phenomena, it can also be used to design distributed systems consisting of many interacting parts, in such a way that the system reaches a desirable form of aggregate behavior without the intervention of any central authority or planner. This approach has many applications in computer science and distributed control theory.

The third part of the book applies the evolutionary framework to study the spread of social and economic innovations. Of particular interest are innovations that exhibit increasing returns: they become more valuable the more people who adopt them. Contemporary examples include social media (Facebook and Linked-In), as well as computer languages and software

(Apple and Microsoft). Other examples with this property are institutional innovations such as forms of contracts and forms of organization.

Although new ways of doing things may offer significant gains in welfare relative to the status quo, they are seldom adopted all at once, but tend to spread gradually as people learn about the innovation from others. The adoption process is often decentralized, undirected, and bottom-up. Unlike in previous chapters, the focus here is on the interim rather than the long-run dynamics of the adoption process. How long does it take between initial discovery and successful adoption by a large proportion of the population? I show how this "lag time" depends on a variety of factors, including the topological structure of the social network through which people interact, the gain in payoff realized by switching from the old to the new way of doing things, and the amount of idiosyncratic variation in individual behaviors. One of the key findings is that, if idiosyncratic perturbations in the adaptive process are small but not vanishingly small, and the payoff gains are sufficiently large, the expected waiting time until a majority have adopted can be bounded independently of the number of agents in the population.

I am greatly indebted to the students and colleagues in economics, game theory, biology, and control theory with whom I have collaborated over the years. They provided many of the crucial insights and lines of inquiry that are described in the following chapters. Most of them appear as co-authors: Itai Arieli, Gurdal Arslan, Yakov Babichenko, Mary Burke, Max Burton-Chellew, Dean Foster, Jason Marden, Jeff Marden, Heinrich Nax, Lucy Pao, Ron Peretz, Jeff Shamma, and Stu West. Others, notably Jean-Paul Carvalho, Sam Jindani, Gabriel Kreindler, and Bary Pradelski, have made major contributions to this line of research that are referenced but could not be included due to space limitations.

Contents

PART I

THE EVOLUTION OF NORMS AND INSTITUTIONS

1
Spontaneous Order

Spontaneous order refers to the emergence of social and economic structure from the accumulation of many individual decisions. It stands in contrast to the notion that order should be imposed from above by a planner or higher authority. The problem with the latter is that, in practice, attempts to design well-functioning social and political institutions often fail to account for the complexities of human motivation and interaction, and the dispersed and imperfect sources of information on which people make decisions. It seems reasonable to suppose that societies that allow individual initiative and discovery to flourish, and that permit order to emerge spontaneously, are more likely to generate new ideas and ways of doing things that benefit society as a whole, as compared to authoritarian or centrally planned regimes that seek to impose order from above. This is not to say that some forms of top-down regulation have no value, but to stress that institutional arrangements that coalesce through a process of discovery and adaptation may often yield superior results.[1]

This concept has deep roots in the history of economic thought; indeed, Adam Smith's "invisible hand" is one of the best-known expressions of it. However, it also applies to many other domains, including the evolution of law and of scientific knowledge. Statute law is analogous to a top-down system that imposes a pre-conceived order on the disposition of particular cases. Case law, by contrast, is built up from an accumulation of many precedents whose interpretation changes over time. Each new judgment is made in light of judgments in previous cases, and thereby influences the resolution of future cases. Constitutional orders exhibit a similar dichotomy. Some, like the US Constitution, explicitly prescribe the relations between different branches of government and the rights of citizens. Others, like the British constitution, are mostly unwritten and evolve from an accretion of precedents, conventions, and understandings that determine the disposition of political and judicial powers.

[1] The term "spontaneous order" was coined by the philosopher and polymath Michael Polanyi in a path-breaking article that presaged the work of Friederich Hayek (Polanyi 1941)

Spontaneous Order. H. Peyton Young, Oxford University Press. © H. Peyton Young (2024).
DOI: 10.1093/oso/9780198892908.003.0001

Scientific knowledge advances in much the same way, namely, through the contributions of dispersed individual researchers who build on previous contributions. Although some scientific projects may be sponsored and directed by the state, taken as a whole science is not a centrally directed activity, but more akin to the evolution of case law in which discoveries bubble up and are judged by others with reference to previous scientific knowledge. The result is a gradual but unplanned advance that orders our way of thinking about the world.[2]

In the social sciences, the concept of spontaneous order and its advantages over planned order figure prominently in the writings of Hume, Smith, and other members of the Scottish Enlightenment. They took a deliberate stance against top-down prescriptions about how to organize political and economic institutions in order to achieve desired social goals. In particular, Adam Smith argued that markets which arise spontaneously through the decentralized choices of many individuals, acting in their own self-interest, may serve the collective benefit more effectively.

In spite of these antecedents, however, the concept of spontaneous order, and more generally the evolutionary point of view in economics, went into abeyance for much of the nineteenth and twentieth centuries. Optimization—both at the individual and institutional level—has become the touchstone of modern economic analysis. Theory is directed toward the design of policies, organizational arrangements, and markets that achieve specific objectives. Individuals within these systems are presumed to be rational and forward-looking, and to optimize their objectives conditional on the optimizing behavior of others. Moreover, the paradigm generally assumes that the entire set-up is common knowledge among the agents, and that their expectations are correctly aligned with the actual behavior of the system. This is a top-down paradigm because it evades the question of how common knowledge and rational expectations become established in the first place, and how such a system comes into equilibrium from out-of-equilibrium conditions.

Admittedly there have been significant countervailing trends, promoted most notably by Menger, Schumpeter, Hayek, von Mises and other members of the Austrian School. Menger, in particular, argued that the emergence of a particular commodity as a means of exchange (money) could be understood in evolutionary terms as the outcome of many individual decisions rather than as the result of edict or decree.[3] It was Hayek, however, who drew attention to

[2] See M. Polanyi (1941, p. 437).

[3] Carl Menger (1963 (1883)). Similarly, Hume (1739) argued that money is a social convention that evolves from the accumulation of precedents.

the Achilles heel of the high rationality position, which is *lack of knowledge*. In an economic system, knowledge is dispersed among many actors, each of whom has partial knowledge of local circumstances, but not knowledge of the whole (let alone common knowledge of the whole). Likewise, the social planner lacks the knowledge necessary to institute an optimal state of affairs from the top down. As Hayek put it,

> The problem of a rational economic order is determined precisely by the fact that the knowledge of the circumstances of which we must make use never exists in concentrated form, but solely as the dispersed bits of incomplete and frequently contradictory knowledge which all the separate individuals possess. ... The problem is thus in no way solved if we can show that all of the facts, if they were known to a single mind ... would uniquely determine a solution; instead we must show how a solution is produced by the interactions of people, each of whom possesses only partial knowledge.[4]

Hayek's main target was the attempt to control economies by central planning and other forms of market intervention. I would argue, however, that Hayek's position applies with equal force to the high rationality doctrines that are now routinely assumed in economic models. In both macro- and microeconomics, it is commonly supposed that the protagonists are rational, forward-looking, and have common knowledge of the structure of the system within which they are embedded. This amounts to a top-down condition on economic order that is imposed by economic theorists rather than economic planners. Missing from their accounts is how a common understanding of the social structure and equilibrium strategies conditional on that structure are arrived at. Where do market-clearing prices come from? How are rational expectations formed? How do players in a repeated game arrive at mutually consistent strategies and beliefs? These aspects of the social and economic order are simply assumed rather than derived from a plausible dynamic adjustment process that pre-supposes no ex ante coordination of players' expectations and behaviors.

In recent years economists have paid increasing attention to these issues, although for the most part they still remain outside mainstream economic analysis.[5] In macroeconomics, for example, there is an emerging body of work showing how self-confirming expectations can arise dynamically through

[4] Friedrich Hayek (1945). For a discussion and critique of Hayek's notion of spontaneous order and its relation to market liberalism see Bowles, Kirman, and Sethi (2017).

[5] An exception is Nelson and Winter (1982), who were among the first to revive the evolutionary approach to modeling economic change.

adaptive learning by boundedly rational agents that are subjected to stochastic shocks, and how the learning process can generate endogenous fluctuations in expectations.[6] There is also a literature at the intersection of economics, sociology, and computer science that shows how complex forms of social order can arise spontaneously from the interactions of many dispersed agents operating with very simple adjustment rules. These "agent-based models" have been applied to a great variety of phenomena, including the emergence of decentralized prices in markets, discriminatory class structures, environmental collapse, and neighborhood segregation, to mention but a few examples.[7]

One limitation of this approach is that the agents' rules of adjustment tend to be imposed in an ad hoc fashion and are not grounded in empirical data on how subjects actually behave in interactive decision-making situations. Another limitation is that the analysis is based mainly on inferences drawn from sample simulation paths, rather than from a rigorous characterization of the long-run properties of the associated dynamical systems.

The approach taken in this book is thematically related to these literatures but differs by providing a rigorous mathematical framework for analyzing the out-of-equilibrium dynamics of economic and social systems, and for characterizing the likelihood with which different outcomes (equilibrium or otherwise) will be realized. The agents' rules of adjustment are founded on game theoretic conceptions of strategic behavior that can be tested empirically in laboratory and field environments.

This framework constitutes a branch of evolutionary game theory, which has its antecedents in the study of competition between species (Maynard-Smith and Price 1973; Maynard-Smith 1974). In the biological setting, there is a large population of biological entities (agents) who engage in pairwise interactions or contests. Each contest has the structure of a two-person game, and each agent is endowed with a strategy for playing this game. A crucial assumption is that agents' strategies are inherited, not chosen, and the payoffs from the pairwise contests determine the rate of reproduction of each strategy. Strategies that do well when paired against randomly drawn opponents in the population tend to increase faster than strategies that do poorly. The object is to characterize the limiting behavior of this dynamical system (known as the replicator dynamic).

[6] See in particular Evans and Honkapohja (1993, 2001), Sargent (1999), Cho, Williams, and Sargent (2002), and Williams (2018). Another strand of the macroeconomics literature analyzes epidemiological models of expectations formation; for a recent survey see C. Carroll and T. Wang (2022).

[7] For comprehensive accounts of agent-based models in the social sciences see J. Epstein and R. Axtell (1996), J. Epstein, (2006), L. Tesfatsion (2006), and D. Helbing (2012).

One limitation of evolutionary models based on the replicator dynamic is the absence of intentionality: agents are *endowed* with strategies, they do not choose them. Although this might seem to be a reasonable model of animal behavior, it is a much less satisfactory model of human behavior. In this book we shall focus on evolutionary models in which actors adjust their behavior by choosing perturbed best replies to their perceived environment. This framework is better suited to modeling the evolution of social and economic phenomena, and has surprisingly powerful implications for the selection of particular forms of social structure. The theoretical framework has four key elements: (i) a large population of agents who interact, either at random or via a given social network; (ii) agents' interactions are defined by a game (either one-shot or repeated); (iii) they adapt their behaviors purposefully based on prior experience and partial knowledge of others' behavior; (iv) their behaviors are subjected to random perturbations that arise from a variety of sources, including payoff shocks, misperceptions of the environment, and experimentation.

In fact, the outlines of this approach (but without the inclusion of stochastic shocks) were suggested as early as 1950 by John Nash in his doctoral dissertation:

> It is unnecessary to assume that the participants have full knowledge of the total structure of the game, or the ability and inclination to go through any complex reasoning processes. But the participants are supposed to accumulate empirical information on the relative advantages of the various pure strategies at their disposal (Nash, 1950).[8]

Unlike classical game theory, this version of game theory does not presume that agents are fully rational, or that they have common knowledge of the interactive structure, nor do they need to be forward-looking (although they might be). They sometimes deviate from best response behavior due to misperceptions, inattention, idiosyncratic utility shocks, or just plain mistakes. The objective is to study the intermediate and long-run dynamics of such systems starting from out-of-equilibrium conditions and to characterize the aggregate patterns or structures that ultimately emerge, whether equilibrium or otherwise. This is the sense in which we use the term "spontaneous order": order in the aggregate arises endogenously from many decentralized interactions of individuals, and not from any external cause. Note that the term is not meant

[8] At the insistence of the editor, this key insight was omitted from the published version of the paper (Nash 1951).

to imply that order arises immediately or even very rapidly; indeed, the process whereby individual adjustments eventually lead to equilibrium may take quite a long time, a topic that we explore in Chapters 11–13.

Beginning in the 1990s, variants of this approach were employed to model the emergence of various forms of equilibrium behavior without invoking high rationality and common knowledge, including risk-dominant equilibria in coordination games (Blume 1993; Kandori, Mailath, and Rob 1993; Young 1993), subgame perfect equilibria in repeated games (Hart 2002), the Nash bargaining solution (Young 1993; Saez-Marti and Weibull 1999; Binmore, Samuelson, and Young 2003), Walrasian equilibrium (Vega-Redondo 1997), and multiple equilibria in social interactions models (Brock and Durlauf 2001). The framework has also been applied to the evolution of diverse institutional structures, including social contracts (Binmore 1994, 1998; Skyrms 1996), neighborhood segregation (Schelling 1971; Young 1998; Zhang 2004), cultural conflict (Carvalho 2017), class discrimination (Axtell, Epstein, and Young 2001), the provision of public goods (Myatt and Wallace 2008), trading relationships in markets (Weisbuch, Kirman, and Herreiner 2000), electoral turnout (Bendor, Diermeier, and Ting 2003), and property rights (Bowles and Choi 2019), to name but a few examples.[9]

The contrast between the evolutionary framework and classical game theory can be illustrated with several examples. Consider the problem of negotiating the terms of a contract between a principal and an agent. In the conventional framework, the outcome is determined by the parties' degree of risk aversion, their discount rates, and their outside options. These parameters are typically assumed to be common knowledge (or at least their distribution is common knowledge), and the equilibrium results from the rational anticipation by each party about what the other is going to agree to. Importantly, the outcome will typically be specific to the characteristics of a particular principal–agent pair. When we look at empirical cases, however, we find that contractual terms often display a remarkable amount of uniformity between different pairs of agents. In the US, real estate agents typically get 6 per cent of the sales price; tort lawyers on continency fee traditionally get 1/3 of the award; tenant farmers almost invariably get a fixed percentage of the crop depending on the region (typically 1/2, 3/5, or 2/3). In short there is a contractual norm that applies to many different principal–agent pairs. An evolutionary model can explain

[9] For book-length treatments of evolutionary game theory and its applications see Jorgen Weibull (1995), Fernando Vega-Redondo (1996), Larry Samuelson (1997), Drew Fudenberg and David Levine (1998), H. Peyton Young (1998), Steven N. Durlauf and H. Peyton Young (2001), Samuel Bowles (2004), and William Sandholm (2010).

how such norms coalesce within a population of bargaining agents having heterogenous characteristics. The key assumption is that in any individual case the parties will be influenced by precedent, that is, by the terms agreed with others in similar circumstances. As will be shown in Chapters 3–5, the resulting feedback loop between precedents, expectations, and future behavior can lead to a bargaining norm within a given population. Moreover, while agents usually choose best responses given their expectations, but sometimes deviate due to random influences, certain norms are more likely to emerge than others. In this account the protagonists are mostly rational, but they lack the detailed knowledge needed to make finely-tuned calculations as is usually assumed in classical game theory. Equilibrium emerges spontaneously from the bottom up, and exhibits much greater homogeneity in behavior than would be predicted by more conventional high-rationality models of bargaining behavior. As will be shown in Chapter 5, the theory also shows why the equilibria of such systems may sometimes consist of localized regional norms rather than a single global norm.

A second example of spontaneous order is the evolution of market structure. Consider a market for perishable goods where buyers and sellers engage in face-to-face bargaining and there is no centralized price-setting mechanism. A pioneering study of this type is Kirman and Vignes's analysis of the Marseille wholesale fish market.[10] Every day suppliers acquire a quantity of fish that they must sell during the day, as the fish will spoil overnight. There are dozens of suppliers and hundreds of prospective retail customers who patronize the market on a regular basis. The supply of particular types of fish varies from one day to the next depending on the nature of the catch. Demand is also variable depending on the needs of the retailers. Retailers engage in face-to-face bargaining with suppliers of each particular type of fish; there are no posted prices and no information about the terms transacted by others. Thus market participants must rely almost entirely on their own experience to navigate the market. This is a game, but it is greatly complicated by variability in the set of players who show up each day, uncertainty in the supply and demand for particular types of fish, and the fact that players have almost no information about the payoffs and strategies of others. Nevertheless, it appears that such markets do achieve a kind of spontaneous order in which stable trading relationships emerge for the majority of customers, while there are some (typically less frequent) customers who continue to shop around and do not settle down with any one supplier.

[10] Kirman and Vignes (1991). See also Weisbuch, Kirman, and Herreiner (2000).

This type of environment is by no means limited to markets for perishable goods. A similar situation occurs in some financial markets, notably the repurchase ("repo") market in which financial institutions borrow from each other on a short-term basis, often just overnight, and post collateral in return (which they subsequently repurchase at a slight premium).[11] The scale of this market is enormous—typically several trillion dollars turn over every day. The participants are highly sophisticated financial institutions, including banks, hedge funds, brokers, and money market funds. As in perishable goods markets, there is an urgency to complete transactions in a timely way: borrowers need to cover their obligations by the end of each trading day, and lenders want to avoid having funds sitting idle overnight. Moreover, in some segments of the market information about transactions carried out by others is not available to participants—they have to shop around to discover the current best deals. In this situation there is a significant benefit to establishing stable trading relationships, as this reduces search costs and the risk that trade will not occur.[12]

A third example of this approach is the dynamic model of segregated neighborhoods proposed by Thomas Schelling (1971). In Schelling's model, agents of two colors are located on the squares of a checkerboard. They have diverse preferences as to the racial composition of their immediate neighborhood: some prefer to live only near people of the same color, others prefer to live in a partially integrated neighborhood, while still others may be indifferent. From time to time agents have the opportunity to move. They are rational in the sense that they move to a preferred location if one is available. However, they are not far-sighted: they do not try to anticipate what other agents are going to do, nor do they consider the ultimate (long-run) consequences of their actions on the composition of the various neighborhoods. Schelling shows that with high probability a segregated pattern emerges in which agents of a given color cluster together. This may occur even though most agents would prefer to live in an integrated neighborhood. Thus, an unintended consequence of their rational actions is often a suboptimal collection of neighborhoods. The key factor behind this result is that an agent's decision to move imposes an externality on his new neighbors (as well as his previous neighbors) because

[11] For a detailed description of the repo market see Kahn et al. (2023).

[12] This is the case in the Tri-Party segment of the market in which borrowers and lenders transact on a bilateral basis, and the Bank of New York Mellon acts as third-party custodian (Han, Nikolaou, and Tase 2022).

it changes the composition of the neighborhood *for them*. These results were given a more rigorous formulation by this author (Young 1998) and Dokumaci and Sandholm (2022); they can also be generalized to include differences in price as well as the racial composition of various neighborhoods (Zhang 2004).

The preceding examples exhibit several key features that distinguish them from more conventional models. These include the following.

1. *Disequilibrium dynamics*. The evolutionary process does not presuppose equilibrium or shared expectations, such as common priors or common knowledge that allow agents to coordinate their behavior. Equilibrium behavior and shared expectations arise (if they arise at all) as a byproduct of agents' decentralized decisions.
2. *Fragmentary knowledge*. Agents are numerous, dispersed, and typically have only fragmentary information about the behavior of others. They may not even know the overall structure of the game in which they are embedded.
3. *Adaptation*. Given their limited information, agents adapt their behavior using simple heuristics such as myopic best response instead of fully rational Bayesian updating.
4. *Stochastic variation*. Individuals' behaviors are not deterministic; they have an idiosyncratic component that results from such factors as experimentation, inattention, unobserved utility shocks, and mistakes.
5. *Long-run selection*. The long-run behavior of the resulting stochastic process can be characterized independently of the starting conditions. The states most likely to emerge in the long run correspond to stable norms or institutions.
6. *Local conformity, global diversity*. When agents interact through a social network, the stable equilibrium patterns tend to reflect local characteristics of the agents, and may vary across different parts of the network. This is a feature of some types of contractual norms such as cropsharing arrangements in agriculture (see Chapter 5).
7. *Speed of convergence*. Both theory and experiments show that under moderate levels of stochasticity in agents' adaptive behavior, convergence may occur quite rapidly. A key determinant is how often agents have the opportunity to revise their strategies. Norms of linguistic usage may evolve very rapidly, for example, whereas norms of divorce and child-bearing evolve much more slowly (see Chapters 12–13).

The following chapters are arranged in three groups that explore different aspects of the theory of spontaneous order. The first group describes the basic analytical framework (stochastic dynamical systems theory), and applies it to the evolution of norms and institutions. Specific examples include bargaining norms, contractual norms, and shared conceptions of what the rules of the game are (see Chapters 2–5).

The second group of chapters focuses on different conceptions of adaptive learning behavior, and contrasts them with standard notions of Bayesian rationality in the game theory literature (see Chapters 6–7). Chapter 8 discusses experiments with laboratory subjects that show how populations of subjects learn to play equilibrium using simple adaptive rules without prior knowledge of the game payoffs or the structure of their interactions. Chapters 9–10 show how agents can be programmed with simple adaptive rules in order to achieve desirable system-wide goals, which has application to the design of systems of distributed control.

The third set of chapters explores the dynamics by which new discoveries (whether social or technological) percolate from the bottom up, and spread in the population of potential adopters. Of particular interest is the rate at which innovations spread, and how the speed of diffusion is affected by the network through which people gain information, the benefits of the innovation relative to the status quo, and the degree of heterogeneity in agents' preferences (see Chapters 11–13).

This volume is not meant to provide a comprehensive account of adaptive learning and evolutionary game theory, which continue to advance at a rapid pace, with applications in economics, sociology, political science, computer science, and control theory. Rather it develops a particular point of view that stands in contrast to much current theorizing in economics, which takes the rules of interaction as given and commonly understood by everyone in the game, assumes a high degree of rationality and forward-looking behavior on the part of the agents, and then attempts to characterize the resulting equilibria. Not only are these assumptions unrealistic, they evade the essential question of how shared expectations and behaviors arise in the first place. The theory described here provides a framework for addressing this question. It does not presume a particular type of bounded rationality on the part of the agents; rather it shows how to characterize the learning dynamics for different classes of adaptive rules. The most likely outcomes of the process depend on the particulars of the rules, as well as on the amount of information available to the protagonists.

One of the key challenges for future research is to bring the theory into closer contact with empirical examples by studying the adaptive rules that agents actually do use in concrete situations, and to ask how long it takes for the system to equilibrate. There is also the potential to harness advances in computing technology to give participants more information about the state of the system without sacrificing their autonomy in how they make use of the information. Of course, providing more information does not necessarily mean that the end result will be superior. A program that tells buyers about the racial composition of different neighborhoods may lead to an outcome that is just as segregated as if they had no such information. For better or worse, new forms of information provision are bound to have a significant impact on the behavior of economic, social, and engineered distributive systems. The outcome will depend not only on the quality of the information itself, but on how people make decisions based on this information. Understanding these effects is one of the most important contemporary issues that can be addressed by the theory of spontaneous order.

References

Axtell, R., Epstein, J. and Young, H. P. (2001), "The emergence of classes in a multi-agent bargaining model," in S. N. Durlauf and H. P. Young, eds., *Social Dynamics* (Cambridge, MA: MIT Press, pp. 191–211).

Bendor, J., Diermeier, D., Ting, M. M. (2003), "A behavioral model of turnout," *American Political Science Review* 97 (2): 261–280.

Binmore, K. (1994), *Game Theory and the Social Contract vol. 1. Playing Fair* (Cambridge, MA: MIT Press).

Binmore, K. (1998), *Game Theory and the Social Contract* vol. 2. Just Playing (Cambridge, MA: MIT Press).

Binmore, K., Samuelson, L., and Young, H.P. (2003), "Equilibrium selection in bargaining models," *Games and Economic Behavior* 45 (2): 296–328.

Blume, L. E. (1993), "The statistical mechanics of strategic interaction," *Games and Economic Behavior* 5 (3): 387–424.

Bowles, S. and J.-K. Choi (2019), "The Neolithic agricultural revolution and the origins of private property," *Journal of Political Economy* 127 (5): 2186–2228.

Bowles, S., Kirman, A., and Sethi, R. (2017), "Retrospectives: Friedrich Hayek and the Market Algorithm," *Journal of Economic Perspectives* 31: 215–230.

Brock, W. A. and Durlauf, S. N. (2001), "Discrete choice with social interactions," *Review of Economic Studies* 68 (2): 235–260.

Carvalho, J.-P. (2017), "Coordination and culture," *Economic Theory* 64: 449–475.

Cho, I.-K., Williams, N., and Sargent, T. J. (2002), "Escaping Nash inflation," *Review of Economic Studies* 69: 1–40.

Dokumaci, E. and Sandholm, W. H. (2022), "Schelling redux: an evolutionary dynamic model of residential segregation," *Journal of Dynamics and Games* 9 (4): 373–403.

Durlauf, S. N. and Young, H. P., eds. (2001), *Social Dynamics* (Cambridge, MA: MIT Press).

Epstein, J. and Axtell, R. (1996), *Growing Artificial Societies: Social Science from the Bottom Up* (Cambridge, MA: MIT Press).

Epstein, J. (2006), *Generative Social Science: Studies in Agent-Based Computational Modelling* (Princeton, NJ: Princeton University Press).
Evans, G. and Honkapohja, S. (1993), "Adaptive forecasts, hysteresis, and endogenous fluctuations," *Economic Review of the Federal Reserve Bank of San Francisco* 1: 3–13.
Evans, G. and Honkapohja, S. (2001), *Learning and Expectations in Macroeconomics* (Princeton: Princeton University Press).
Foster, D. and Young, H. P. (1991), "Cooperation in the short and in the long run," *Games and Economic Behavior* 3 (1): 145–156.
Fudenberg, D. and Levine, D. (1998), *Theory of Learning in Games* (Cambridge, MA: MIT Press).
Han, S, Nikolaou, K., and Tase, M. (2022), "Trading relationships in secured markets: evidence from triparty repos," *Journal of Banking and Finance* 139: 106486.
Hart, S. (2002), "Evolutionary dynamics and backward induction," *Games and Economic Behavior* 41 (2): 227–264.
Hayek, F. (1945), "The use of knowledge in society," *American Economic Review* 35 (4): 519–530.
Helbing, D. ed. (2012), *Social Self-Organization: Agent-Based Simulations and Experiments to Study Emergent Social Behavior* (Springer).
Hume, D. (1739), *A Treatise of Human Nature* (London: John Noon).
Kahn, R. J., McCormick, M., Nguyen, V., Paddrik, M., and Young, H. P. (2023), "Anatomy of the repo rate spikes in September 2019," Washington DC: Office of Financial Research, Working Paper 23-04.
Kirman, A. and Vignes, A. (1991), "Price dispersion. Theoretical considerations and empirical evidence from the Marseille fish market," in K. J. Arrow (ed.), *Issues in Contemporary Economics* (London: Macmillan).
Maynard-Smith, J. and Price, G. R. (1973), "The logic of animal conflict," *Nature* 246: 15–18.
Kandori, M., Mailath, G., and Rob, R. (1993), "Learning, mutation, and long-run equilibria in games," *Econometrica* 61: 29–56.
Maynard-Smith, J. (1974), "The theory of games and the evolution of animal conflicts," *J. Theoretical Biology* 47 (1): 209–221.
Menger, C. (1963), *Problems of Economics and Sociology.* (Urbana, IL: University of Illinois Press). Originally published in 1883 as *Untersuchungen uber die Methode der Sozialwissenschaften und der Politischen Oekonomie* (Leipzig: Duncker & Humblot).
Myatt, D. P. and Wallace, C. (2008), "When does one bad apple spoil the barrel? An evolutionary analysis of collective action," *Review of Economic Studies* 75 (2): 499–527.
Nash, J. F. (1950), *Non Cooperative Games*, Ph.D. thesis, Princeton University.
Nash, J. F. (1951), "Equilibrium Points in n-Person Games," *PNAS* 36 (1): 48–49.
Nelson, R. R. and Winter, S. G. (1982), *An Evolutionary Theory of Economic Change*, (New Haven, CT: Harvard University Press).
Polanyi, M. (1941), "The growth of thought in society," *Economica* New Series 8 (32): 428–456.
Saez-Marti, M. and Weibull, J. (1999), "Clever agents in Young's bargaining model," *Journal of Economic Theory* 59: 145–168.
Samuelson, L. (1997), *Evolutionary Games and Equilibrium Selection* (Cambridge, MA: MIT Press).
Sandholm, W. (2010), *Population Games and Evolutionary Dynamics* (Cambridge, MA: MIT Press).

Sargent, T. J. (1999), *The Conquest of American Inflation* (Princeton, NJ: Princeton University Press).
Schelling, T. C. (1971), "Dynamic models of segregation," *Journal of Mathematical Sociology* 1: 143–186.
Skyrms, B. (1996), *Evolution of the Social Contract* (Cambridge, MA: Cambridge University Press).
Tesfatsion, L. (2006), "Agent-based computational economics: a constructive approach to economic theory," in Tesfatsion, L. and Judd, K. L, *Handbook of Computational Economics*, vol. 2: 831–880 (North-Holland).
Vega-Redondo, F. (1996), *Evolution, Games, and Economic Behavior* (Oxford: Oxford University Press).
Vega-Redondo, F. (1997), "The evolution of Walrasian behavior," *Econometrica* 65: 375–384.
Weibull, J. (1995), *Evolutionary Game Theory* (Cambridge, MA: MIT Press).
Weisbuch, G., Kirman, A., and Herreiner, D. (2000), "Market organization and trading relationships," *The Economic Journal* 110: 411–436.
Williams, N. (2018), "Escape dynamics in learning models," *Review of Economic Studies* 86: 882–912.
Young, H. P. (1993), "The evolution of conventions," *Econometrica* 61: 57–84.
Young, H. P. (1998), *Individual Strategy and Social Structure: An Evolutionary Theory of Institutions* (Princeton, NJ: Princeton University Press).
Zhang, J. (2004), "A dynamic model of residential segregation," *Journal of Mathematical Sociology* 28: 147–170.

Preface to Chapter 2
Evolutionary Dynamics with Persistent Perturbations

During my time at the University of Maryland I taught a graduate course in game theory that was based in the mathematics department. It attracted students from a wide variety of subjects, including statistics, computer science, and economics. This was the heyday of high rationality models in game theory; accordingly, the focus of the course was on equilibria in repeated games and how they are sustained by perfectly rational, forward-looking agents. In games with multiple equilibria, the standard approach was to invoke various refinement criteria such as such as perfect equilibrium, sequential equilibrium, the intuitive criterion, and so on.

I could not, however, disguise my dissatisfaction with this approach: it seemed completely at odds with how players actually navigate repeated interactions in real-world settings. Moreover, the implementation of refinements criteria requires a great deal of knowledge about the game parameters. Is it reasonable to assume that players know their opponents' utility functions or the distribution thereof? Do they know that their opponents know these things? Can they be sure their opponents will behave rationally? An even more fundamental problem is that equilibrium is treated as something that has already established itself. But how does equilibrium (of any sort) come about, starting from out-of-equilibrium conditions? In short, what are the dynamics that lead to equilibrium?

To address these questions, I began to explore alternative approaches to equilibrium selection that are founded on bounded rationality, limited knowledge, and are explicit about the dynamics of adjustment. It turned out that prior work in this vein had been done mainly by biologists. In biological settings, species interact and compete for resources using inherited strategies, which they pass on to their descendants. These *replicator dynamics* are very natural because the payoffs can be interpreted as the species' reproductive success rates. There is no need to invoke notions of "rationality": in the long run some species' strategies are more reproductively fit than others given the strategies being used by everyone else.

There remains, however, the question of which equilibrium, among many possible ones, is most likely to be realized. To address this problem, Maynard-Smith and Price introduced the concept of an *evolutionary stable strategy* (ESS).[1] The essential idea is that, given a sufficiently small, one-time deviation from an equilibrium, the replicator dynamics should revert to it. In biological applications such deviations can arise for many reasons, including the appearance of mutant strategies, variability in matchings between agents, variability in reproductive success rates, and so forth. Such deviations are continually jostling the dynamical system. The problem with ESS as a stability concept is that a succession of shocks can kick the process out of the basin of attraction of the ESS and into an entirely new regime. In other words, the system may not have time to revert to the ESS before another deviation occurs. In general, the time it takes for the process to return to a particular ESS (assuming it ever does return) depends on the global structure of the dynamical system. The challenge is to characterize the asymptotic behavior of such a system when it is subjected to *recurring* stochastic shocks.

Fortunately, the mathematical tools for addressing this issue were right at hand, because one of the pioneers in stochastic dynamical systems theory, Mark Freidlin, was a member of the Maryland mathematics department. Together with Dean Foster (one of the students in the game theory class) we showed how the Freidlin–Wentzell framework could be used to characterize the probability of different equilibrium (and disequilibrium) regimes when the evolutionary process is governed by a perturbed version of the replicator dynamics.[2] Specifically, we considered the replicator dynamic perturbed by a continuous-time, continuous-space Wiener process, and studied its asymptotic behavior for smaller and smaller levels of variance. We also assumed that mutations and immigration effectively keep the population proportions bounded away from zero, thus ensuring the existence of an ergodic distribution. A key result is that the ergodic distribution becomes increasingly concentrated on a small set of states—sometimes a unique state—as the stochastic component tends to zero. We called this the *stochastically stable set*.[3]

We then showed that this idea can be applied to the problem of equilibrium selection. This notion of selection differs from traditional ones, however,

[1] J. Maynard-Smith and G. R. Price (1973), "The logic of animal conflict," *Nature* 246: 15–18; J. Maynard-Smith (1974), "The theory of games and the evolution of animal conflicts," *J. Theoretical Biology* 47 (1): 209–221.

[2] M. Freidlin and A. D. Wentzell (1984), *Random Perturbations of Dynamical Systems*, Springer Verlag.

[3] D. Foster and H. P. Young (1990), "Stochastic evolutionary game dynamics," *Theoretical Population Biology* 38: 219–232.

because the stochastic process will typically not converge to an equilibrium, or even to a neighborhood of an equilibrium: it is constantly in motion. Nevertheless, selection occurs in the sense that, in the long run, some equilibria are *more likely* to be visited than others when the variance of the white-noise process is sufficiently small. The intuitive idea is that the relative stability of different equilibria depends crucially on the "depth" of the basins of attraction in which they are embedded. Furthermore, the stochastic process may select a limit cycle (or other disequilibrium configuration) even when equilibria exist. This can happen, for example, when the equilibrium states lie in relatively shallow basins of attraction compared to the basin of the limit cycle.

One limitation of the continuous set-up is that it does not specify the sources of stochasticity in individual behaviors. It turns out that there are finite-state, discrete-time versions of the framework in which the behavior of individuals can be modeled explicitly; moreover they are easier to solve analytically. Chapter 2 develops such a framework for finite n-person games played by a finite population of agents who interact globally. The framework can encompass many different forms of adaptive behavior (including the replicator dynamics), but the chapter focuses on the case of perturbed best response dynamics. Agents interact pairwise and (usually) choose best responses based on partial knowledge of recent precedents, but occasionally they choose non-best responses due to such factors as utility shocks, inattention, experimentation, or just plain "mistakes."

This set-up defines a finite-state Markov chain that consists of two parts: there is a "primary" transition probability matrix determined by the adjustment process without perturbations, and superimposed on the primary process there is a "secondary" process defined by the transition probability matrix induced by the perturbations. The combination of the two processes constitutes a *perturbed Markov chain.*

The concept of stochastic stability is readily adapted to this framework: we wish to identify the minimal set of states whose long-run probability approaches one when the perturbations have probability approaching zero. As we show in Chapter 2, these *stochastically stable states* can be found by representing the states as vertices of a directed graph, whose edges are weighted by the probability of transiting between states, and invoking the Markov Chain Tree Theorem to solve for the ergodic distribution.

A particularly interesting case arises when the underlying game has the following property: from any initial configuration of actions, there exists a best-reply path (one agent moving at a time) to a strict Nash equilibrium.

Such a game is said to be *weakly acyclic* in best replies. Weakly acyclic games include coordination games and potential games, but they are considerably more general. Chapter 2 shows that if agents update their actions by myopic best responses to incomplete samples of recent precedents, and if the game is weakly acyclic in best replies, the process converges with probability one to a strict Nash equilibrium, which we may interpret as a social convention. In other words, equilibrium at the population level arises spontaneously through the decentralized choices of individuals who have only fragmentary information about the behavior of others. When this process is buffeted by deviations from best-response behavior of the types described above, we obtain a perturbed Markov process, and can solve for the stochastically stable equilibria.

A particularly transparent class of examples are 2×2 coordination games. When such a game is played by a large population of interacting agents, we would like to know whether their behaviors eventually come close to one of the two coordination equilibria, and if so which of them is most likely to occur over the long run. In our 1990 article, Dean Foster and I considered this situation in the context of the replicator dynamic perturbed by a continuous stochastic process. Subsequent articles treated the case of perturbed best response dynamics in a large but finite population.[4] As in the continuous model, the dynamics oscillate between states in which most (but not all) players are playing one of the two equilibria. When the perturbation probability decreases, the dynamics spend a longer and longer proportion of the time near one of them (unless the payoffs are such that the two equilibria are tied). It turns out that, under various specifications of the perturbation distribution, the risk-dominant equilibrium will be selected. The proof amounts to showing that the basin of attraction of the risk-dominant equilibrium is more difficult to escape than the complementary basin.[5]

Chapter 2 shows that when we move beyond the framework of 2×2 coordination games, however, the analysis of stochastic stability becomes more complex. For example, the following is a symmetric coordination game in which neither the risk-dominant equilibrium (A) nor the Pareto dominant equilibrium (C) is selected under a best response process perturbed by independent random mutations.

[4] M. Kandori, G. Mailath, and R. Rob (1993), "Learning, mutation, and long-run equilibria in games," *Econometrica* 61 (1): 29–56; L. E. Blume (1993), "The statistical mechanics of strategic interaction," *Games and Economic Behavior* 5 (3): 387–424; and H. P. Young (1993), "The evolution of conventions," *Econometrica* 61: 57–84.

[5] The perturbation distributions for which this statement holds are characterized by L. E. Blume (2003), "How noise matters," *Games and Economic Behavior* 44 (2): 251–271.

	A	B	C
A	6, 6	0, 5	0, 0
B	0, 5	7, 7	5, 5
C	0, 0	5, 5	8, 8

The method for solving this and more complex examples is described in Chapter 2. Moreover, the Appendix of Chapter 2 develops a very general framework for characterizing the asymptotic behavior of various types of adaptive dynamics (not just best-response dynamics) that are buffeted by random perturbations. The perturbations need not be "mindless" mutations, but may reflect deliberate experimentation with alternative strategies that form an integral part of the adaptive process itself. The theory also applies to situations where agents are embedded in a social network and interact solely with their neighbors instead of with the population at large (see Chapters 12 and 13).

One criticism that has been leveled at the stochastic evolutionary approach to equilibrium selection is that the process may take too long to be a meaningful predictor of actual behavior. If the population of agents is large and fully mixed, and if mutations occur independently with some small probability $\varepsilon > 0$, the expected time it takes to reach a neighborhood of the stochastically stable equilibrium grows exponentially in $1/\varepsilon$ as well as in the population size Although this statement is correct as such, the claim that evolutionary dynamics *in general* take an excessively long time to reach equilibrium is not warranted. Experiments with human subjects tend to support a behavioral model in which deviations from best response actually occur with fairly substantial probability; moreover, deviations do not occur with uniform probability but tend to be more likely when they do not involve large payoff losses. Furthermore, when agents are embedded in a social network, changes in behavior can occur quite rapidly at a local level, and then spread through the population more quickly than if the population were fully mixed. The combination of these features can induce change surprisingly rapidly, as will be shown in the final chapter.

2

The Evolution of Conventions

1 Introduction

A CONVENTION IS A PATTERN of behavior that is customary, expected, and self-enforcing.[1] Everyone conforms, everyone expects others to conform, and everyone wants to conform given that everyone else conforms.[2] Familiar examples include driving on the right when others drive on the right, going to lunch at noon if others go at noon, accepting dollar bills in payment for goods if others accept them, and so forth. Conventions need not be symmetric. Men conventionally propose to women. Sailboats on the port tack yield the right-of-way to sailboats on the starboard tack. In some regions, tenant farmers customarily get one-third of the harvest and landlords get two-thirds, whereas in other regions the reverse convention holds.[3] For each role in such asymmetric interactions there is a customary and expected behavior, and everyone prefers to follow the behavior expected of him provided that others follow the behavior expected of them. Under these circumstances we say that people follow a convention. A convention is an equilibrium that everyone expects. But how do expectations become established when there is more than one equilibrium?

One explanation is that some equilibria are a priori more reasonable than others. A deductive theory of this type has been proposed by Harsanyi and Selten (1988). A second explanation, proposed by Schelling (1960), is that agents focus their attention on one equilibrium because it is more prominent or conspicuous than the others. Yet a third explanation is that, over time, expectations converge on one equilibrium through positive feedback effects. Suppose that a game is played repeatedly, either by the same or different agents. Past plays

Originally published as H. P. Young, "The Evolution of Conventions," Econometrica, Vol. 61, No. 1, 1993, pp. 57–84. Reproduced with permission

[1] This work was supported in part by the Santa Fe Institute. I am indebted to Dean Foster, David Canning, Michael Cohen, Michihiro Kandori, David Lane, George Mailath, and Rafael Rob for stimulating conversations on this topic, and to the referees for several constructive suggestions. The usual caveat applies.

[2] See Lewis (1967).

[3] See Bardhan (1984) for an account of sharecropping patterns in India.

Spontaneous Order. H. Peyton Young, Oxford University Press. © H. Peyton Young (2024).
DOI: 10.1093/oso/9780198892908.003.0002

have a feedback effect on the expectations and behaviors of those playing the game now because people pay attention to precedent. Eventually, one equilibrium becomes entrenched as the conventional one, not because it is inherently prominent or focal, but because the dynamics of the process happen to select it.

This evolutionary explanation for the origin of conventions has been suggested in a variety of papers, but the precise dynamics of the process by which expectations and behaviors evolve has not been clearly spelled out.[4] In particular it is not clear whether it works. Does the process converge to an equilibrium, and if so, are all equilibria equally likely to be selected? We shall show that the process converges in an asymptotic sense provided that the underlying game has an acyclic best reply structure, and provided there is sufficient stochastic variability in the players' responses. In this case, society is at or close to a Nash equilibrium most of the time. Not all Nash equilibria are equally likely to be selected, however. In fact, typically only *one* Nash equilibrium will be observed with high probability in the long run. Such an equilibrium is said to be *stochastically stable* (Foster and Young (1990)). Building on work of Freidlin and Wentzell (1984) and Kandori, Mailath, and Rob (1993), we shall show how to compute the stochastically stable equilibria by solving a series of shortest path problems in a graph. This is an application of a more general result (proved in the Appendix) that characterizes graph-theoretically the stochastically stable communication classes of a perturbed finite-state Markov process.

2 Outline of the Model

We consider a fixed n-person game that is played once each period. The players are drawn at random from a large, finite population of individuals. Each player chooses an optimal strategy based on his beliefs about his environment, which he takes to be stationary. He forms his beliefs by looking at what other agents have done in the recent past. Since gathering information is costly, however, each player knows only a small portion of the history, that is, he bases his current actions on a *sample* of plays from recent time periods. We shall also assume that the players occasionally experiment with different strategies, or simply make mistakes.

The strategies that the agents choose in the current period are recorded and the game is played again in the next period by another random draw of n agents

[4] See, for example, Lewis (1967), Axelrod (1986), Sugden (1986), Bicchieri (1990), and Warneryd (1990).

from the fixed population. Each of these agents takes a random sample of previous plays and reacts accordingly. Actions in earlier periods therefore have a feedback effect on actions by agents in later periods. To emphasize that convergence in this model has nothing to do with learning at the individual level, we can assume that after an agent plays the game once he dies and is replaced by a naive agent of the same type (same sample size and same utility function) but no prior information. Thus each time an agent plays he starts afresh and must ask around to find out what is going on. This assumption is not necessary from a mathematical point of view, but it underscores the fact that we are ignoring learning and reputation effects.

The adaptive dynamics described above define a Markov chain whose states are the histories of play truncated to a finite number of periods. It is similar to fictitious play in that agents choose best replies to other agents' past actions. In fictitious play, however, agents base their decisions on the entire history of actions by other agents. Here we assume that agents base their decisions on *limited* information about actions of other agents in the *recent* past, and they do not always optimize. These assumptions seem less fictitious than fictitious play; hence we call this process *adaptive play.*

For general n-person games, adaptive play need not converge to a Nash equilibrium, either pure or mixed, as we shall show below by example. Nevertheless, there is an important class of games for which it does converge. These games have the property that, from any initial choice of strategies, there exists a sequence of best replies that leads to a strict, pure strategy Nash equilibrium. This class includes, but is substantially more general than, coordination games and common interest games. For these *weakly acylic* games, adaptive play converges with probability one to a pure strategy Nash equilibrium *provided* that the samples are sufficiently incomplete and the players never make mistakes. Incompleteness creates enough stochastic variability to prevent the process from becoming stuck in suboptimal cycles. Finite memory allows past miscoordinations to be forgotten eventually. Once a given equilibrium has been played for as long as anyone can remember, then this equilibrium becomes entrenched as the "conventional" way of playing the game. It is an absorbing state of the process. One cannot say in advance, of course, which equilibrium will become the conventional one, since this depends on the vagaries of the process and on the initial state. What can be said is that some equilibrium will eventually be selected with probability one, and it will *not* be a mixed strategy equilibrium.

If the players occasionally experiment or make mistakes, however, then more can be said. In this case the process has no absorbing states; rather, it has a stationary distribution that describes the relative frequency with which different states are observed in the long run. We shall show that, if the probability of mistakes is small, then this stationary distribution is concentrated around a particular subset of pure strategy Nash equilibria. In fact, typically it puts almost all the weight on exactly one equilibrium. This *stochastically stable equilibrium* will be observed with probability close to one when the noise is very small.

This concept differs in an important respect from other notions of equilibrium stability, such as evolutionarily stable strategies.[5] An evolutionary stable strategy is a strategy (or frequency distribution of strategies) that is restored after a small one-time shock to the system. A stochastically stable distribution is a distribution that is restored repeatedly when the evolutionary process is constantly buffeted by small random shocks. This concept was first defined for general evolutionary processes by Foster and Young (1990). Subsequently it was applied to a discrete model of equilibrium selection in a pioneering paper by Kandori, Mailath, and Rob (1993). They consider an evolutionary learning process defined on symmetric 2×2 games. In each period every player plays every other. Successful strategies are adopted with higher probability than unsuccessful ones, and there is a small probability that players make mistakes. Kandori, Mailath, and Rob show that this stochastic process selects the risk dominant Nash equilibrium when the mistake probability is small.[6] The techniques of analysis in both papers are similar, and build on the theory of perturbed dynamical systems developed by Freidlin and Wentzell (1984).

Using somewhat different methods, Canning (1992) studies a general class of learning models in which agents adapt their behavior to the current state and occasionally make mistakes. He shows that, under certain regularity conditions, the stationary distribution of the perturbed process converges to a stationary distribution of the unperturbed one. In this chapter we shall show how to characterize the support of the limiting stationary distribution (the "stochastically stable set") by solving a series of shortest path problems in a graph. Very often the support consists of a single absorbing state, in which case we obtain a complete characterization of the limiting distribution, and a unique stochastically stable equilibrium. We then apply this result to compute the stochastically stable equilibria of adaptive play. For 2×2 coordination

[5] Models of equilibrium selection based on the concept of evolutionarily stable strategy (ESS) include Axelrod (1984), Fudenberg and Maskin (1990), Crawford (1991), Samuelson (1991a), and Samuelson and Zhang (1992). For other models of evolutionary dynamics in games, see Samuelson (1988), Nachbar (1990), and Friedman (1991).

[6] Models of equilibrium selection based on the concept of stochastic stability include Young and Foster (1991), Fudenberg and Harris (1992), Kandori and Rob (1992), and Samuelson (1991b). For other stochastic selection models see Evans and Honkapohja (1995a, 1995b), and Kirman (1995).

games we show that the risk dominant equilibrium is the unique stochastically stable equilibrium. In coordination games with more than two strategies, the stochastically stable equilibrium may be neither risk dominant nor Pareto optimal, as we show by example. Although there appears to be no simple formula that characterizes the stable equilibria in the general case, they may be computed by a general algorithm that is efficient to implement.

3 Adaptive Play

Let Γ be an n-person game in strategic form, and let S_i be the finite set of strategies available to player i. Let N be a finite population of individuals that is partitioned into n nonempty classes $C_1, C_2, \ldots, C_n$. Each member of C_i is a candidate to play role i in the game. For example, C_1 is the class of men, C_2 is the class of women, and the game is Battle of the Sexes. We shall assume that all individuals in class i have the same utility function $u_i(s)$ for strategy-tuples $s = (s_1, s_2, \ldots, s_n) \in \prod S_i$, which we shall identify with outcomes.

Let $t = 1, 2, \ldots,$ denote successive time periods. The game G is played once each period. In period t, one individual is drawn at random from each of the n classes and is assigned to play the appropriate role in the game. It will be convenient to refer to the individual playing role i as player "i" even though the identity of this individual may change from one period to the next. Player i chooses a pure strategy $s_i(t)$ from his strategy space according to a rule that will be defined below. The strategy-tuple $\mathbf{s}(t) = (s_1(t), s_2(t), \ldots, s_n(t))$ is recorded and will be referred to as the *play at time t*. The *history* of plays up to time t is the sequence $h(t) = (\mathbf{s}(1), \mathbf{s}(2), \ldots, \mathbf{s}(t))$. We assume that the histories are anonymous: it does not matter *who* played a given strategy in a given period, only that it was played by someone.

The players decide how to choose their strategies as follows. Fix integers k and m such that $1 \leq k \leq m$. In period $t + 1$ $(t \geq m)$ each player inspects k plays drawn without replacement from the most recent m periods $t, t-1, t-2, \ldots, t-m+1$. The draws are independent for the various players. One way to think about the sampling procedure is that each player "asks around" to find out how the game was played in recent periods. He stops when he has learned about k different plays within the last m periods (say because he has reached his capacity to retain information). Another way of thinking about the sampling procedure is that each agent passively hears about certain precedents, and k is the number of precedents that come to the agent's attention. In either case, the fraction k/m measures the *completeness* of the agents' information relative to the surviving precedents. It is not necessary to assume that every subset of k precedents out of the last m is *equally likely* to constitute an agent's information. For example, an agent might be more likely to hear about recent precedents than more dated ones. It is enough to assume that every subset of k has a positive probability of being agent i's information for every i.[7]

Assume for the sake of generality that the first m plays are randomly selected. Thus we can think of the sampling process as beginning in period $t = m + 1$ from some arbitrary initial sequence of m plays $h(m) = (s(1), s(2),\ldots, s(m))$. We then obtain a finite Markov chain on the state space H consisting of all sequences of length m drawn from $\prod S_i$, beginning with some arbitrary "initial" state $h(m)$.

Successor *A* successor *of a state $h \in H$ is any state $h' \in H$ obtained by deleting the left-most element of h and adjoining a new right-most element.*

The process moves from the current state h to a successor state h' in each period according to the following transition rule. For each $s \in S_i$, let $p_i(s|h)$ be the probability that agent i chooses s. We assume that $p_i(\cdot)$ is a *best-reply distribution* in the sense that $p_i(s|h) > 0$ if and only if there exists a sample of size k to which s is i's best reply, and that $p_i(s|h)$ is independent of t. If s is the right-most element of h, the probability of moving from h to h' is

$$P^0_{hh'} = \prod_{i=1,n} p_i(s_i \mid h). \quad (1)$$

$P^0{}_{hh'} = 0$ if h' is not a successor of h. We call the process P^0 *adaptive play with memory m and sample size k.*

4 Convergence of Adaptive Play When there are no Mistakes

Let us begin by observing that h is an absorbing state of this process if and only if it consists of a strict pure strategy Nash equilibrium played m times in succession. Suppose, indeed, that $h = (s^1,\ldots, s^m)$ is an absorbing state. For each agent i let s_i be i's best reply to some subset of k plays drawn from h, and let $s = (s_1,\ldots, s_n)$. By assumption, there is a positive probability of moving from h to $h' = (s^2,\ldots,s^m, s)$ in one period. Since h is absorbing, $h = h'$ and hence $s^1 = s^2$. Continuing in this fashion we conclude that $s^1 = s^2 = \ldots = s^m = s$. Hence

[7] The model can be modified by assuming that different agents have different sample sizes. An application of this type is described in Young (1993).

$h = (s, s, \ldots, s)$. By construction, s_i is a best reply to some sample of k elements from h. Hence s_i is a best reply to $\mathbf{s}_{-i}$ for each i. It must also be a unique best reply to $\mathbf{s}_{-i}$, because otherwise the process could move to a successor that is different from h. So $\mathbf{s}$ is a strict, pure strategy Nash equilibrium. Conversely, any state h consisting of m repetitions of a strict, pure strategy Nash equilibrium is clearly an absorbing state. Such a state will be called a *convention.*

If adaptive play converges to an absorbing state, then clearly the game must have a strict Nash equilibrium in pure strategies. The existence of such an equilibrium is not a sufficient condition for convergence, however. Consider the following variation of an example due to Shapley:

Example 1

	a	b	c	d
a	2,1	0,0	1,2	−1,−1
b	1,2	2,1	0,0	−1,−1
c	0,0	1,2	2,1	−1,−1
d	−1,−1	−1,−1	−1,−1	3,3

Here d is a best response to itself, but it is not a best response to any mixture of a, b, and c. If the initial state does not involve d, then adaptive play (like fictitious play) cycles. Consider, for example, the case where $m = 2$ and $k = 1$. Let the first two plays be $\begin{pmatrix} a \\ a \end{pmatrix}$ and $\begin{pmatrix} a \\ c \end{pmatrix}$. In period 3, Column will sample one of Row's previous two choices (both a) and react by playing c. Row will sample one of Column's previous two choices (a or c) with equal probability and react by playing a or c. So the next play will be $\begin{pmatrix} a \\ c \end{pmatrix}$ or $\begin{pmatrix} c \\ c \end{pmatrix}$ with equal probability. The process therefore moves from state $\begin{bmatrix} a & a \\ a & c \end{bmatrix}$ to state $\begin{bmatrix} a & a \\ c & c \end{bmatrix}$ with probability one-half, and to state $\begin{bmatrix} a & c \\ c & c \end{bmatrix}$ with probability one-half. The subsequent transitions form a cycle of length six imbedded within a cycle of length twelve, as shown in Figure 2.1. This cycle constitutes a recurrent communication class of the Markov process defined by (1).

When cycling is built into the best reply structure of the game, as in the above example, we cannot expect adaptive play to converge. Nevertheless, there are many games that do not have a cyclic best-reply structure. Consider

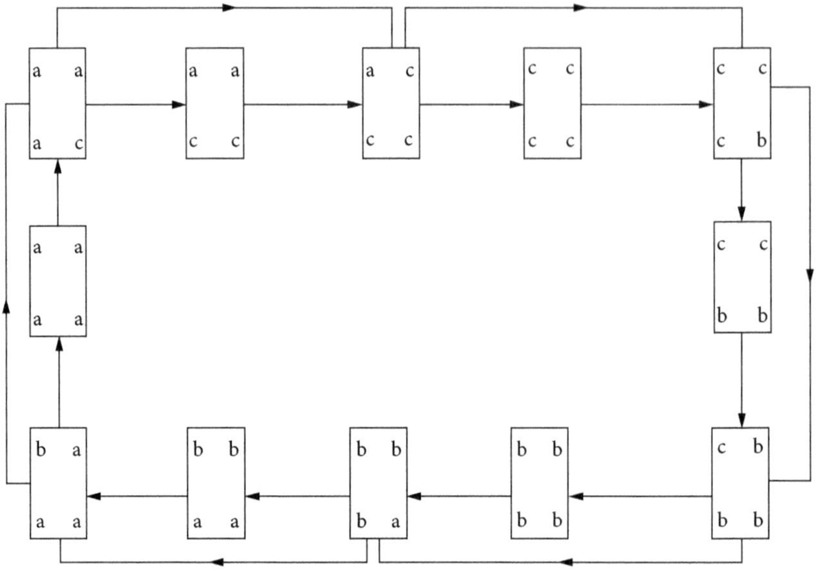

Figure 2.1 A recurrent communication class of adaptive play with $m = 2$, $k = 1$.

a two-person coordination game in which both agents have the same number of strategies, and each agent strictly prefers to play his jth strategy if and only if the other agent plays his jth strategy for every j. Clearly there is no cycling problem here: once one of them chooses a pure strategy and the other responds optimally, then they have achieved a coordination equilibrium.

To take another example, suppose that the agents have *common interests*: for every two strategy tuples $\boldsymbol{s}$ and $\boldsymbol{s}'$, either everyone prefers $\boldsymbol{s}$ to $\boldsymbol{s}'$ or everyone prefers $\boldsymbol{s}'$ to $\boldsymbol{s}$. Assume further that no two strategy-tuples are payoff equivalent. Given an arbitrary strategy-tuple $\boldsymbol{s}$ that is not a Nash equilibrium, there exists some agent i who can do better by playing s'_i instead of s_i. Let $\boldsymbol{s}' = (s'_i, \boldsymbol{s}_{-i})$. If $\boldsymbol{s}'$ is not Nash equilibrium, there is some agent j who can do better by playing s_j'' instead of s'_j. Let $\boldsymbol{s}'' = (s''_j, \boldsymbol{s}'_{-j})$, and so forth. At each step of this adjustment process everyone's utility increases, so the process cannot cycle and it must end at a pure strategy Nash equilibrium. Moreover, this equilibrium must be strict because of the assumption that no two strategy-tuples are payoff equivalent and the players have common interests.

This construction can be generalized as follows. Let Γ be an n-person game in normal form on a finite strategy space $\prod S_i$. Define the *best-reply graph* of Γ as follows: each vertex is an n-tuple of strategies $\boldsymbol{s} \in \prod S_i$, and for every two vertices $\boldsymbol{s}$ and $\boldsymbol{s}'$ there is a directed edge $\boldsymbol{s} \to \boldsymbol{s}'$ if and only if $\boldsymbol{s} \neq \boldsymbol{s}'$ and there exists exactly one agent i such that s'_i is a best reply to $\boldsymbol{s}_{-i}$ and $\boldsymbol{s}'_{-i} = \boldsymbol{s}_{-i}$.

Acyclic Game *A game Γ is* acyclic *if its best reply graph contains no directed cycles. It is* weakly acyclic *if, from any initial vertex* **s**, *there exists a directed path to some vertex* **s*** *from which there is no exiting edge (a* sink*).*

Every sink of the best reply graph is clearly a strict Nash equilibrium in pure strategies. So a game is weakly acyclic if, and only if, from every strategy-tuple there exists a finite sequence of best replies by one agent at a time that ends in a strict, pure strategy Nash equilibrium. We shall show that, for this class of games, adaptive play converges with probability one provided that sampling is sufficiently incomplete and the players do not make mistakes.

Let Γ be a weakly acyclic n-person game. For each strategy-tuple $\boldsymbol{s}$, let $L(\boldsymbol{s})$ be the length of a shortest directed path in the best reply graph from $\boldsymbol{s}$ to a strict Nash equilibrium, and let $L_\Gamma = \max_s L(\boldsymbol{s})$.

Theorem 1 *Let Γ be a weakly acyclic n-person game. If $k \le m/(L_\Gamma + 2)$, then adaptive play converges almost surely to a convention.*

Proof Fix k and m, where $k \le m/(L_\Gamma + 2)$. We shall show that there exists a positive integer $\boldsymbol{M}$, and a positive probability p, such that from any state h, the probability is at least p that adaptive play converges within M periods to a convention. M and p are time-independent and state-independent. Hence the probability of *not* reaching a convention after at least rM periods is at most $(1 - p)^r$, which goes to zero as $r \to \infty$.

Let $h = (\boldsymbol{s}(t - m + 1), \ldots, \boldsymbol{s}(t))$ be the state in period $t \ge m$. In period $t + 1$ there is a positive probability that each of the n agents samples the last k plays in h, namely, $(\boldsymbol{s}(t - k + 1), \ldots, \boldsymbol{s}(t)) = \eta$. There is also a positive probability that, from periods $t + 1$ to $t + k$ inclusive, every agent draws the sample η every time. Finally, there is a positive probability that, if an agent has a choice of several best replies to η, then he will choose the same one k times in succession. Thus there is a positive probability of a run $(\boldsymbol{s}, \boldsymbol{s}, \ldots, \boldsymbol{s})$ from periods $t + 1$ to $t + k$ inclusive. Note that this argument depends on the agents' memory being at least $2k - 1$, since otherwise they could not choose the sample η in period $t + k$.

Suppose that $\boldsymbol{s}$ happens to be a strict Nash equilibrium. There is a positive probability that, from periods $t + k + 1$ through $t + m$, each agent will sample only the last k plays, in which case the unique best response of each agent i is s_i. So they play $\boldsymbol{s}$ for $m - k$ more periods. At this point an absorbing state has been reached, and they continue to play $\boldsymbol{s}$ forever.

Suppose instead that $\boldsymbol{s}$ is not a strict Nash equilibrium. Since Γ is weakly acyclic, there exists a directed path $\boldsymbol{s}, \boldsymbol{s}', \ldots, \boldsymbol{s}^r$ in the best reply graph such that s^r is a strict Nash equilibrium. The first edge on this path is $s \to s'$. Let i be the index such that $s'_{-i} = s_{-i}$ and s'_i is a best reply to s_{-i}. Consider the event in which agent i samples from the run of s established in periods $t + 1$ to $t + k$ and responds by playing s'_i, while every agent $j \neq i$ draws the sample $\eta = (s(t - k + 1), \ldots, s(t))$. By assumption, a best response of every agent j to this sample is s_j. These events occur together with positive probability, and there is a positive probability that they occur in every period from $t + k + 1$ to $t + 2k$, assuming that $m \geq 3k - 1$. The result is a run of $s' = (s'_i, s_{-i})$ for k periods in succession.

Continuing in this fashion, we see that there is a positive probability of obtaining a run of s, followed by a run of s'... followed eventually by a run of s^r. Each run is of length k, and the run of s^r occurs from period $t + kr + 1$ to $t + kr + k$. To reach this point may require that some agent look back $kr + 2k - 1$ periods, namely, from period $t + kr + k$ to period $t - k + 1$. This is possible because of the assumption that $k \leq m/(L_\Gamma + 2)$.

After this, the process can converge to the absorbing state $(s^r, s^r, \ldots, s^r)$ by period $t + kr + m$ if each agent samples the previous k plays from periods $t + kr + k + 1$ to $t + kr + m$ inclusive.

Since $r \leq L_\Gamma$, we have established that, given an initial state h, there is a probability $p_h > 0$ of converging to an absorbing state within $M = kL_\Gamma + m$ periods. Letting $p = \min_{h \in H} p_h > 0$, it follows that from any initial state the process converges with probability at least p to an absorbing state within at most M periods. This completes the proof.

We do not claim that the bound $k \leq m/(L_\Gamma + 2)\ m/(L_\Gamma + 2)$ is best possible for all weakly acyclic games. Without incomplete sampling, however, the process may not converge to a pure Nash equilibrium. Consider the following version of "Battle of the Sexes":

Example 2

	Yield	Not Yield
Yield	0, 0	$1, \sqrt{2}$
Not Yield	$\sqrt{2}, 1$	0, 0

Let $k = m$, so that both players sample the same m plays in each period. Consider any initial sequence of m plays in which the players have always miscoordinated, that is, they both yielded or they both failed to yield in each period. Let f be the frequency with which they yielded in this sequence. In the next period, Row yields if and only if $1 - f > f\sqrt{2}$, and Column does the same. So they miscoordinate again. (Since f is rational, the inequality is always strict so we never have to consider ties.) Thus, if they begin in a state of perfect miscoordination, then they miscoordinate forever.

The same holds if memory is unbounded, as in fictitious play: if they miscoordinate on the first move, then they will continue to miscoordinate forever. In this case the frequency distributions converge to the mixed Nash equilibrium of the game in which both players yield with probability $1/(1 + \sqrt{2})$. Note, however, that their *behaviors* do not converge to the behavior specified by the mixed Nash equilibrium, because in each period their moves are correlated: either both yield or both do not yield. So even in 2×2 games fictitious play need not converge in a behavioral sense. This problem does not arise in adaptive play, because when it converges to an absorbing state, this state must correspond to a Nash equilibrium in pure strategies. Hence the behaviors also converge.

The feature of adaptive play that allows it to break out of suboptimal cycles is incomplete sampling, which introduces stochastic variation into the players' responses. There is a possibility that they will coordinate by chance, and if they do so often enough the process eventually locks in to a pure strategy Nash equilibrium. This equilibrium then becomes the conventional way of playing the game, because for as long as anyone can remember, the game has always been played in this way. Therefore sampling does not matter any more, because no matter what samples the agents take, their optimal responses will be to play the equilibrium that is already in place.

5 Adaptive Play with Mistakes

Theorem 1 relies on the assumption that, while agents base their decisions on limited information, they always choose a best response given their information. This assumption is clearly unrealistic. Agents sometimes make mistakes; they may also experiment with nonoptimal responses. In this case the stochastic process does not converge to an absorbing state, because it has no absorbing states. Mistakes constantly perturb the process away from equilibrium. If we assume, however, that all mistakes are possible and that the mistake probabilities are time-independent, then the process does have a unique stationary distribution. Hence we can study its asymptotic behavior. When the probability of mistakes is small, we shall show that this stationary distribution is concentrated around a particular convention (or, in the event of ties, a subset of conventions). These are the *stochastically stable* conventions—the ones that will be observed with positive probability in the long run when the noise is small but nonvanishing. We therefore obtain a theory of equilibrium selection. As we shall see below, it yields (except in the 2×2 case) quite different answers than the theory of Harsanyi and Selten (1988).

Our model of mistakes generalizes an approach pioneered by Canning (1992) and Kandori, Mailath, and Rob (1993). Fix the sample size k and memory m. Suppose that, in each time period, there is some small probability $\varepsilon\lambda_i > 0$ that player i experiments by choosing a strategy randomly from S_i instead of optimizing based on a sample of size k. The ratio λ_i / λ_j is the relative probability with which a player of type i experiments as compared to a player of type j. The factor ε determines the probability with which players in general experiment. The event that i experiments is assumed to be independent of the event that j experiments for every $i \neq j$. For every i, let $q_i(s|h)$ be the conditional probability that i chooses $\boldsymbol{s} \in \boldsymbol{S}_i$ given that i experiments and the process is in state h, where $\sum_{s \in S_i} q_i(s \mid h) = 1$ for every i and h. We shall assume that $q_i(s|h)$ is independent of t, and that $q_i(s|h) > 0$ for all $s \in S_i$. The latter assumption is made for ease of exposition; similar results hold provided the $q_i(\cdot)$ have enough support that every state is reachable from every other state in a finite number of periods by agents who experiment.

A priori we do not know the distributions $\boldsymbol{q} = (q_1(\cdot), q_2(\cdot), \ldots, q_n(\cdot))$ or the relative probabilities of experimentation $\lambda = (\lambda_1, \lambda_2, \ldots, \lambda_n)$. It turns out, however, that this does not matter. If the overall probability of experimentation ε is small, and if the agents experiment independently of one another, then the selected equilibria are *independent* of $\boldsymbol{q}$ and λ.

The perturbed process may be described as follows. Suppose that the process is in state h at time t. Let J be a subset of j players, $1 \leq j \leq n$. The probability is $\varepsilon^j \left(\Pi_{j \in J} \lambda_j\right) \left(\Pi_{j \notin J} (1 - \varepsilon\lambda_j)\right)$ that exactly the players in J experiment and the others do not. Conditional on this event, the transition probability of moving from h to h' is

$Q^J_{hh'} = \prod_{j \in J} q_j(s_j \mid h) \prod_{j \notin J} p_j(s_j \mid h)$ if h' is a successor of h and s is the rightmost element of h';

$Q^J_{hh'} = 0$ if h' is not a successor of h.

If no agent experiments, then the transition probability of moving from h to h' in one period is $P^0_{hh'}$ as defined in (1). This event has probability $\Pi_{i=1,n}$ $(1 - \varepsilon\lambda_i)$. The perturbed Markov process therefore has the transition function:

$$P^{\varepsilon}_{hh'} = \left(\prod_{i=1,n} (1 - \varepsilon\lambda_i) \right) P^0_{hh'} + \sum_{J \subseteq N, J \neq \phi} \varepsilon^{|J|} \left(\prod_{j \in J} \lambda_j \right) \left(\prod_{j \notin J} (1 - \varepsilon\lambda_j) \right) Q^J_{hh'}. \quad (2)$$

The process P^{ε} will be called *adaptive play with memory m, sample size k, experimentation probabilities* $\varepsilon\lambda_i$ *and experimentation distributions* q_i. Note that P^0 is the process defined in (1), which we shall refer to as the *unperturbed* process.

6 Asymptotic Behavior of Adaptive Play

We shall now characterize the asymptotic behavior of process (2) when the overall probability of experimenting ε is close to zero. Let h and h' be two distinct states. If P^{ε} is in state h at time t, there is a positive probability that all players will experiment for m periods in succession. Thus there is a positive probability that the process arrives at state h' at time $t + m$, so P^{ε} is irreducible. It is aperiodic because the process can move from h to h in exactly m periods, and also in exactly $m + 1$ periods. Hence P^{ε} has a unique stationary distribution μ^{ε} satisfying the equation $\mu^{\varepsilon}P^{\varepsilon} = \mu^{\varepsilon}$. The process is strongly ergodic, and (with probability one) $\mu^{\varepsilon}{}_h$ is the cumulative relative frequency with which state h will be observed when the process runs for a very long time. It is also the probability that h will be observed at any given time t, provided that t is sufficiently large.

The following concept was introduced by Foster and Young (1990).

Stochastic Stability *A state* $h \in H$ *is* stochastically stable *relative to the process* P^{ε} *if* $lim_{\varepsilon\to 0}\mu^{\varepsilon}{}_h > 0$.

Over the long run, states that are not stochastically stable will be observed infrequently compared to states that are, provided that the probability of mistakes ε is small. If there is a *unique* stochastically stable state, then it will be observed almost all of the time when ε is small.

We shall now show how to compute the stochastically stable states of adaptive play for a general n-person game. In the process we shall show that these states are essentially *independent* of the particular mistake distributions q_i and the mistake probabilities λ_i. Then we shall specialize to the case of weakly acyclic games, and show that if $k \leq m/(L_{\Gamma} + 2)$, every stochastically stable state is a convention, and if k and m are sufficiently large, then typically it is unique.

Mistake *Let* h' *be a successor of* h *and let* **s** *be the right-most element of* h'. *A* mistake *in the transition* $h \to h'$ *is a component* s_i *of* **s** *that is not an optimal response by agent* i *to any sample of size* k *from* h.

A mistake can only arise if a player experiments, but an experimental choice need not be a mistake, since it could (by chance) be an optimal choice.

Resistance *For any two states h, h′ the* resistance $r(h, h')$ *is the total number of mistakes involved in the transition* $h \to h'$ *if* h' *is a successor of h; otherwise* $r(h, h') = \infty$.

Let us now view the state space H as the vertices of a directed graph. For every pair of states A, h' insert a directed edge $h \to h'$ if $r(h, h')$ is finite, and let $r(h, h')$ be its "weight" or "resistance." The edges of zero resistance correspond to the transitions that occur with positive probability under P^0. Let $H_1, H_2, \ldots, H_J$ be the *recurrent communication classes* of P^ε. These classes are disjoint, and they are characterized by the following three properties: (i) from every state there is a path of zero resistance to at least one of the classes H_i; (ii) within each class H_i there is a path of zero resistance from every state to every other; (iii) every edge exiting from H_i has positive resistance.

Given any two distinct classes H_i and H_j consider all directed paths that begin in H_i and end in H_j. There is at least one such path, because the perturbed process P^ε is irreducible. Among all such paths, find one with *least total resistance*, and let this resistance be denoted by r_{ij}. Clearly $r_{ij} \geq 0$. Computing r_{ij} amounts to solving a shortest path problem in a directed graph, for which there exist very efficient algorithms. Note that r_{ij} is independent of which vertex in H_i is the starting point and which vertex in H_j is the termination point, because every two states *within* the same class are accessible from each other by paths of zero resistance.

Now define a new directed graph G as follows: there is one vertex i for each recurrent communication class H_i, and for every distinct $1 \leq i, j \leq J$ the directed edge (i, j) has "weight" or *resistance* r_{ij}. The following concept is due to Friedlin and Wentzell (1984).

i-TREE: An i-tree *in G is a spanning tree such that from every vertex* $j \neq i$ *there is a unique path directed from j to i.*

For every vertex i let $\mathcal{T}_i$ be the set of all i-trees on G. The *resistance* of an i-tree $\tau \in \mathcal{T}_i$ is the sum of the resistances of its edges,

$$r(\tau) = \sum_{(i,j) \in \tau} r_{ij} \,. \tag{3}$$

Stochastic Potential *The* stochastic potential *of the recurrent class* H_i *is the least resistance among all i-trees:*

$$\gamma_i = \min_{\tau \in \mathcal{T}_i} r(\tau) \,. \tag{4}$$

Computing γ_i for a given set of weights r_{ij} is a standard problem in combinatorial optimization known as the *arborescence problem*. There exist algorithms for solving it in the order of $|J|^2$ steps (see Chu and Liu (1965), Edmonds (1967), and Tarjan (1977)). Since there are $|J|$ vertices in G, and one arborescence problem must be solved for each, the potential function can be computed in $O(|J|^3)$ steps.

Note that the numbers r_{ij} depend only on the *number* of mistakes in making various transitions, not on the relative probability with which specific mistakes are made. Hence the potential function is independent of the parameters λ_i and q_i. This is important, for in applications one would rarely know the relative probabilities of various mistakes, only that they are possible. What matters is that the probability of mistakes is small, that all mistakes are possible, and the agents make them independently of one another.

Theorem 2 *Let Γ be an n-person game on a finite strategy space. The stochastically stable states of adaptive play P^ε are the states contained in the recurrent communication classes of P^0 with minimum stochastic potential. These states are independent of the experimentation probabilities λ_i and the experimentation distributions q_i so long as they have full support.*

Corollary *If Γ is weakly acyclic and $k \leq m/(L_\Gamma + 2)$, the stochastically stable states of adaptive play are the convention(s) of minimum stochastic potential.*

Theorem 2 follows from a general theorem on perturbed Markov processes that we prove in the Appendix. The corollary is a direct consequence of Theorems 1 and 2.

The stochastically stable states are computed in three steps. First we identify the recurrent communication classes of the process P^0 without mistakes. For a general n-person game these classes can be quite complex. If the game is weakly acyclic and the sampling is sufficiently incomplete, however, then Theorem 1 tells us that the recurrent classes correspond one-to-one with the strict pure strategy Nash equilibria, which are easy to identify. The second step is to compute the least resistance in moving from every recurrent class to every other. In theory this involves solving a series of shortest path problems, but in practice the computation can often be made directly from the payoff matrix of the game. The third and final step is to construct a complete directed graph with these resistances as weights, and to find the arborescence having least weight. This identifies the stochastically stable convention(s), which is unique except in the event of ties. In the remainder of the chapter we shall illustrate

Table 2.1 A succession of k' mistakes by row causes the process to converge to h_2.

Period	1	$2 \ldots m$	$m+1$	$m+2 \ldots$	$m+k'$	$m+k'+1$	...	$m+k'+k$	$m+k'+k+1\ldots$
Row	1	1...1	2*	2*	2*	1	...	1(2)	2
Column	1	1...1	1	1	1	2		2	2

Note: 2* Denotes a mistaken choice of 2, 1(2) an optimal choice of either 1 or 2.

the technique for 2×2 and 3×3 matrix games, and show how the stochastically stable equilibria relate to standard concepts of equilibrium selection such as risk dominance.

7 The 2×2 Case

Let Γ be a 2×2 matrix game with two strict Nash equilibria in pure strategies. It is clear that Γ is acyclic and $L_\Gamma = 1$. Without loss of generality we may write Γ in the form

	1	2
1	a_{11}, b_{11}	a_{12}, b_{12}
2	a_{21}, b_{21}	a_{22}, b_{22}

where $a_{11} > a_{21}, b_{11} > b_{12}, a_{22} > a_{12}$, and $b_{22} > b_{21}$. The strict, pure strategy Nash equilibria are (1,1) and (2,2). Theorem 1 implies that, if $k \leq m/3$, adaptive play without mistakes has two absorbing states: h_1 = ((1,1), (1,1),. . ., (1,1)) and h_2 = ((2,2), (2,2),. . ., (2,2)). To determine which of these states is stochastically stable, we must compute the path of least resistance from h_1 to h_2, and the path of least resistance from h_2 to h_1.

Let h_1 be the state at time $t = m$. To go from h_1 to h_2 requires that at least one player choose strategy 2 by mistake. Moreover, he must choose strategy 2 so often that the other's optimal reply is also strategy 2 for at least one sample of size k, for otherwise the process cannot lock in to the absorbing state h_2.

Suppose, for example, that Row chooses 2 by mistake from periods $t = m + 1$ to $t = m + k'$ inclusive, where $k' \leq k$. From then on Row makes no further mistakes. These choices are marked 2* in Table 2.1.

If Column draws a sample that includes these k' choices of 2, as well as $k - k'$ choices of 1, then Column's best reply is 2 provided that

$$\left(1 - k'/k\right) b_{12} + \left(k'/k\right) b_{22} \geq \left(1 - k'/k\right) b_{11} + \left(k'/k\right) b_{21} \ ;$$

that is,

$$k' \geq \frac{b_{11} - b_{12}}{b_{11} - b_{12} - b_{21} + b_{22}} k \,. \tag{5}$$

If equality holds in (5) then strategy 2 is among Column's best replies, so Column will play it with positive probability.

Suppose that (5) holds and that Column's sample happens to include Row's mistakes in every period from $m + k' + 1$ to $m + k' + k$ inclusive. This event has positive probability provided that m is sufficiently large relative to k. (It suffices that $m \geq 2k$.) Then Column's best reply is to play 2 from periods $m + k' + 1$ to $m + k' + k$ and none of these choices are mistakes. In period $m + k' + k + 1$ Row may sample Column's choices of 2, while Column samples Row's choices of 2, in which case their best replies are to play 2. In the next period there is again a positive probability that both sample enough choices of 2 to want to play 2 again, and so forth. So with positive probability the process converges to the absorbing state h_2 with no further mistakes. In other words, k' mistakes are sufficient to move the process from h_1 to h_2 provided that k' satisfies (5) and m/k is large enough.

Similarly, the process converges with positive probability to h_2 if Column chooses 2 by mistake k'' times, where

$$k'' \geq \frac{a_{11} - a_{21}}{a_{11} - a_{12} - a_{21} + a_{22}} k \,.$$

Let

$$R_1 = \min\left\{\frac{a_{11} - a_{21}}{a_{11} - a_{12} - a_{21} + a_{22}}, \frac{b_{11} - b_{12}}{b_{11} - b_{12} - b_{21} + b_{22}}\right\}.$$

For every real number x, let $[x]$ denote the least integer greater than or equal to x. We have just shown that the resistance in going from h_1 to h_2 is $[R_1 k]$. A similar argument shows that the resistance in going from h_2 to h_1 is $[R_2 k]$ where

$$R_2 = \min\left\{\frac{a_{22} - a_{12}}{a_{11} - a_{12} - a_{21} + a_{22}}, \frac{b_{22} - b_{21}}{b_{11} - b_{12} - b_{21} + b_{22}}\right\}.$$

If $R_1 > R_2$, then (1,1) *risk dominates* (2,2) in the terminology of Harsanyi and Selten (1988). The pair (1,1) *weakly risk dominates* (2,2) if $R_1 \geq R_2$. If $R_1 > R_2$, then the unique stochastically stable convention is h_1 for all sufficiently large values of k and m/k. If $R_1 = R_2$, then both h_1 and h_2 are stochastically stable conventions for all sufficiently large values of k and m/k.

Note that the discrimination of the process grows with the sample size, because the players can only respond to frequency distributions that involve integers between 0 and k. For all sufficiently large k, the process can discriminate any difference in resistance between the two equilibria. This leads to the following definition.

Generic Stability *A strict pure strategy Nash equilibrium is* generically stable *if the associated convention is stochastically stable for all sufficiently large k and m such that $k \leq m/(L_\Gamma + 2)$.*

Theorem 3 *Let Γ be a 2×2 matrix game with two strict Nash equilibria in pure strategies. The generically stable equilibria are the weakly risk dominant Nash equilibria.*

Kandori, Mailath, and Rob (1993) obtain a similar result for symmetric games using a somewhat different dynamic adjustment process. In their model, there is a single homogeneous population of N individuals who play a symmetric 2×2 game. At each time period t, z_t is the current number of players who have "adopted" strategy 1, and $N - z_t$ is the number who have adopted strategy 2. For $i = 1,2$ let $\pi_i(z_t)$ be the total payoff to strategy i in state z_t when every player is matched once against every other. The dynamical assumption is that players adopt successful strategies more often than unsuccessful strategies. That is, there is a deterministic dynamic $z_{t+1} = f(z_t)$ such that

$$\text{for all } 0 \leq z_t < N, \quad z_{t+1} \geq z_t \quad \text{if and only if } \pi_i(z_t) \geq \pi_2(z_t).$$

In addition, there is a small probability ε that each player will switch from playing strategy 1 to strategy 2 or vice versa, where the switches are independent across players. Using similar techniques to those developed here, Kandori, Mailath, and Rob show that the resulting irreducible, aperiodic process $P(N, \varepsilon)$ has a stationary distribution that, for all sufficiently large N, puts all of the probability on the risk dominant equilibria as ε goes to zero.

8 The 3×3 Case

When the agents have three or more strategies, there is no simple formula analogous to risk dominance that identifies the stochastically stable equilbria. First, the path of least resistance must be computed from every equilibrium to every other. Then a minimum arborescence problem must be solved for each equilibrium. We shall illustrate by solving an example.

Example 3

	1	2	3
1	6, 6	0, 5	0, 0
2	5, 0	7, 7	5, 5
3	0, 0	5, 5	8, 8

The pairs (i, i) are the strict, pure strategy Nash equilibria, $i = 1, 2, 3$. Let h_i denote the convention in which (i, i) is played m times in succession. Theorem 1 says that these are the absorbing states of the unperturbed process provided that $k \leq m/3$. Let us compute the path of least resistance from every convention to every other.

Suppose that the perturbed process is in state h_1. To exit to h_2 or h_3, one agent must choose a sufficient number of 2s or 3s (or both) to cause the other agent to choose 2 or 3. Since the game is symmetric, it does not matter which player makes the mistakes and which player reacts. Assume that the Column player chooses 2 at least $k'' = [(1/8)k]$ times in succession. If Row samples these choices (plus $k - k''$ choices of strategy 1), then Row's best reply is also strategy 2. At this point there is a positive probability that the process will converge to h_2 with no further mistakes. (This relies on the assumption that $k \leq m/3$.) It may be checked that this is the least number of mistakes to go from h_1 to h_2 by any route. Thus the least resistant path is *direct* in the sense that it only involves strategies 1 and 2.

Not all paths of least resistance are direct, however. For example, suppose that the process is in state h_3 and we want to exit to state h_2. The direct route is for one player to choose strategy 2 by mistake at least $[(3/5)k]$ times, which causes the other player to reply with strategy 2. But if one player chooses strategy 1 by mistake at least $[(3/8)k]$ times (and at most $(5/6)k$ times), then the best reply of the other player is strategy 2. Thus the indirect route has lower resistance when k is large enough.

Another example of an indirect route involves the transition from h_1 to h_3. The direct route requires some player to make $[(5/8)k]$ mistaken choices of strategy 3, since otherwise strategy 1 or 2 is a better reply. But there is a path of less resistance, namely, go first to h_2 and then to h_3. The total resistance of this path is $[(1/8)k] + [(2/5)k]$, which is approximately $(21/40)k$ when k is large. The resistances between every pair of equilibria are shown in Figure 2.2.

For each vertex h_i there are three h_i-trees, and the h_i-tree of least resistance determines the stochastic potential of h_i, as shown in Figure 2.3.

It is readily seen that the least resistant tree is rooted at h_2 and has total resistance $[(1/8)k] + [(3/8)k]$. For all sufficiently large k this is the unique tree of least resistance, so h_2 is the unique generically stable convention. The risk

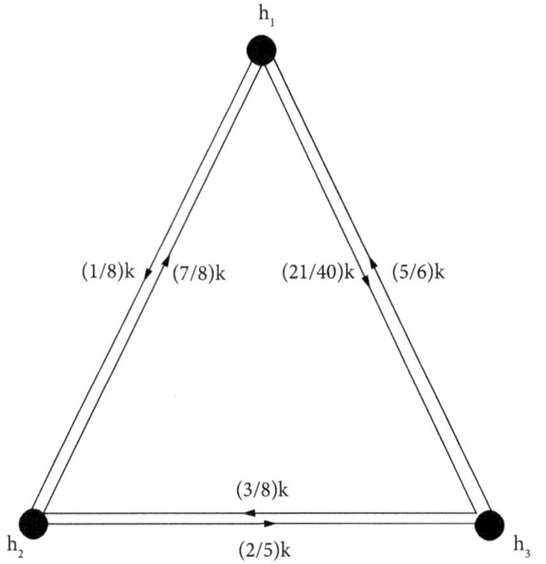

Figure 2.2 Pairwise resistances (unrounded) for the pure strategy equilibria of Example 3.

dominant equilibrium, however, is strategy 3. It is also the Pareto optimal equilibrium. To verify risk dominance, one checks that strategy 3 risk dominates strategy 2 in the subgame restricted to strategies 2 and 3, and that strategy 3 risk dominates strategy 1 in the subgame restricted to strategies 1 and 3. (See Harsanyi and Selten (1988).)

There are essentially two respects in which risk dominance and stochastic stability differ. First, they employ different notions of resistance. Strategy i risk dominates j if it requires fewer mistakes to go from j to i than from i to j within the subgame consisting of just these two strategies (i.e., by a direct route). Stochastic stability requires us to look at all transitions from i to j, including those that involve other strategies. The second distinction is that risk dominance is only defined when there is one strategy that risk dominates *every* other in pairwise comparisons. Such a strategy may not exist because of cycles in the pairwise comparisons. Stochastic stability, by contrast, relies on a global criterion (the resistance of spanning trees) to compare the stability of different strategies. The difference between the two concepts may therefore be summarized as follows: risk dominance selects the equilibrium that is easiest to flow into from every other equilibrium considered in isolation (assuming such an equilibrium exists). Stochastic stability selects the equilibrium that is easiest to flow into from all other states combined, including both equilibrium and nonequilibrium states.

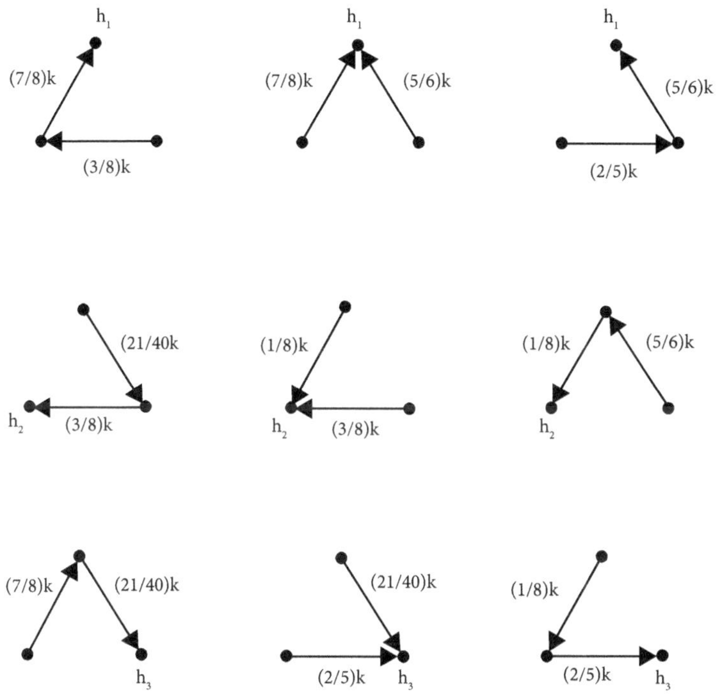

Figure 2.3 The nine h_i-trees ($i = 1, 2,3$). Resistances are not rounded.

9 Conclusion

In this chapter we have shown how an equilibrium can evolve in a game that is played repeatedly by different agents. The model is similar to fictitious play in that agents' expectations are shaped by precedent. It differs in that agents base their choices on an incomplete knowledge of recent precedents and they occasionally make mistakes. These assumptions seem more natural than the deterministic dynamics of fictitious play, so we can justify them on the grounds of realism. They also play an important technical role: by introducing noise into the dynamic adjustment process, they select among pure strategy Nash equilibria for weakly acyclic games, and among more complex regimes for general n-person games. Unlike other evolutionary models, these perturbations are not one-shot affairs but form an integral part of the dynamics. By incorporating noise directly into the dynamics, one is led to a different criterion of equilibrium selection than classical notions like evolutionary stability and risk dominance.

Several questions remain to be explored. One is the sensitivity of the equilibrium selected to the way in which the model is specified. We have shown that the stochastically stable equilibria are invariant with respect to the distribution of the perturbations so long as they are independent across players, have positive support, and are stationary. In addition, we showed that for 2×2 games the stochastically stable equilibria are independent of m and k so long as m/k and k are sufficiently large. It is not clear whether this result holds for weakly acyclic games in general, although we know of no examples in which it fails to hold.

A second question is whether the stochastically stable equilibria can be characterized more succinctly. The algorithm described in the Appendix shows how to compute the stochastically stable equilibria in a wide class of dynamical models with perturbations, but it does not describe these equilibria in terms of a simple formula. There is no reason to think that such a formula exists in the general case. In specific classes of games, however, one can sometimes exploit the payoff structure to obtain more specific answers. Kandori and Rob (1992) explore this issue for pure coordination games and differentiated-product oligopoly games.

Another class of games where one can obtain an explicit formula for the stochastically stable equilibria is the bargaining problem. Consider the two-person Nash demand game in which each player demands a share of a fixed pie. They get their shares if the shares sum to 1 or less; otherwise they get nothing. The strategy spaces can be made discrete by assuming that the shares are rounded to a large, fixed number of decimal places. Let two disjoint populations of agents play this game adaptively, where all members of one population have utility function u and the others have utility function v. Then the stochastically stable equilibria are close to the Nash solution. (See Young (1993).)

A third issue is whether and how the stochastically stable equilibria change when the agents are allowed more decision-making scope. For example, what happens when the agents learn as they play the game over time, or make inferences about the others' decision rules, or choose optimal sample sizes based on their costs of gathering information? These additions complicate both the state space and the stochastic process. They may also require more common knowledge on the part of the agents. In any event, if the agents make mistakes infrequently and independently of each other, then the stochastically stable states can still be analyzed using the general techniques developed here. We have deliberately chosen to focus on the case where agents do *not* learn in order to show that convergence to equilibrium can occur with no common knowledge and with only a minimum degree of rationality on the part of the agents. Society can "learn" even when its members do not.

Appendix

Here we shall prove a general result on finite Markov chains of which Theorem 2 is a special case. This result amounts to a finite version of results obtained by Freidlin and Wentzell for continuous diffusion processes (see Freidlin and Wentzell (1984, pp. 186–187)). Let P^0 be a stationary Markov chain defined on a finite state space X. Suppose that this process is subjected to a small perturbation or noise. By this we mean that with high probability the process follows the transition function P^0, but with small probability certain transitions occur that could *not* have occurred via P^0. We shall assume that the perturbed process can be modeled as a stationary Markov chain on X with transition function P^ε, where ε is a scalar parameter that measures the overall level of noise, ε takes on all values in some interval $(0, a]$, and the following conditions hold for all $x, y \in X$:

$$P^\varepsilon \text{is aperiodic and irreducible for all } \varepsilon \in (0, a], \tag{6}$$

$$\lim_{\varepsilon \to 0} P^\varepsilon_{xy} = P^0_{xy} \tag{7}$$

$$P^\varepsilon_{xy} > 0 \text{ for some } \varepsilon \text{ implies } \exists r \geq 0 \text{ s.t. } 0 < \lim_{\varepsilon \to 0} \varepsilon^{-r} P^\varepsilon_{xy} < \infty. \tag{8}$$

Condition (6) implies that the perturbed process has a unique stationary distribution μ^ε for every $\varepsilon \in (0, a]$. Condition (7) says that the perturbed process converges to the unperturbed one. Condition (8) says that the transition $x \to y$ is either impossible in the perturbed process P^ε for all $\varepsilon \in (0, a]$, or P^ε_{xy} is of order ε^r for some unique real number $r \geq 0$ as ε becomes small. In the latter case we set $r(x, y) = r$ and call $r(x, y)$ the *resistance* of the transition $x \to y$. By virtue of (7), $r(x, y) = 0$ if and only if $P^0_{xy} > 0$. Thus the transitions of zero resistance are the same as the feasible transitions under P^0. Any family of Markov processes P^ε satisfying (6)–(8) will be called a *regular perturbation* of P^0.

The family P^ε defined by (2) in the text is a regular perturbation of the process P^ε defined in (1), and the resistance of a one-period transition is the minimum number of mistakes required to make it.

By hypothesis, the perturbed process P^ε is aperiodic and irreducible, so it has a unique stationary distribution μ^ε for every $\varepsilon > 0$. The unperturbed process, by contrast, may have many stationary distributions. We are going to show that $\lim_{\varepsilon \to 0} \mu^\varepsilon = \mu^0$, where μ^0 is one of the stationary distributions of P^0. Thus the perturbations effectively *select* among the stationary distributions of P^0. The support of the stationary distribution μ^0 is a subset of the recurrent communication classes of P^0. Thus the perturbations effectively select among the recurrent communication classes. In fact, the perturbations typically select *exactly one* communication class of P^0. The selected class (or classes) are computed by finding a path of least resistance from every recurrent communication class to every other. Hence the selected classes depend only on the resistances $r(x, y)$, that is, only on the *order of magnitude* of the various perturbations. This is important in applications, where the general form of the perturbations may be known, but not their precise values.

To characterize the limiting distribution μ^0, we shall define two directed graphs. The first graph G has vertex set X (the set of all states) and there is a directed edge from state x to state y if and only if the one-period transition $x \to y$ has positive probability under P^ε for all sufficiently small $\varepsilon > 0$. In this event, let $r(x, y)$, as defined by (8), be the weight or *resistance* of the directed edge (x, y).

Let the recurrent communication classes of P^0 be denoted by $X_1, \ldots, X_J$. These classes can be characterized within the graph G as follows: (i) from every vertex there exists a path of zero resistance to at least one of the classes X_i; (ii) for every two vertices x and y within the same class X_i there is a path of zero resistance from x to y and vice versa; (iii) every edge from a vertex in X_i to a vertex not in X_i has positive resistance.

For every $i \neq j$ let r_{ij} be the *least resistance* among all directed paths that begin in X_i and end in X_j. (For this purpose it is sufficient to fix any two states $x \in X_i$ and $y \in X_j$ and find a least-resistant path from x to y.) This is well-defined because there is at least one path from every class to every other, by virtue of the assumption that P^ε is irreducible when $\varepsilon > 0$.

Now define a second graph $\mathcal{G}$ as follows: the vertices are the indices $\{1,2,\ldots, J\}$, and for each pair (i, j) there is a directed edge from i to j with weight r_{ij}. $\mathcal{G}$ is normally much smaller than G; in fact it may have only a few vertices. In the case of a 2×2 coordination game, for example, adaptive play has only two recurrent classes (assuming $k \leq m/3$), so $\mathcal{G}$ has exactly two vertices.

Define a *j-tree* in $\mathcal{G}$ to be a spanning subtree of $\mathcal{G}$ such that for every vertex $i \neq j$ there exists exactly one directed path from i to j. Let $\mathcal{T}_j$ be the set of all j-trees in $\mathcal{G}$. For each j find a j-tree of least total resistance, and let this resistance be denoted by γ_j. γ_j is the *stochastic potential* of the class X_j.

We shall now prove the following result, of which Theorem 2 is a special case.

Theorem 4 *Let P^0 be a stationary Markov process on the finite state space X with recurrent communication classes $X_1, \ldots, X_J$. Let P^ε be a regular perturbation of P^0, and let μ^ε be its unique stationary distribution for every small positive ε. Then*

(i) *as $\varepsilon \to 0$, μ^ε converges to a stationary distribution μ^0 of P^0.*
(ii) *x is stochastically stable ($\mu^0_x > 0$) if and only if x is contained in a recurrent class X_j that minimizes γ_j.*

Freidlin and Wentzell (1984, pp. 186–187) prove an analogous result when the unperturbed process is a deterministic dynamical system described by a differential equation on a manifold, and the perturbed process is a family of diffusion processes whose drift converges to that of the unperturbed process as the diffusion goes to zero. The minimum resistance between two states is found by integrating a certain "action functional" along all continuous paths from one state to the other and taking the infimum over all such paths. (See Freidlin and Wentzell (1984, p. 161).) However, their result requires numerous regularity conditions not needed here (Freidlin and Wentzell 1984, pp. 155 and 169), and in any event the characterization given in Theorem 4 is much simpler analytically.

The proof will be divided into two lemmas. In the first, we establish statement (i) and show how to characterize the stochastically stable states by solving a series of arborescence problems in the graph G. In the second lemma we show that this characterization can be reduced to the simpler problem of solving arborescence problems in the (typically) much smaller graph $\mathcal{G}$.

We begin by characterizing the stochastically stable states in terms of the graph G, whose vertex set is the whole state space X. For any vertex z of G a *z-tree* T is a spanning tree in G such that, for every vertex $x \neq z$, there exists a unique directed path from x to z. Let $\mathcal{T}_z$ be the set of all z-trees in G and define

$$\gamma(z) = \min_{T \in \mathcal{T}_z} \sum_{[x,y] \in T} r(x, y) . \tag{9}$$

The following result generalizes Theorem 1 in Kandori, Mailath, and Rob (1993), and utilizes a lemma on Markov chains due to Freidlin and Wentzell.[8]

[8] Using different methods, Canning (1992) proves that μ^ε converges to a stationary distribution μ^0 of P^0 under more general conditions, but he does not characterize the support of the limiting distribution.

Lemma 1: *Let P^{ε} be a regular perturbation of P^0 and let μ^{ε} be its stationary distribution. Then $\lim_{\varepsilon\to 0}\mu^{\varepsilon} = \mu^0$ exists and μ^0* is a stationary distribution of P^0. Moreover, $\mu_x^0 > 0$ if and only if $\gamma(x) \le \gamma(y)$ *for all* y *in* X.

Proof: Let P' be the transition function of any aperiodic, irreducible, stationary Markov process defined on the finite space X. For each $z \in X$, define the number

$$p'_z = \sum_{T\in\mathcal{T}_z} \prod_{(x,y)\in T} P'_{xy},$$

where p'_z is positive because P' is irreducible. Let

$$\mu'_z = p'_z / \left(\sum_{x\in X} p'_x\right) > 0.$$

It may then be verified that $\mu' P' = \mu'$, from which it follows that μ' is the unique stationary distribution of, . (See Freidlin and Wentzell (1984, Chapter 6, Lemma 3.1).)

Now let us apply this result to the process P^{ε} hypothesized in Lemma 1. Let

$$p_z^{\varepsilon} = \sum_{T\in\mathcal{T}_z} \prod_{(x,y)\in T} P_{xy}^{\varepsilon}. \tag{10}$$

By the above result, the stationary distribution of P^{ε} is given by the formula

$$\mu_z^{\varepsilon} = p_z^{\varepsilon} / \sum_{x\in X} p_x^{\varepsilon}. \tag{11}$$

Define $\gamma(z)$ as in (9) and let $\gamma* = \min_z \gamma(z)$. We are going to show that $\mu^{\varepsilon} > 0$ if and only if $\gamma(z) = \gamma*$. Choose a z-tree T with resistance $\gamma(z)$ and consider the identity

$$\varepsilon^{-\gamma^*} \prod_{(x,y)\in T} P_{xy}^{\varepsilon} = \varepsilon^{r(T)-\gamma^*} \prod_{(x,y)\in T} \varepsilon^{-r(x,y)} P_{xy}^{\varepsilon}. \tag{12}$$

By (8)

$$\lim_{\varepsilon\to 0} \varepsilon^{-r(x,y)} P_{xy}^{\varepsilon} > 0 \text{ for every } (x,y) \in T. \tag{13}$$

If $r(T) = \gamma(z) > \gamma^*$, it follows from (12) and (13) that

$$\lim_{\varepsilon\to 0} \varepsilon^{-\gamma^*} \prod_{(x,y)\in T} P_{xy}^{\varepsilon} = 0,$$

so

$$\lim_{\varepsilon\to 0} \varepsilon^{-\gamma^*} p_z^\varepsilon = 0.$$

Similarly, if $r(T) = \gamma(z) = \gamma^*$ we obtain

$$\lim_{\varepsilon\to 0} \varepsilon^{-\gamma^*} p_z^\varepsilon > 0. \tag{14}$$

From (13), (14), and the identity

$$\mu_z^\varepsilon = \varepsilon^{-\gamma^*} p_z^\varepsilon / \sum_{x\in X} \varepsilon^{-\gamma^*} p_x^\varepsilon,$$

it follows that

$$\lim_{\varepsilon\to 0} \mu_z^\varepsilon = 0 \text{ if } \quad \gamma(z) > \gamma^*$$

and

$$\lim_{\varepsilon\to 0} \mu_z^\varepsilon > 0 \text{ if } \quad \gamma(z) = \gamma^*.$$

Thus we have shown that $\lim_{\varepsilon\to 0} \mu^\varepsilon$ exists, and its support is the set of states z that minimize $\gamma(z)$. Since μ^ε satisfies the equation $\mu^\varepsilon P^\varepsilon = \mu^\varepsilon$ for every $\varepsilon > 0$, it follows from assumption (7) that $\mu^0 P^0 = \mu^0$. Hence μ^0 is a stationary distribution of P^0. This completes the proof of Lemma 1.

Since μ^0 is a stationary distribution of P^0, ${\mu^0}_z = 0$ for every state z that is not recurrent under P^0. To find the stochastically stable states, it therefore suffices to compute the potential function only on the recurrent states.

It is easy to see that all states in the same recurrent communication class have the same potential. Suppose, in fact, that x is in X_j and T is an x-tree in G with potential $\gamma(x)$. Let y be some other vertex in X_j. Choose a path of zero resistance from x to y. The union of this path and T contains a y-tree T' that has the same resistance as T. From this it follows that the potential of y is no larger than the potential of x, and a symmetric argument shows that it is no smaller. Hence they have the same potential.

We shall now show that the potential on each recurrent class X_j is precisely γ_j, namely, the least resistance among all j-trees in the graph $\mathcal{G}$. Thus the potential may be computed simply by solving an arborescence problem on the reduced graph $\mathcal{G}$.

Lemma 2: $\gamma(x) = \gamma_j$is the stochastic potential of all states $x \in X_j$.

Proof: Fix one state x_j in each recurrent class X_j. We shall show first that $\gamma(x_j) \leq \gamma_j$; then we shall show the reverse inequality.

Fix a class X_j and a j-tree τ in $\mathcal{G}$ whose resistance $r(\tau)$ equals $\gamma_{j'}$. For every $i \neq j$, there exists exactly one outgoing edge $(i, i) \in \tau$. In the graph G (whose vertices are the states) choose a directed path $D_{ii'}$ from x_i to $x_{i'}$ having resistance $r_{ii'}$. Now choose a directed subtree T_i that spans the vertex set X_i such that from every vertex in X_i there is a unique directed path to x_i. Since X_i is a communication class of P^0, T_i can be chosen so that it has zero resistance.

Let E be the union of all of the edges in the trees T_i and all of the edges in the directed paths $D_{ii'}$ where $(i, i') \in \tau$. By construction, E contains at least one directed path from every vertex in X to the fixed vertex x_j. Therefore it contains a subset of edges that form an x_j-tree T in G. The resistance of T is clearly less than or equal to the sum of the resistances of the paths $D_{ii'}$, so it is less than or equal to $r(\tau)$. Thus $\gamma(x_j) \leq \gamma_j = r(\tau)$ as claimed.

To show that $\gamma(x_j) \geq \gamma_j$, fix a class j and a least-resistant x_j-tree T among all x_j-trees on the vertex set X. Label each of the specially chosen vertices x_i by the class "i" to which it belongs. These will be called *special vertices.* A *junction* in T is any vertex y with at least two incoming T-edges. If the junction y is not a special vertex, label it "i" if there exists a path of zero resistance from y to X_i. (There exists at least one such class because they are the recurrent classes of P^0; if there are several such classes choose any one of them as label.) Every labeled vertex is either a special vertex or a junction (or both), and the label identifies a class to which there is a path of zero resistance.

Define the *special predecessors* of a state $x \in X$ to be the special vertices x_i that strictly precede x in the fixed tree T (i.e., such that there is a path from x_i to x in T) and such that there is no other special vertex x_j on the path from x_i to x.

If x_i is a special predecessor of a labeled vertex x, then the unique path in the tree from x_i to x has resistance at least r_{ik}, where k is the label of x. (15)

Property (15) clearly holds for the tree T because any path from the special vertex x_i to a vertex labeled "k" can be extended by a zero resistance path to the class X_k, and the total path must have resistance at least r_{ik}. We shall now perform certain operations on the tree T that preserve property (15), and that bring it into a form that is more or less congruent to a j-tree in $\mathcal{G}$. We shall do this by successively eliminating all junctions that are not special vertices.

Suppose that T contains a junction y that is not a special vertex, and let its label be "k." We distinguish two cases, depending on whether the special vertex x_k is or is not a predecessor of y in the tree.

> *Case 1*: If x_k is not a predecessor of y in the tree (see Figures 2.4a and 2.4b), cut off the subtree consisting of all edges and vertices that precede y and glue them onto the tree at the vertex x_k.
> *Case 2*: If x_k is a predecessor of y (see Figures 2.5a and 2.5b), cut off the subtree consisting of all edges and vertices that precede y (*except* for the path from x_k to y and all of its predecessors) and glue them onto x_k.

Both of these operations preserve property (15) because x_k and y have the same label. Each operation reduces by one the number of junctions that are not special vertices. Thus we eventually obtain an x_j-tree T^* in which every junction is a special vertex, property (15) is satisfied, and T^* has the same resistance as the original tree T.

Now construct a j-tree τ on the vertex set J as follows. For every two classes i and i' put the directed edge (i, i') in τ if and only if x_i is a special predecessor of $x_{i'}$ in T^*. By construction, τ forms a j-tree. Let $D_{ii}{}^*$ be the unique path in T^* from x_i to $x_{i'}$. By property (15) its resistance is at least $r_{ii'}$. The paths $D_{ii}{}^*$ are edge-disjoint because every junction is one of the special vertices. Since T^* contains their union, the resistance of T^* is at least $\sum_{(i,i')\in\tau} r_{ii'}$. But $\sum_{(i,i')\in\tau} r_{ii'}$ is the resistance of τ. Hence $\gamma(x_j) = r(T^*) \geq \sum_{(i,i')\in\tau} r_{ii'} = r(\tau) \geq \gamma_j$ as claimed.

This completes the proof of Lemma 2, which, together with Lemma 1, establishes Theorem 4.

Manuscript received November, 1989; final revision received March, 1992.

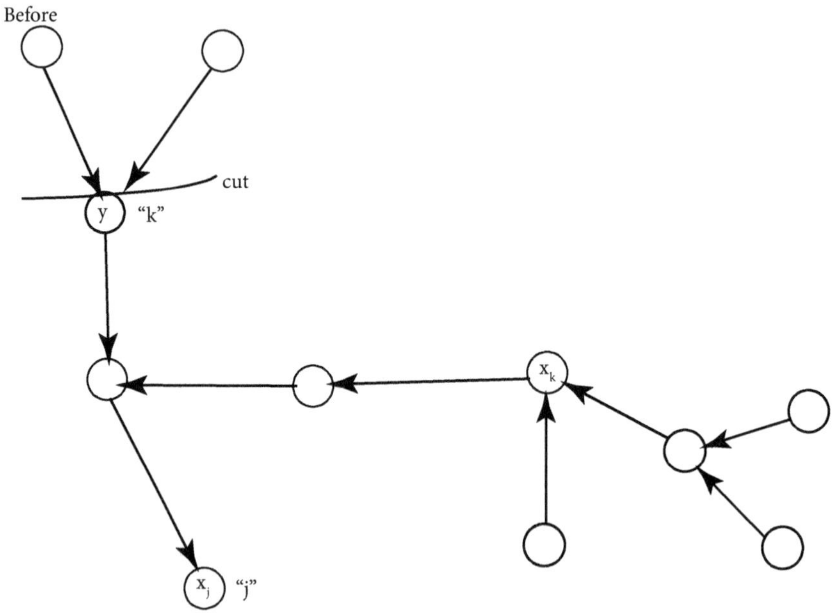

Figure 2.4a Case 1 surgery: before.

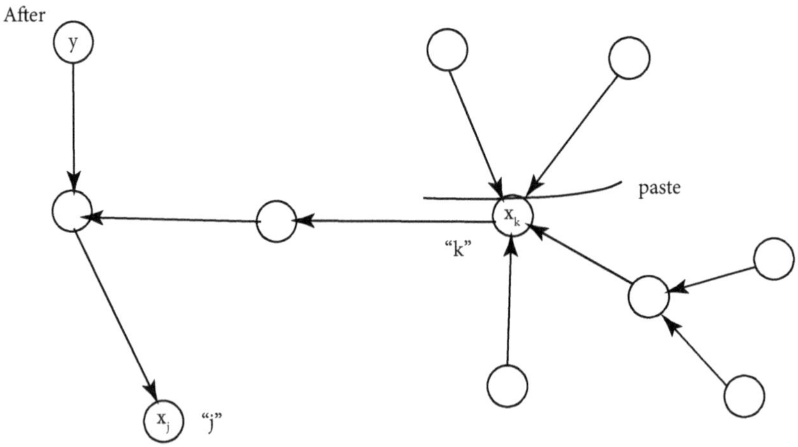

Figure 2.4b Case 1 surgery: after.

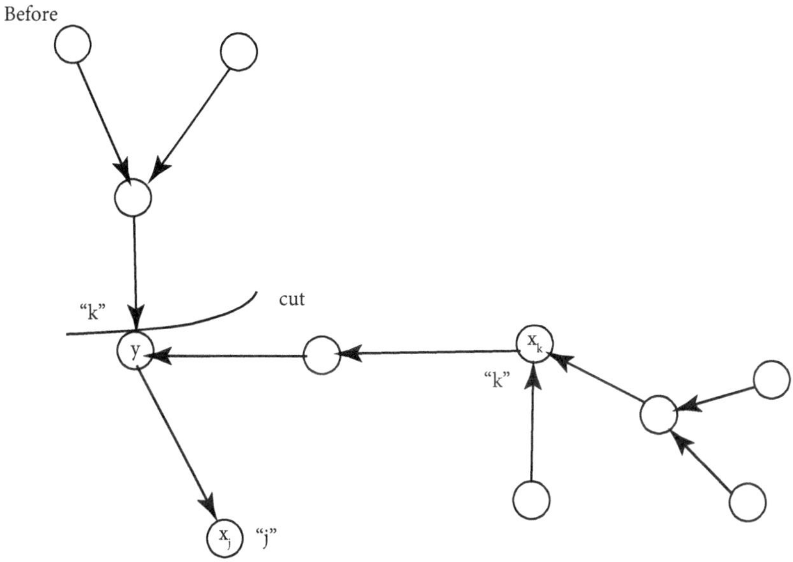

Figure 2.5a Case 2 surgery: before.

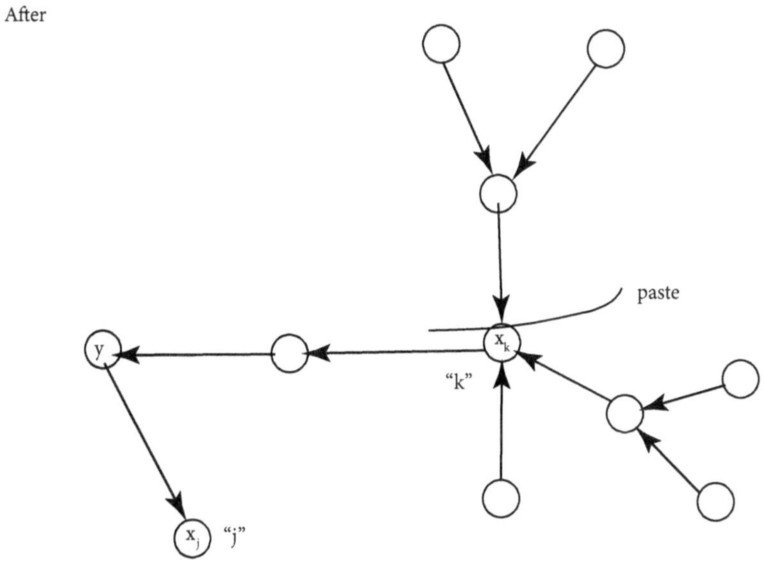

Figure 2.5b Case 2 surgery: after.

References

Axelrod, Robert (1984), *The Evolution of Cooperation.* New York: Basic Books.
Axelrod, Robert. (1986), "An evolutionary approach to norms," *American Political Science Review* 80: 1097–1111.
Bardhan, Pranab (1984), *Land, Labor, and Rural Poverty.* New York: Columbia University Press.
Bicchieri, Christina (1990), "Norms of cooperation," *Ethics* 100: 838–861.
Canning, David (1992), "Average behavior in learning models," *Journal of Economic Theory* 57: 442–472.
Chu, Y. J. and T. H. Liu (1965), "On the shortest arborescence of directed graphs," *Scientia Sinica* 14: 1390–1400.
Crawford, Vincent (1991), "An evolutionary interpretation of Van Huyck, Battalio, and Beil's experimental results on coordination," *Games and Economic Behavior* 3: 25–59.
Edmonds, Jack (1967), "Optimum branchings," *Journal of Research of the National Bureau of Standards* 71B: 233–240.
Evans, George and S. Honkapohja (1995a), "Increasing social returns, learning, and bifurcation phenomena," in Alan Kirman and Mark Salmon (eds.), *Learning and Rationality in Economics.* New York: Basil Blackwell.
Evans, George and S. Honkapohja (1995b), "Exceptional stability and adaptive learning: An introduction" in Alan Kirman and Mark Salmon (eds.), *Learning and Rationality in Economics.* New York: Basil Blackwell,.
Foster, Dean and H. Peyton Young (1990): "Stochastic evolutionary game dynamics," *Theoretical Population Biology* 38: 219–232.
Freidlin, Mark and Alexander Wentzell (1984), *Random Perturbations of Dynamical Systems.* New York: Springer Verlag.
Friedman, Daniel (1991), "Evolutionary games in economics," *Econometrica* 59: 637–666.
Fudenberg, Drew and C. Harris (1992), "Evolutionary dynamics in games with aggregate shocks," *Journal of Economic Theory* 57: 420–441.
Fudenberg, Drew and Eric Maskin (1990), "Evolution and cooperation in noisy repeated games," *American Economic Review* 80: 274–279.
Harsanyi, John and Reinhard Selten (1988), *A General Theory of Equilibrium Selection in Games.* Cambridge, MA: MIT Press.
Kandori, Michihiro and Rafael Rob (1992), "Evolution of Equilibria in the Long Run: A General Theory and Applications," manuscript, Princeton University.
Kandori, Michihiro, George Mailath, and Rafael Rob (1993), "Learning, mutation, and long run equilibria in games," *Econometrica* 61: 29–56.
Kirman, Alan (1995), "Ants, rationality, and recruitment," in Alan Kirman and Mark Salmon (eds.), *Learning and Rationality in Economics*,. London: Basil Blackwell.
Lewis, David (1967), *Convention: A Philosophical Study.* Cambridge, MA: Harvard University Press.
Nachbar, John H. (1990), "Evolutionary selection dynamics in games: Convergence and limit properties," *International Journal of Game Theory*: 19, 59–89.
Samuelson, Larry (1988), "Evolutionary Foundations of solution concepts for finite, 2-player, normal-form games," in M. Y. Vardi (eds.), *Theoretical Aspects of Reasoning About Knowledge.* Los Altos: Morgan Kauffman.
Samuelson, Larry (1991a), "Limit evolutionarily stable strategies in two-player, normal form games," *Games and Economic Behavior* 3: 110–128.

Samuelson, Larry (1991b), "How to tremble if you must," manuscript, University of Wisconsin.
Samuelson, Larry, and J. Zhang (1992), "Evolutionary Stability in Asymmetric Games," *Journal of Economic Theory* 57: 363–391.
Schelling, Thomas C. (1960), *The Strategy of Conflict.* Cambridge, MA: Harvard University Press.
Sugden, Robert (1986), *The Evolutions of Rights, Cooperation and Welfare.* New York: Basil Blackwell.
Tarjan, R. E. (1977), "Finding optimum branchings," *Networks* 7: 25–35.
Warneryd, Karl (1990), "Conventions," *Constitutional Political Economy* 1: 83–107.
Young, H. Peyton (1993), "An evolutionary model of bargaining," *Journal of Economic Theory*, 59, 145–168.
Young, H. Peyton and Dean Foster (1991), "Cooperation in the short and in the long run," *Games and Economic Behavior* 3: 145–156.

Preface to Chapter 3
The Spontaneous Emergence of Bargaining Norms

Conventional bargaining theory predicts that the outcome of a bargain depends on the bargainers' degrees of risk aversion, their levels of impatience, and their alternatives to an agreement if negotiations break down. In its modern incarnation these predictions are derived from a model in which the agents are imagined to play an infinitely repeated game in which they alternate making offers and counteroffers. I have already remarked in Chapter 1 on the demanding assumptions needed to characterize the equilibrium outcome of such a model, including common knowledge of the agents' utility functions, their degrees of impatience, and common belief in their rationality. In practice, however, it seems highly implausible that a protagonist would know the other party's level of risk aversion or degree of impatience. They probably do not even have an accurate idea of the distribution of these characteristics in the population.

Leaving aside the implausibility of this scenario, the alternating offers model also predicts that when agents have different levels of risk aversion and impatience (as seems highly likely in practice), the equilibrium outcomes will vary from one pair of bargainers to another. In practice, however, bargains between principals and agents often exhibit a striking degree of uniformity within a given profession and geographical region. This pattern has been documented in a wide variety of settings, including real estate commissions, fund managers' bonuses, lawyers' contingency fees, agricultural cropsharing contracts, and many others (see Chapter 5). Such bargaining norms call for an explanation as they are at odds with the predictions of standard bargaining theory.

Chapter 3 shows how bargaining norms can arise "spontaneously" in a population of interacting agents. The model is an application of the stochastic dynamic framework set out in Chapter 2. Suppose that there are two classes of bargainers who are randomly matched to negotiate their shares of the value created by their cooperative effort. Suppose that with high probability each chooses a best response, based on a random sample of past demands that were made by the other side. (This is a way of representing "partial knowledge" in the sense of Hayek.) Imagine that with some small probability they make idiosyncratic demands that depart from best response behavior. Chapter 3 demonstrates that this process settles into a pattern in which most of the time most pairs of bargainers follow an established norm: every principal demands some fixed share x, every agent demands the complementary share $1 - x$, each expects the other side to make these demands, and hence their demands are in equilibrium. Chapter 3 also shows that, if it should happen that all principals have the same utility function $u(x)$, all agents have the same utility function $v(1 - x)$, and if everyone collects the same amount of information (has the same sample size), then the stochastically stable outcome is the Nash bargaining solution, namely, the division that maximizes $u(x)v(1 - x)$.[1]

There are several key differences between this framework and the standard alternating offers model. First, there is no need to assume that players know the utility functions and time preferences of those with whom they are bargaining. It is enough to know the previous demands of (possibly different) agents in that role, and to choose a best response with high probability. Second, even if the population is completely heterogeneous with respect to the agents' utility functions, time preferences, and sample sizes, the equilibrium outcome will be a social norm that applies to everyone. The norm depends on the *distribution* of these characteristics in the population, but unlike standard models of incomplete information, bargainers do not need to *know* the distribution of these characteristics. They simply adapt their behavior to the observed behavior of others, and this process leads to an outcome that depends on the underlying (but unknown) characteristics of others. In this sense the evolutionary model is much less demanding with respect to the information required of the protagonists than the standard alternating offers model.

In the version of the model just described, players are assumed to be myopic and to naively choose best responses given the observed behavior of their opponents. What happens if players are more sophisticated and think ahead? Is the Nash bargaining solution still selected? Suppose, for example, that some "clever" agents try to *anticipate* how much their opponents are going to demand and choose a best response accordingly (subject to idiosyncratic perturbations).[2] To be concrete, suppose there are two distinct populations of bargainers, the As and the Bs. In each period one A – B pair is chosen at random. Each knows a random sample of past demands by the opposite side.

[1] Variants of this framework are treated by K. Binmore, L. S. Samuelson, and H. P. Young (2003), "Equilibrium selection in bargaining models," *Games and Economic Behavior* 45 (2): 296–328.

[2] The following account is based on M. Saez-Marti and J. Weibull (1999), "Clever agents in Young's evolutionary bargaining model," *Journal of Economic Theory* 86 (2): 268–279.

A naive A-player will choose a best response to a sample distribution of prior demands by the Bs. A sophisticated B-player will anticipate that this is the plan of the naive A-players, which leads her to sample the prior demands of members of her *own* population, and to choose a best response to the own-population sample (subject to perturbations). Saez-Marti and Weibull (1999) show that if players in both populations have the same sample size, but one of the populations (say the Bs) has a positive share of both sophisticated and unsophisticated players, then the outcome is the same as before: the increased sophistication of players on one side does not increase their share of the pie.

More generally, the apparatus of stochastic stability can be adapted to games where some of the agents are not just myopic best-responders; they best-respond to the *anticipated* moves of their opponents. There is, of course, the question of how many steps ahead the agents try to anticipate: do they anticipate that the others will anticipate? Do they anticipate that the others will anticipate that the others will anticipate? These differences in degree of sophistication are captured by the notion of level-k reasoning: level-1 agents naively best respond to the current behavior of others without anticipating that their behavior may change, level-2 agents best-respond to the anticipated best responses of level-1 agents, and so forth. When agents' choices are subjected to stochastic perturbations, one can study the ergodic distribution of the process as the stochastic perturbations tend to zero, under various assumptions about the degree of sophistication among the agents.[3]

More generally, the evolutionary framework developed in this book does not pre-suppose that agents have low rationality or are not forward-looking; it can incorporate different amounts of rationality, different amounts of information, different time preferences and other characteristics that are distributed heterogeneously in the population. Unlike conventional game theory models, there is no need to assume that the distribution of these characteristics is known to the players themselves. The process of adjustment leads to outcomes that depend on the underlying distribution of these characteristics, but the outcomes arise spontaneously without the players knowing anything about the distribution that is driving the process.

[3] D. Myatt and C. Wallace (2002), "Sophisticated play by idiosyncratic agents," *Journal of Evolutionary Economics* 13: 319–345; A. Matros (2003), "Clever agents in adaptive learning," *Journal of Economic Theory* 111: 110–124; H. H. Nax and J. Newton (2022), "Deep and shallow thinking in the long run," *Theoretical Economics* 17: 1501–1527.

3

An Evolutionary Model of Bargaining

1 Introduction

Classical bargaining theory holds that bargains between rational agents depend solely on the agents' *utilities* for the outcomes, not the outcomes themselves. As Nash put it: "What the actual courses of action are among which the individuals must choose is not regarded as essential information ... Only the attitudes (like or dislike) of the two individuals towards the ultimate results are considered" (Nash 1950). The assumption that only utility matters is implicit in almost all treatments of the bargaining problem. These fall into three broad categories. The axiomatic approach originally introduced by Nash seeks to deduce a solution from plausible first principles (Nash 1950; Raiffa 1953; Kalai and Smorodinsky 1975). The noncooperative approach (also introduced by Nash) views the outcome as the equilibrium of a one-shot game in which the players make certain demands, which they get provided that the demands do not exceed the amount available (Nash 1953; Binmore 1987). The third approach, introduced by Stahl (1972) and Rubinstein (1982), views the outcome as the subgame perfect equilibrium of a game in which the players alternate in making offers to one another.

In all three approaches the outcome is highly sensitive to the utility functions of the parties. It is also assumed that these utility functions are common knowledge, or at least that their distribution is common knowledge (Harsanyi and Selten 1972; Binmore 1987)]. Yet it is common knowledge that utility functions are often *not* common knowledge. When the utility functions are not known, bargainers may try to coordinate by appealing to custom, that is, by relying on what is usual and expected in the given situation. A well-known example is fifty–fifty division. Fifty–fifty is customary in many everyday bargains, and it is frequently observed in laboratory settings. Curiously, it is also quite common even when agents are in quite asymmetric positions, as in sharecropping agreements between tenants and landlords (Bardhan 1984).

In this chapter we propose to explain how such customs arise, and why some (like fifty–fifty) are more likely than others. The essential idea is that bargainers' expectations are shaped by precedent. Each bargainer has limited

Spontaneous Order. H. Peyton Young, Oxford University Press. © H. Peyton Young (2024).
DOI: 10.1093/oso/9780198892908.003.0003

information about what other bargainers have demanded in past encounters. Each chooses a best reply assuming that these precedents are a reasonable predictor of what other agents are going to demand now. There is no common knowledge. When agents look at different precedents or the precedents conflict, their responses will very likely be uncoordinated. It may happen, however, that a succession of bargainers coordinate *by chance* on the same rule over a period of time. This establishes a set of common precedents and common expectations. People begin to expect that similar bargains will be solved by this rule in the future. Under circumstances that we shall describe below, this positive feedback loop eventually drives society toward a fixed division. It becomes the norm, not because it is inherently "focal," but because early chance events happened to favor it and now everyone expects it.

Of course, a given rule does not remain entrenched forever. Changing economic circumstances may eventually cause the shares to shift in favor of one party or another. Furthermore, there will inevitably be some variability in bargaining outcomes due to differences in agents' behavior and beliefs. When these stochastic influences are included, the long-run behavior of the evolutionary process becomes more complex. In particular, it has no absorbing states. A rule of division will remain in use for a time, after which it will be displaced by another rule. These displacements are irregular and infrequent, but they tend to occur quite rapidly when they do occur. The process then settles down to another period of relative tranquility in which a fixed rule is in force. We cannot predict which rule will be in use at any given time because of the stochastic nature of the process. Nevertheless we can say that some rules are more likely to be observed than others over the long run. In fact, when the stochastic perturbations are very small, usually one rule will be observed almost all of the time. This *stochastically stable* rule can be calculated explicitly. It depends on the agents' utility functions, but it does not require the agents to know anything about the *others'* utility functions.

The results may be summarized as follows. When each class of agents is *homogeneous*—all members of the same class have the same utility function and the same amount of information—the long-run stable division is the one that corresponds to the asymmetric Nash solution. The asymmetry arises from informational differences between the two classes, where the class with more information (the larger sample size) has the advantage. This division does not result from calculations involving other agents' utility functions, but from its inherent stability *given* their utility functions.

When each class is *heterogeneous*, the stable division depends on the types of agents represented in each class, but not on their relative frequency in the population. It therefore generalizes the Nash solution to mixed populations in a substantially different way than Harsanyi and Selten's bargaining solution with incomplete information (Harsanyi and Selten 1972). When there is even a small amount of mixing between the classes, the heterogeneity within each class no longer matters and the unique stable division is fifty–fifty. This suggests yet another reason (in addition to the usual focal point argument) why fifty–fifty is observed so often in practice, even when bargainers are in asymmetric positions.

2 The Model

Consider two finite, nonempty classes of individuals A and B. To be concrete we shall think of A as the class of landlords and B as the class of tenants. In each period, one landlord meets one tenant and they bargain over their shares of the crop. The share of the landlord will generically be denoted by x and the share of the tenant by y. For technical reasons we are going to assume that there are a finite number of feasible divisions. Let p be a positive integer, and let D be the set of all p-place decimal fractions that are positive and less than or equal to one. D is the set of *feasible* demands, and $\delta = 10^{-p}$ is the precision of the demands.

The bargaining process has the following structure. In each period $t = 1, 2, \ldots$, one landlord α is drawn at random from the class A and one tenant β is drawn at random from the class B. They then play the *Nash demand game*: α demands some fraction $x \in D$, β demands some fraction $y \in D$, and they get their demands if $x + y \leq 1$; otherwise they get nothing.

Let the demands in period t be denoted by (x_t, y_t). The complete *history* up to and including period t is the sequence $(x_1, y_1), (x_2, y_2), \ldots, (x_t, y_t)$.

Suppose that agents α and β are chosen in period $t + 1$. Assume that neither party has prior knowledge or beliefs about the utility function of the other side, or about the distribution of the utility functions in the general population. To decide what to do they rely on the precedents that they happen to have heard about. Formally we may suppose that agent α draws a random sample of k_α items from the last m records $\mathbf{s} = ((x_{t-m+1}, y_{t-m+1}), \ldots, (x_t, y_t))$. The ratio k_α/m is a measure of x's *information*. We may think of k_α/m as the extent to which α is informed or "gets around." It is not the result of a process in which α sets out to gather an optimal amount of information. Agent α chooses an optimal reply to the observed frequency distribution of the demands made by tenants in the sample that he knows about. More precisely, α forms the cumulative probability distribution $F(y)$ of the demands y_j made by tenants in his sample:

$$\forall y \in D,\ F(y) = h/k_x \quad \text{iff there are exactly } h \text{ demands } y_j \text{ in the sample such that } y_j \le y.$$

F is a random variable that depends on the particular sample that α happened to draw at time $t + 1$.

Agent α is assumed to have a von Neumann Morgenstern utility function $u_\alpha(x)$, where $u_\alpha(x)$ is concave (not necessarily strictly concave) and strictly increasing in α's own share. The utility function is normalized so that $u_\alpha(0) = 0$. We shall assume that $u_\alpha(x)$ is defined for all $x \in [0, 1]$, because later on we shall want to treat the level of precision δ as a variable.

If α demands x and the tenant demands y, then α gets x if $y \le 1 - x$ and zero otherwise. Thus α believes his *expected payoff* (given the observed cumulative distribution F) is:

$$u_\alpha(x)\,F(1-x) + u_\alpha(0)\,(1 - F(1-x)) = u_\alpha(x)\,F(1-x).$$

We assume that α chooses an amount x_{t+1} in period $t + 1$ that maximizes his expected payoff:

$$x_{t+1} = \text{argmax}\ u_\alpha(x)\,F(1-x). \tag{1}$$

If there are several values of x that maximize (1), then α chooses each of them with positive probability.

The tenant behaves similarly: β draws k_β items at random from the last m plays, and chooses an optimal response to the observed distribution of demands made by the landlords in his sample. All best replies are chosen with positive probability. The probability distribution over best replies may depend on the sample, but we assume it is stationary over time. The process is therefore a variation of fictitious play in which each agent reacts to a sample of the opponents' *recent* moves, rather than to all of the opponents' past moves. We shall call this process *adaptive play* (see Chapter 2 (Young 1993)).

We make no claim that this is how perfectly rational agents would play the game if they were aware of the process in which they are embedded. We are not assuming, however, that the agents in this model are perfectly rational or fully informed. They are merely *sensible*: they possess some information about the world around them, and they choose best replies based on this information. Note that we do not assume that agents learn through repeated plays of the game. We can assume, in fact, that after two individuals play they die and are replaced by agents of the same *type* as their forebears. Each has the same information-gathering propensity and the same utility function as its parent but not the parent's information. Thus each time an agent plays he starts afresh and must ask around to find out what is going on. This assumption is not necessary from a mathematical point of view, but it underscores the fact that "learning" need not occur at the individual level for the process to converge at the social level.

The agents' response rules determine a stationary Markov chain. The *state space* S consists of all sequences $\mathbf{s}$ of length m whose elements are pairs $(x, y) \in D \times D$. Let $p_\alpha(x|\mathbf{s})$ be the conditional probability that α demands x given that the state is $\mathbf{s}$, and let $p_\beta(y|\mathbf{s})$ be the conditional probability that β demands y given that the state is $\mathbf{s}$. We assume that p_α is a *best reply distribution*, that is, $p_\alpha(x|\mathbf{s}) > 0$ if and only if x is a best reply by α to a sample of size k_α drawn from $\mathbf{s}$. Similarly $p_\beta(y|\mathbf{s}) > 0$ if and only if y is a best reply by β to a sample of size k_β drawn from $\mathbf{s}$. The samples need not be drawn with uniform probability. For example, recent samples might be more likely than older samples. All that matters is that any appropriate-sized sample *could* be drawn, any best reply to such a sample *could* be made, and the probability of these events (conditional on the state) is stationary.

The process begins at time $t = m$ in some arbitrary initial state $\mathbf{s}^0$, that is, some arbitrarily chosen sequence of m pairs from $D \times D$. (This is clearly more general than assuming that the process begins at time $t = 1$ with a randomly chosen pair (x, y).) In each subsequent period one pair of agents $(\alpha, \beta) \in A \times B$ is drawn at random. Every pair has a positive probability $\pi(\alpha, \beta) > 0$ of being drawn, though it is not necessarily the same probability for all pairs. Given a state $\mathbf{s} = ((x_{t-m+1}, y_{t-m+1}), \ldots, (x_t, y_t))$ we say that $\mathbf{s}'$ is a successor of $\mathbf{s}$ if it has the form $\mathbf{s}' = ((x_{t-m+2}, y_{t-m+2}), \ldots, (x_t, y_t), (x_{t+1}, y_{t+1}))$. If the process is in state $\mathbf{s}$ at time t, then it moves to the successor state $\mathbf{s}'$ at time $t + 1$ with transition probability

$$P_{\mathbf{ss}'} = \sum_{\alpha \in A} \sum_{\beta \in B} \pi(\alpha, \beta)\, p_x(x_{t+1}|\mathbf{s})\, p_\beta(y_{t+1}|\mathbf{s}). \qquad (2)$$

If $\mathbf{s}'$ is not a successor of $\mathbf{s}$, then $P_{\mathbf{ss}'} = 0$. This Markov process will be called the *evolutionary bargaining process* with precision δ, memory m, information parameters $\{k_\alpha/m, k_\beta/m\}$ and best reply distributions $\{p_\alpha, p_\beta\}$.

3 Convergence of the Evolutionary Bargaining Process

Convention *A state* $\mathbf{s}$ *is a* convention *if it consists of some fixed division* $(x, 1 - x)$ *repeated* m *times in succession, where* $x \in D$ *and* $0 < x < 1$. *We shall denote this convention by* $\boldsymbol{x}$.

We claim that every convention is an absorbing state of the Markov process P defined by (2). Suppose, indeed, that at time $t \geq m$ the process is in the convention $\boldsymbol{x}$. Let $\boldsymbol{\alpha}$ and β be an arbitrary landlord and tenant who bargain in period $t + 1$. In every sample drawn by the landlord, the previous tenants always demanded $1 - x$, so the landlord's unique optimal reply is $x > 0$. Similarly, in every sample drawn by the tenant, the previous landlords always demanded x, so the tenant's unique optimal reply is $1 - x > 0$. (This argument relies on the assumption that $0 < x < 1$.) Thus the optimal demands in period $t + 1$ are $(x, 1 - x)$, so the state in period $t + 1$ is the same as it was in period t.

We now show that, if the information in each class is sufficiently *incomplete*, then the process P converges with probability one to a convention.

Theorem 1 *If at least one agent in each class samples at most half of the surviving records, then from any initial state the evolutionary bargaining process converges almost surely to a convention.*

Proof We shall prove that there exists a positive integer N and a positive probability p (both independent of t) such that, from any states, the probability of converging to a convention within N steps is at least p. Thus the probability of not reaching a convention within rN steps is at most $(1 - p)^r$, which goes to zero as $r \to \infty$. Hence the set of all sample paths that do not land in an absorbing state have probability zero.

Fix two agents α and β, where α has minimum information among all agents in A and β has minimum information among all agents in B. Let α have information $k/m \leq 1/2$ and let β have information $k^*/m \leq 1/2$. We may assume without loss of generality that $k \geq k^*$. Suppose that at time $t \geq m$ the process is in state $\mathbf{s} = ((x_{t-m+1}, y_{t-m+1}), \ldots, (x_t, y_t))$. There is a positive probability that agents α and β (or agents just like them) will be selected to play the Nash demand game in *every* period from $t + 1$ to $t + k$ inclusive. There is also a positive probability that α will draw the specific sample $\sigma = ((x_{t-k+1}, y_{t-k+1})$ $\ldots, (x_t, y_t))$ in each of these periods, and that β will draw the (possibly smaller) sample $\sigma^* = ((x_{t-k^*+1}, y_{t-k^*+1}), \ldots, (x_t, y_t))$ in each of these periods. Let x and y be best replies to these samples by α and β, respectively. Then there is a positive probability that α will demand x, and that β will demand y in every period from $t + 1$ to $t + k$ inclusive. Hence there is a positive probability of obtaining a *run* $\rho = ((x, y), \ldots, (x, y))$ from periods $t + 1$ to t + k inclusive.

From periods $t + k + 1$ to $t + 2k$ there is a positive probability that two agents like α and β will be selected to play every time and that both of them will always sample from the run p. (Thus α draws the whole sample ρ, and β draws a subsample of k^* items from p.) Their unique best replies are $(1 - y, 1 - x)$, so we obtain a run $\rho' = ((1 - y, 1 - x), \ldots, (1 - y, 1 - x))$ from periods $t + k + 1$ to $t + 2k$ inclusive.

In period $t + 2k + 1$, there is a positive probability that a landlord like α will be chosen again and that she draws the sample ρ, and also that a tenant like β will be chosen again and that he draws the sample ρ^* consisting of the last k^* items in ρ'. Their best replies are then $(1 - y, y)$. Note that this requires α to look back $2k$ periods, and β to look back k^* periods, both of which are possible because of the assumption that $k, k^* \leq m/2$.

In period $t + 2k + 2$, however, the oldest record in ρ may have disappeared (if $m = 2k$). Nevertheless, α can still draw a sample of size k in which the other side always demanded y, namely, the most recent $k - 1$ plays in ρ and the play in the preceding period $t + 2k + 1$. Meanwhile β can still look at the sample ρ^* because none of these records has disappeared yet. So there is a positive probability that the best replies in period $t + 2k + 2$ will again be $(1 - y, y)$. Continuing in this manner, we see that there is a positive probability of obtaining a run of $(1 - y, y)$ for m periods in succession, at which point a convention has been reached.

It follows that, from a given initial state **s**, there is a positive probability (possibly depending on **s**) of reaching a convention within $2k + m$ periods. Since the number of states is finite, there is a positive probability ρ (independent of the initial state) of reaching a convention within $2k + m$ periods. This concludes the proof of Theorem 1.

The analog of Theorem 1 holds for a more general class of games (Young 1993), though a lower bound than $m/2$ may be needed to guarantee convergence (depending on the structure of the game). We do not claim that $m/2$ is the best bound possible in the present case, but without incomplete sampling the result fails. Consider the following example. Let $\delta = 0.1$ and let all agents in A have utility function u and all agents in B have utility function v, where u and v are any concave functions on D such that:

$x =$	0	0.3	0.5	0.7	1
$u(x) =$	0	0.35	$0.4\sqrt{2}$	0.75	1
$v(x) =$	0	0.50	$0.5\sqrt{2}$	0.90	1

Suppose that both agents have complete information (they draw complete samples) and suppose that all previous pairs of demands were either (0.7,0.5) or (0.5, 0.3). Thus they miscoordinated every time. At time t, let f_t be the relative frequency with which the pair (0.7, 0.5) was played in the preceding m periods, and let $1 - f_t$ be the relative frequency with which (0.5, 0.3) was played in the preceding m periods. The landlord's best reply will be either 0.5 or 0.7 depending on the value of f_t. It will be 0.7 if

$$\left(1 - f_t\right)\ u(0.7)\ \geq\ u(0.5),$$

and 0.5 if the inequality runs the other way. Similarly, the tenant's best reply is 0.5 if

$$\left(1 - f_t\right)\ v(0.5)\ \geq\ v(0.3),$$

and 0.3 if the inequality runs the other way. Substituting for the values of u and v we find that these two inequalities are the same, and hold if and only if $f_t < 1 - 1/\sqrt{2}$. (The inequality is always strict because f_t is rational.) Thus the demands in period $t + 1$ are either (0.7, 0.5) or (0.5, 0.3). It follows that if the agents begin in any state where they have always miscoordinated, then they continue to miscoordinate forever.

4 Mistakes and Experimentation

The evolutionary bargaining process defined in (2) is based on the assumption that agents always choose best replies given their information. A more realistic assumption is that agents sometimes make mistakes or experiment with other choices. These random choices or "trembles" keep the process constantly in motion and test the viability of different regimes, much like mutations in biological models of evolution.

Let $\varepsilon\lambda_\alpha$ be the probability that agent α experiments in any given period, and let $q_\alpha(x|\mathbf{s})$ be the conditional probability that α chooses the reply x, given that α is experimenting and that the current state is $\mathbf{s}$. Similarly define $\varepsilon\lambda_\beta$ and $q_\beta(y|\mathbf{s})$ for every $\beta \in B$. We assume that $q_\alpha(x|\mathbf{s})$ and $q_\beta(y|\mathbf{s})$ have full support, that is, all choices are possible in any given state. The parameters $\lambda_\alpha > 0$ and $\lambda_\beta > 0$ define the *relative* probabilities with which particular agents experiment, and the factor ε determines the *absolute* probability with which agents in general experiment. (This model generalizes an approach introduced by Canning (1991) and Kandori, Mailath, and Rob 1993.)

These random choices yield a stationary Markov chain on S in which the probability of moving from state $\mathbf{s} = ((x_{t-m+1}, y_{t-m+1}), \ldots, (x_t, y_t))$ at time $t \geq m$ to the successor state $\mathbf{s}' = ((x_{t-m+2}, y_{t-m+2}), \ldots, (x_{t+1}, y_{t+1}))$ at time $t+1$ is

$$
\begin{aligned}
P^{\varepsilon}_{\mathbf{ss}'} = \sum_{\alpha \in A} \sum_{\beta \in B} \pi(\alpha, \beta) \big[& (1 - \varepsilon\lambda_\alpha)(1 - \varepsilon\lambda_\beta)\, p_\alpha(x|\mathbf{s})\, p_\beta(y|\mathbf{s}) \\
& + \varepsilon\lambda_\alpha(1 - \varepsilon\lambda_\beta)\; q_\alpha(x|\mathbf{s})\; p_\beta(y|\mathbf{s}) + \varepsilon\lambda_\beta(1 - \varepsilon\lambda_\alpha)\; p_\alpha(x|\mathbf{s})\; q_\beta(y|\mathbf{s}) \\
& + \varepsilon^2\lambda_\alpha\lambda_\beta\; q_\alpha(x|\mathbf{s})\; q_\beta(y|\mathbf{s}) \big].
\end{aligned}
\tag{3}
$$

$P^{\varepsilon}_{\mathbf{ss}'} = 0$ if s′ is not a successor of **s**. P^{ε} is the perturbed evolutionary bargaining process. Note that the *unperturbed* process P^0 is the same as the process P defined in (2).

Before showing how to compute the stable conventions in particular cases, let us first consider the qualitative behavior of the evolutionary bargaining process when the noise level ε is small. If agents never experiment ($\varepsilon = 0$), then Theorem 1 tells us that the process converges to a convention provided that some agent in each class samples at most half the surviving records. These are the absorbing states of the unperturbed process. If the agents occasionally experiment, then the process has no absorbing states. It still gravitates toward them, but never comes to rest because of the random fluctuations caused by experimentation. Nevertheless, the process is at or near an absorbing state (i.e., a convention) most of the time. Occasionally, a large accumulation of mistakes pushes the process away from a given convention and into the basin of attraction of another convention. The likelihood of such an event is small, but if we wait long enough, it is bound to occur. When it does, the process tends to gravitate toward the new convention. If it reaches it, the process is likely to stay there for a while before another displacement occurs. This pattern of relative stability punctuated by episodes of instability continues indefinitely. Some conventions, however, are harder to displace than others. The ones that are hardest to displace are the ones that will be observed most often over the long run. When the noise is very small, in fact, the convention(s) that are hardest to displace will be observed *almost all of the time* over the long run.

Let us now make these observations more precise by employing the theoretical framework in the Appendix to Chapter 2 (Young 1993). The perturbed process P^{ε} is irreducible because of the assumption that the experimental distributions q_α, q_β have full support. Hence P^{ε} has a unique stationary distribution μ^{ε}. P^{ε} is also strongly ergodic, so for every $\mathbf{s} \in S$, $\mu^{\varepsilon}{}_{\mathbf{s}}$ is (with probability one) the relative frequency with which state **s** will be observed in the first t periods as $t \to \infty$.

Stochastically Stable Convention *A convention* s *is stochastically* stable *if* $\lim_{\varepsilon \to 0} \mu^{\varepsilon}_{s}$ *exists and is positive. It is* strongly stable *if* $\lim_{\varepsilon \to 0} \mu^{\varepsilon}_{s} = 1$.

Over the long run, stable conventions will be observed much more frequently than unstable conventions when the probability ε of the perturbations is small. A strongly stable convention will be observed almost *all* of the time when ε is small.[1]

We now describe the technique for computing the stable conventions, which always exist. Moreover, we show that generically there exists a strongly stable convention, which of course must be unique if it exists.

Mistake Let $\mathbf{s} = ((x^1, y^1), (x^2, y^2), \ldots, (x^m, y^m))$ be some state and let $\mathbf{s}' = ((x^2, y^2), \ldots, (x^m, y^m), (x, y))$ be a successor of $\mathbf{s}$. x is a *mistake* in the transition $\mathbf{s} \to \mathbf{s}'$ if, for every landlord α, x is not a best reply by α to any sample of size k_α drawn from $\mathbf{s}$. Similarly, y is a mistake if, for every tenant β, y is not a best reply by β to any sample of size k_β drawn from $\mathbf{s}$.

Resistance If s′ is a successor of $\mathbf{s}$, the *resistance* $r(\mathbf{s}, \mathbf{s}')$ of the one-period transition $\mathbf{s} \to \mathbf{s}'$ is the minimum number of mistakes involved in the transition. Clearly $r(\mathbf{s}, \mathbf{s}') = 0, 1,$ *or* 2. For every two states $\mathbf{s}^1$ and $\mathbf{s}^2$, the resistance $r(\mathbf{s}^1, \mathbf{s}^2)$ is the least total number of mistakes in any sequence of one-period transitions that leads from $\mathbf{s}^1$ *to* $\mathbf{s}^2$.

To compute $r(\mathbf{s}^1, \mathbf{s}^2)$ one solves a shortest path problem in a directed graph, namely, the graph in which every state is a vertex, there is an edge directed from $\mathbf{s}$ to $\mathbf{s}'$ if and only if $\mathbf{s}'$ is a successor of s, and its weight is $r(\mathbf{s}, \mathbf{s}')$.

Now define another graph $\mathcal{G}$ as follows. There is one vertex for each convention $\mathbf{x}$, and a directed edge from every vertex to every other. The "weight" or resistance of the directed edge $\mathbf{x} \to \mathbf{x}'$ is the resistance $r(\mathbf{x}, \mathbf{x}')$ of moving from the convention $\mathbf{x}$ to the convention $\mathbf{x}'$.

$\mathbf{x}$-*tree.* An $\mathbf{x}$-*tree* is a collection of edges in $\mathcal{G}$ such that, from every vertex $\mathbf{x}' \neq \mathbf{x}$ there is a unique directed path to $\mathbf{x}$, and there are no cycles. Let $\mathcal{T}_\mathbf{x}$ be the set of all $\mathbf{x}$-trees.

Stochastic Potential. The *stochastic potential* of the convention $\mathbf{x}$ is the least resistance among all $\mathbf{x}$-trees:

$$\gamma(\mathbf{x}) = \min_{T \in \mathcal{T}_\mathbf{x}} \sum_{(\mathbf{x}', \mathbf{x}'') \in T} r(\mathbf{x}', \mathbf{x}'')$$

[1] The origins of this concept are discussed in Chapter 1.

The following is a special case of Theorem 4 in Chapter 2 (Young 1993).

Theorem 2 *The sequence of stationary distributions* μ^ε *converges to a stationary distribution* μ^0 *of* P^0 *as* $\varepsilon \to 0$. *Moreover,* **s** *is stochastically stable* ($\mu^0{}_s > 0$) *if and only if* **s** = **x** *is a convention and* $y(\mathbf{x})$ *has minimum stochastic potential among all conventions.*

In the remainder of the chapter we shall characterize the stable conventions under various assumptions about the composition of the bargaining populations.

5 Homogeneous Populations and the Nash Solution

Let a be a rational fraction, $0 < a < 1$, and let $u(x)$ be a concave utility function defined for all $x \in [0, 1]$. An agent is of *type* (a, u) if he always samples the fraction a of the last m records and his utility function is u. Let A and B be the two classes of agents, and suppose that each class is *homogeneous*: all members of A are of the same type (a, u), and all members of B are of the same type (b, v). Since the classes A and B are treated symmetrically in the model, there is no loss of generality in assuming that $a \geq b$. We shall also assume that m is chosen so that ma and mb are integers, such an m is "admissible".

Generic Stability. A division $(x, 1 - x)$ is *generically stable* for types (a, u) and (b, v) if the associated convention **x** is stochastically stable for all admissible m.

Asymmetric Nash Bargaining Solution. Given types (a, u) and (b, v), the *asymmetric Nash bargaining solution* is the unique division $(x, 1 - x)$ that maximizes

$$u^a(x)\, v^b(1 - x) \quad \text{subject to } 0 \leq x \leq 1. \tag{4}$$

When $a = b$ this reduces to the ordinary Nash bargaining solution.

Theorem 3 *Let A and B be homogeneous populations composed respectively of types* (a, u) *and* (b, v), *where* $a, b \leq 1/2$. *For every precision* $\delta > 0$ *there exists at least one and at most two generically stable divisions, and as* $\delta \to 0$ *they converge to the asymmetric Nash bargaining solution.*

Stated informally, this result says that if memory is sufficiently large and the precision is sufficiently small, most people will be using the same rule of division most of the time, and this rule will be close to the asymmetric Nash solution. In particular, agents who are risk averse or poorly informed will fare worse than those who are not, all else being equal. For example, if the landlords are risk neutral and the tenants are risk averse, and if both have the same amount of information ($a = b$), then most of the time the landlords will be getting the larger share. If both sides are risk neutral, but the landlords have more information ($a > b$), then again the landlords will get the larger share most of the time. The reason is that agents with more information are less likely to respond to mistakes by the other side, so they are steadier. Similarly, agents who are less risk averse are more likely to take chances, so they are more demanding.

The proof of Theorem 3 proceeds by a series of lemmas that provide a sharp estimate of the stable conventions for *any* choice of parameters δ and m.

Fix the precision $\delta = 10^{-p}$ and let $D^0 = \{x \in D : \ \delta \le x \le 1 - \delta\}$. Fix $a,\ b \le 1/2$ and let m be such that ma and mb are integers. For every real number r let $\lceil r \rceil$ denote the least integer greater than or equal to r. Also, let $r \wedge r'$ denote the minimum of r and r'.

x-*Basin*. For every $x \in D^0$ the **x**-*basin* is the set of all states from which the unperturbed process P^0 converges to the convention **x** with positive probability.

Lemma 1 *For every $x \in D^0$ the minimum resistance to moving from **x** to a state in some other basin is $\lceil mr_\delta(x) \rceil$, where*

$$r_\delta(x) = a\,(1 - u(x-\delta)/u(x)) \wedge b(1 - v(1-x-\delta)/v(1-x)) \wedge bv(1-x)/v(1-\delta). \quad (5)$$

Before proving Lemma 1, we give the intuition behind it and show how it leads to a proof of Theorem 3. The three terms in expression (5) arise in the following way. To displace an established convention **x** requires one side or the other to demand something other than the conventional amount. Suppose that the tenants demand just a little more than they should, say δ more. The landlords will resist this, and the amount of their resistance is the relative loss in utility that they would suffer by giving up a δ-increment of their current share x, times their sample size. This yields the first term $a(1 - u(x - \delta)/u(x))$. Similarly, if the landlords demand δ more than the conventional amount the tenants will resist, and the extent of their resistance is the relative loss in utility that they would suffer by giving up δ, times their sample size. This yields the second term $b(1 - v(1 - x - \delta)/v(1 - x))$.

A third possibility is that some landlord demands *less* than the conventional amount by mistake. This would be a "silly" mistake, of course, but it could happen. The silliest mistake of all would be to demand only δ. (Recall that we exclude demands of 0, because it makes little sense to "demand" nothing.) If the landlords make enough mistakes of this sort, the tenants will switch to $1 - \delta$, and their resistance to switching is given by the third term, $bv(1 - x)/v\,(1 - \delta)$. Of course, the tenants might also make silly mistakes. If they demand δ often enough, the landlords will switch to $1 - \delta$. It can be shown, however, that when $a \geq b$ the resulting term $au(x)/u(1 - \delta)$ is never strictly smaller than the term $b(1 - v(1 - x - \delta)/v(1 - x))$ for any concave utility functions u and v. Hence mistakes of the second type have no effect on the long-run stability of various divisions.

The function $r_\delta(x)$ is the minimum of three monotone functions. The first is strictly decreasing in x, the second is strictly increasing in x, and the third is strictly decreasing in x. Hence, as x increases, $r_\delta(x)$ is first strictly increasing and then strictly decreasing. On the subset D^0 it therefore achieves its maximum at one, or at most two, values. We shall show that these are the generically stable divisions, and that, as δ approaches 0, they converge to the asymmetric Nash solution (Lemmas 2 and 3 below).

Example 1 Let all agents in A sample 1/3 of the surviving records and have utility function $u(x) = \sqrt{x}$. Let all agents in B sample 1/10 of the surviving records and have utility function $v(y) = y$. The asymmetric Nash solution is (5/8, 3/8). Let $\delta = 0.1$. Then $r_\delta(x)$ is the minimum of the following three functions

$$f_1(x) = (1/3)(1 - u(x - \delta)/u(x)) = (1/3)(1 - \sqrt{(1 - 0.1/x)})$$
$$f_2(x) = (1/10)(1 - v(1 - x - \delta)/v(1 - x)) = 0.01/(1 - x)$$
$$f_3(x) = (1/10)v(1 - x)/v(1 - \delta) = (1/9)(1 - x).$$

The function $r_\delta(x) = min\{f_1(x), f_2(x), f_3(x)\}$ is graphed in Figure 3.1. It achieves its maximum on D^0 at the value $x_\delta = 0.6$, so by Theorem 3, (0.6, 0.4) is the generically stable division.

Proof of Lemma 1. Let the types (a, u) and (b, v) be given, where $a,\ b \leq 1/2$. Assume without loss of generality that $a \geq b$, and let $k = ma$. Suppose that the process is at the convention $\boldsymbol{x}$, where $x \in D^0 = \{x \in D : \ \delta \leq x \leq 1 - \delta\}$ Let π be a path of least resistance from $\boldsymbol{x}$ to a state that is in some other basin. Clearly, π must pass through some state $\mathbf{s}$ such that:

(i) some landlord's best reply to a sample from $\mathbf{s}$ is different from x, and/or

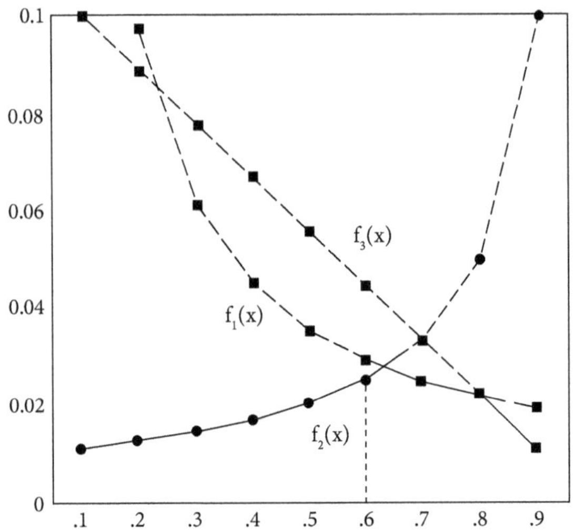

Figure 3.1 The function $r_\delta(x) = \min\{f_1(x), f_2(x), f_3(x)\}$ for Example 1.

(ii) some tenant's best reply to a sample from **s** is different from $1 - x$.

Let **s** be the first such state. Without loss of generality we can assume that some landlord (say α) has a best reply $x' \neq x$ to a sample σ of size k drawn from **s**. By choice of **s**, every tenant's demand in **s** that differs from $1 - x$ must have been a mistake. Let $i \leq k$ be the total number of mistakes by tenants in the sample σ that the landlord drew.

Now construct another sequence σ' from σ by replacing each mistaken demand of the tenants by the demand $1 - x'$. Thus in σ' there are exactly i instances of $1 - x'$ on the tenants' side, and $k - i$ instances of $1 - x$.

Landlord α's best reply to σ' is still x', because his best reply to σ was x', and the optimality of x' is not compromised when the tenants demand $1 - x'$ more often and other things less often.

We now construct an alternative path π' from **x** to the **x**$'$-basin in which the total number of mistakes is i, from which it follows that π' is also a path of least resistance. Beginning at the convention **x**, let a succession of tenants mistakenly demand $1 - x'$ i times in a row. With positive probability the landlord α' drawn in the next period will sample the most recent k plays. This sample σ'' consists of i instances of $1 - x'$ and $k - i$ instances of $1 - x$. The relative frequency of $1 - x'$ is the same in σ' and σ''. Since α' and α have the same utility function, and x' is the best reply by α to σ', it follows that x' is also the best reply by α' to σ''.

With positive probability the landlords will sample σ'' for k periods in succession and reply with x' each time. This establishes a run of x' (by the landlords) without introducing more mistakes. After this there is a positive probability that a succession of k tenants will sample from this run and demand $1 - x'$ for k periods in a row. From here it is clear (as in the proof of Theorem 1) that the process converges with positive probability to the convention $\mathbf{x}'$ with no further mistakes.

Thus we have constructed an alternative least-resistant path from the convention $\mathbf{x}$ to the convention $\mathbf{x}'$ in which the only mistakes are an initial succession of i mistaken demands $1 - x'$. To compute the least number of mistakes necessary to exit from the $\mathbf{x}$-basin, it therefore suffices to consider, for every $x' \neq x$, the least number of initial mistakes $1 - x'$ by tenants that would cause a landlord to reply with $x' \neq x$, and the least number of initial mistakes x' by landlords that would cause a tenant to reply with $1 - x'$, and to take the smaller of these two numbers.

Choose an arbitrary $x' \neq x$. We distinguish two cases: $x' < x$ and $x' > x$.

Case 1: $x' < x$.

Let the process be in the convention $\mathbf{x}$. Suppose that the landlords make i successive demands of x' that cause some tenant's best reply to switch to $1 - x'$ instead of $1 - x$. We can assume that $i \leq mb$, which is the tenant's sample size. When a tenant samples these i mistaken demands x', together with $mb - i$ of the previous "conventional" demands x, he switches to $1 - x'$ provided that $(i/mb)\, v(1 - x') \geq v(1 - x)$, that is

$$i \geq mbv(1 - x)/v\left(1 - x'\right).$$

Over all feasible $x' < x$ the minimum value of i occurs when $x' = \delta$ and

$$i = \lceil mbv(1 - x)/v(1 - \delta)\rceil. \tag{6}$$

The other possibility is that the tenants make the mistakes. Suppose that a succession of $j \leq k = ma$ tenants demand $1 - x'$ by mistake instead of $1 - x$. (Note that these are only "mistakes" in the sense that they are not best replies to the existing set of precedents. The tenants would obviously like to get $1 - x'$ rather than $1 - x$ if they thought they could get away with it.) If some landlord samples the j mistaken demands of $1 - x'$, together with $k - j$ of the previous conventional demands $1 - x$, then he switches to x' provided that $u(x') \geq (1 - j/k)\, u(x)$, that is,

$$j \geq k(1 - u(x')/u(x)) = ma(1 - u(x')/u(x)).$$

Over all feasible $x' < x$, the minimum such j occurs when $x' = x - \delta$. Hence

$$j = \lceil ma(1 - u(x - \delta)/u(x)) \rceil. \tag{7}$$

Case 2: $x' > x$.
An analogous argument shows that, if the landlords make i mistaken demands of x', then some tenant switches provided that

$$i \geq mb(1 - v(1 - x')/v(1 - x)).$$

The minimum occurs when $x' = x + \delta$:

$$i = \lceil mb(1 - v(1 - x - \delta)/v(1 - x)) \rceil. \tag{8}$$

If the tenants make j mistaken demands of $1 - x'$, then some landlord switches provided that

$$j \geq mau(x)/u(x').$$

Since all demands are positive, the minimum occurs when $1 - x' = \delta$, that is,

$$j = \lceil mau(x)/u(1 - \delta) \rceil. \tag{9}$$

Combining (6)–(9) it follows that the least number of mistakes to exit from the **x**-basin is $\lceil mr_\delta(x) \rceil$, where

$$r_\delta(x) = a(1 - u(x - \delta)/u(x)) \wedge b(1 - v(1 - x - \delta)/v(1 - x))$$
$$\wedge bv(1 - x)/v(1 - \delta) \wedge au(x)/u(1 - \delta).$$

We claim that the last term of this expression is at least as large as the second term for all $x \in D^0$, that is,

$$au(x)/u(1 - \delta) \geq b(1 - v(1 - x - \delta)/v(1 - x)). \tag{10}$$

Indeed, since v is concave, we have

$$(v(1 - x) - v(1 - x - \delta))/\delta \leq (v(1 - x) - v(0))/(1 - x),$$

because the rate of loss of utility in going from $1 - x$ to $1 - x - \delta$ is no more than the rate of loss of utility in going from $1 - x$ to 0. Since $v(0) = 0$ this implies that

$$1 - v(1 - x - \delta)/v(1 - x) \leq \delta/(1 - x). \tag{11}$$

Similarly, since u is concave, we have

$$(u(1-\delta)-u(0))/(1-\delta) \;\leq\; (u(x)-u(0))/x,$$

from which it follows that

$$u(x)/u(1-\delta) \;\geq\; x/(1-\delta). \tag{12}$$

It may be checked that $x/(1-\delta) \geq \delta/(1-x)$ whenever $\delta \leq x \leq 1-\delta$. Combining this with (11) and (12) we therefore have

$$u(x)/u(1-\delta) \;\geq\; x/(1-\delta) \;\geq\; \delta/(1-x) \;\geq\; 1-v(1-x-\delta)/v(1-x).$$

This, together with the assumption that $a \geq b$, proves (10). Hence the rightmost term in $r_\delta(x)$ can be omitted, so $r_\delta(x)$ is given by (5). This completes the proof of Lemma 1.

Lemma 2 *A division $(x, 1-x)$ is generically stable if and only if x maximizes the function $r_\delta(x)$ on D^0.*

Proof The function $r_\delta(x)$ is first strictly increasing in x and then strictly decreasing in x. It therefore achieves its maximum on D^0 either at a unique value x_δ or at two adjacent values x_δ and $x_\delta+\delta$. Suppose first that the maximum is achieved at a unique value x_δ. Construct a directed tree T_δ having root $\mathbf{x}_\delta$ as follows:

(i) for every $x \in D^0$ such that $x < x_\delta$, put the directed edge $(\mathbf{x}, \mathbf{x}+\delta)$ in T_δ and let its weight be $\lceil mb(1-v(1-x-\delta)/v(1-x))\rceil$;
(ii) for every $x \in D^0$ such that $x > x_\delta$ and $a(1-u(x-\delta)/u(x)) \leq bv(1-x)/v(1-\delta)$, put the directed edge $(\mathbf{x}, \mathbf{x}-\delta)$ in T_δ and let its weight be $\lceil ma(1-u(x-\delta)/u(x))\rceil$;
(iii) for every $x \in D^0$ such that $x > x_\delta$ and $a(1-u(x-\delta)/u(x)) > bv(1-x)/v(1-\delta)$, put the directed edge $(\mathbf{x}, \boldsymbol{\delta})$ in T_δ and let its weight be $\lceil mbv(1-x)/v(1-\delta)\rceil$.

Then T_δ has the following structure. Every node $\mathbf{x}$ except $\mathbf{x}_\delta$ has a unique outgoing edge. If $\mathbf{x}$ lies to the *left* of $\mathbf{x}_\delta$ the outgoing edge is directed to $\mathbf{x}$'s immediate neighbor on the right (case *i*). If x lies to the *right* of $\mathbf{x}_\delta$ the outgoing edge is directed *either* to $\mathbf{x}$'s immediate neighbor to the left (in case ii) *or* to the node $\boldsymbol{\delta}$ (in case iii). From every vertex other than $\mathbf{x}_\delta$ there exists a unique path

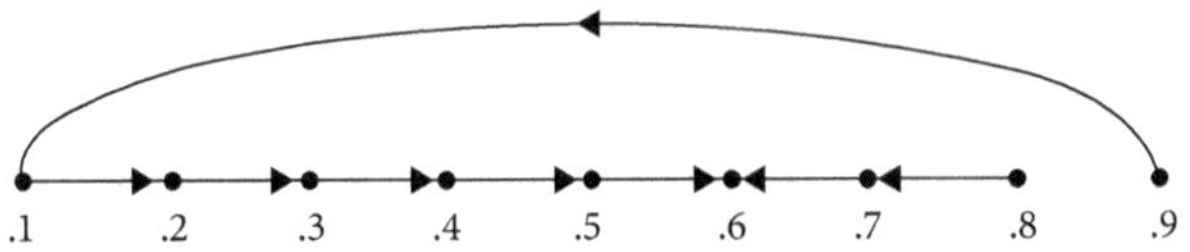

Figure 3.2 The $\mathbf{x}_\delta$-tree of least resistance for Example 1 ($x_\delta = 0.6$).

to $\mathbf{x}_\delta$. Thus T_δ is an $\mathbf{x}_\delta$-tree. Since T_δ was constructed by choosing at each node the outgoing edge with least resistance, it is clear that T_δ has least resistance among all $\mathbf{x}_\delta$-trees. Figure 3.2 illustrates the structure of T_δ for Example 1.

We claim that T_δ has least resistance among *all* $\mathbf{x}$-trees. To prove this, let T be an $\mathbf{x}$-tree for some $x \neq x_\delta$. In T, $\mathbf{x}_\delta$ has a unique outgoing edge e. Let r be the resistance of e. By Lemma 1, $\lceil mr_\delta(x_0)\rceil$ is the minimum resistance among *all* possible edges directed away from $\mathbf{x}_\delta$, so $r \geq \lceil mr_\delta(x_0)\rceil$. By construction, $\lceil mr_\delta(x)\rceil$ is the resistance of the unique edge exiting from $\mathbf{x}$ in the tree T_δ. Further, for every node x' different from both $\mathbf{x}_\delta$ and $\mathbf{x}$, the resistance of the unique outgoing edge from $\boldsymbol{x}'$ in T_δ is no greater than the resistance of the unique outgoing edge from $\boldsymbol{x}'$ in T. Hence

$$r(T) \geq r(T_\delta) + r - \lceil mr_\delta(x)\rceil.$$

Since

$$r \geq \lceil mr_\delta(x_\delta)\rceil \geq \lceil mr_\delta(x)\rceil,$$

it follows that $r(T) \geq r(T_\delta)$ as claimed. Moreover, since x_δ is the unique maximum of r_δ, we have $\lceil mr_\delta(x_\delta)\rceil > [mr_\delta(x)]$ for all sufficiently large m, so $r(T) > r(T_\delta)$. It follows from Theorem 2 that $\mathbf{x}_\delta$ is a stochastically stable convention for all m, and it is the unique stochastically stable convention for all sufficiently large m.

If r_δ is maximized at two neighboring values $\mathbf{x}_\delta$ and $\mathbf{x}_\delta + \delta$, construct a tree for each of them as above. A similar argument shows that these are the $\mathbf{x}$-trees of least resistance for all sufficiently large m. Hence $\mathbf{x}_\delta$ and $\mathbf{x}_\delta + \delta$ are the unique generically stable conventions. This concludes the proof of Lemma 2.

Lemma 3 *The maxima of the function $r_\delta(x)$ converge to the asymmetric Nash solution as $\delta \to 0$.*

Proof For each δ, view $r_\delta(x)$ as a function defined on the whole interval $[0, 1]$. Let x'_δ be the *unique* real value at which the maximum is achieved. It is clear that $|x_\delta - x'_\delta| \leq \delta$, where x_δ maximizes $r_\delta(x)$ on D^0. To prove Lemma 3, it suffices to show that the sequence $\{x'_\delta\}$ converges to the asymmetric Nash solution as $\delta \to 0$.

For each precision δ define $f_\delta(x) = r_\delta(x)/\delta$. Clearly x'_δ also maximizes $f_\delta(x)$ on [0, 1]. As δ goes to zero, the third term of $f_\delta(x)$ becomes large while the other two terms are bounded. Hence the third term may be ignored in the limit (since $f_\delta(x)$ is the minimum of the three terms). Let $u^-(x)$ denote the left-hand derivative of u at x and let $u^+(x)$ denote the right-hand derivative of u at x (both of which exist). Then for each x in (0, 1) we have

$$\lim_{\delta \to 0} f_\delta(x) = au \quad (x)/u(x) \wedge bv^-(1-x)/v(1-x) = f(x).$$

Since u and v are concave and bounded, the function $f(x)$ defined in the above expression is strictly quasiconcave and upper-semicontinuous. Hence it attains its maximum at a unique value x^*. We claim that x^* is the asymmetric Nash bargaining solution. To see this, suppose first that $u(x)$ and $v(y)$ are differentiable. Then $u'(x) = du(x)/dx$ and $v'(y) = dv(y)/dy$ are continuous and nonincreasing. Thus $au'(x)/u(x)$ is continuous and strictly decreasing, while $bv'(1-x)/v(1-x)$ is continuous and strictly increasing. The maximum of $f(x)$ occurs where the curves $au'(x)/u(x)$ and $bv'(1-x)/v(1-x)$ cross, that is, at the unique value $x^* \in (0, 1)$ such that

$$au'(x^*)/u(x^*) = bv'(1-x^*)/v(1-x^*).$$

This is a necessary and sufficient condition for x^* to be the unique point where the strictly concave function $a \ln u(x) + b \ln v(1-x)$ is maximized. It follows that x^* maximizes $u^a(x)\, v^b(1-x)$, so it is the asymmetric Nash solution.

Suppose now that u and v are not differentiable. Nevertheless they are subdifferentiable. For each $x \in (0, 1)$ consider the set of all lines in R^2 that pass through the point $(x, u(x))$ and are tangent to the curve $u(x)$. The slopes of these lines are the *subgradients* of u at x, and the set of them is the *subdifferential* at x, denoted by $\partial u(x)$ (Rockafellar 1970). In general we have $\delta u(x) = \{r : u^-(x) \geq r \geq u^+(x)\}$. Similarly define $\partial v(y)$ for every $y \in (0, 1)$.

Suppose that f attains its maximum at x^*. Then for all small $\varepsilon > 0$

$$au^-(x^*)/u(x^*) \geq f(x^*) \geq au^+(x^* + \varepsilon)/u(x^* + \varepsilon).$$

It is a standard result that $u^+(x^* + \varepsilon)/u(x^* + \varepsilon) \to u^+(x^*)/u(x^*)$ as $\varepsilon \to 0$ (Rockafellar 1970, 228). Hence

$$au^-(x^*)/u(x^*) \geq f(x^*) \geq au^+(x^*)/u(x^*),$$

and therefore,

$$f(x^*) \in a\partial u(x^*)/u(x^*). \tag{13}$$

Similarly, conclude that

$$f(x^*) \in b\partial v(1-x^*)/v(1-x^*). \tag{14}$$

Consider the strictly concave function $F(x) = a \ln u(x) + b \ln v(1 - x)$. Then (Rockafellar 1970, 223)

$$\begin{aligned}\partial F(x) &= \partial(a \ln u(x)) + \partial(b \ln v(1-x)) \\ &= a\partial u(x)/u(x) - b\partial v(1-x)/v(1-x).\end{aligned} \tag{15}$$

In particular this holds for x^*. From this and (13)–(14) it follows that $0 \in \partial F(x^*)$. Therefore x^* is the unique minimum of F (see Rockafellar 1970, 264), so by definition x^* is the asymmetric Nash solution.

The only point that remains to be verified is that, if x^*_δ maximizes $f_\delta(x)$ for every δ, then the sequence $\{x^*_\delta\}$ converges to x^*. Suppose, by way of contradiction, that x' is an accumulation point of the sequence $\{x^*_\delta\}$ that differs from x^*. For notational simplicity we shall assume that $\{x^*_\delta\}$ actually converges to x'. For each δ let

$$C_\delta = \{x \in [0,1] : f_\delta(x) \geq f_\delta(x^*)\}.$$

C_δ is convex, closed, and contains both x^* and x^*_δ. Hence it contains the interval I_δ spanned by x^* and x^*_δ. Since x^*_δ converges to x', the intervals I_δ converge to the interval I spanned by x' and x^*. Fix some z strictly between x^* and x'. For all sufficiently small δ, $z \in I_\delta$ and hence $z \in C_\delta$. It follows by definition of C_δ that $f_\delta(z) \geq f_\delta(x^*)$ for all δ. Hence $f(z) \geq f(x^*)$ because f_δ converges pointwise to f. This is a contradiction, however, because x^* uniquely maximizes f. This completes the proof of Lemma 3, and Theorem 3 follows at once.

The above proof shows that Theorem 3 holds under a variety of other assumptions about how agents make mistakes. For example, suppose that agents only make "small" mistakes. In other words, imagine that an agent first samples the past and determines the best reply, but that occasionally he demands a little bit more (or a little bit less) than he should – say δ more or δ less. Then the expression for the resistance function r_δ in (5) is the same except that the third term $bv(1-x)/v(1-\delta)$ is deleted. Since in any event this term is not relevant when δ goes to zero (because it is strictly larger than the first two terms), the proofs of Lemmas 2 and 3 go through much as before. Hence the generically stable division(s) are close to the asymmetric Nash solution when δ is close to zero.

Theorem 3 differs in several important respects from other ways of deriving the Nash solution based on perturbations. In his 1953 paper, for example, Nash suggested that his bargaining solution is the unique limiting outcome of the Nash equilibria of a perturbed version of the Nash demand game in which there is a small uncertainty in the payoff function (Nash 1953)). A more rigorous argument along these lines was provided by Binmore (1987). Carlsson (1991) investigated variations of the Nash demand game in which agents' demands are modified by a small error term, and showed that the Pareto optimal Nash equilibria of this noncooperative game converge to the Nash solution as the error term becomes vanishingly small.

These models of equilibrium stability differ from the present one in that they presuppose that the outcome of the perturbed game *is* a Nash equilibrium. To justify this approach, it must be assumed (among other things) that the properties of the game are common knowledge among the agents. In our model, there is no common knowledge and no assumption that a Nash equilibrium will be played. Instead, equilibrium emerges (in an asymptotic sense) as a consequence of the long-run dynamics, without the players being aware of it.

6 Heterogeneous Populations and A Generalization of the Nash Solution

Theorem 3 shows how the Nash bargaining solution can be selected by a dynamic process that does not require any common knowledge on the part of the agents. This conclusion is based, however, on the special assumption that all agents in the same class have the same utility function. Moreover, the result is very sensitive to this assumption. If just *one* agent has a utility function that differs from the norm in his or her class, the long-run outcome of the process may be completely different from the one described in Theorem 3. The reason is that the long-run behavior of the process is determined by those who would suffer the *greatest proportional loss in utility* from a given reduction in share. If there is just one landlord, for example, who is more risk averse than the others in this sense, then his choices will determine the long-run behavior of the process despite there being many other landlords who are less risk averse.

Let us therefore consider the more general situation in which each population of agents contains various types of individuals. Let T_A be the finite set of types represented in class A, and T_B the set of types represented in class B. Let a^* be the least a among all (a, u) in T_A and let b^* be the least b among all

(b, v) in T_β. We may assume without loss of generality that $a^* \geq b^*$. We shall also assume that $a^*, b^* \leq 1/2$. Then, just as in the proof of Lemma 1, it follows that the minimum resistance to moving from the convention $\boldsymbol{x}$ into the basin of attraction of some other convention is $[mR_\delta(x)]$, where

$$R_\delta(x) = \min_{(a,u) \in T_A} a(1 - u(x - \delta)/u(x)) \wedge \min_{(b,v) \in T_B} b(1 - v(1 - x - \delta)/v(1 - x))$$
$$\wedge \min_{(b,v) \in T_B} bv(1 - x)/v(1 - \delta).$$

As before we are able to omit the terms of form $au(x)/u(1 - \delta)$, because for *every* type $(a, u) \in T_A$ we have

$$b^*(1 - v^*(1 - x - \delta)/v^*(1 - x)) \leq au(x)/u(1 - \delta),$$

where (b^*, v^*) is a type with minimum information in T_B.

It is easy to see that $R_\delta(x)$ is unimodal. An argument analogous to the proof of Lemma 2 shows that $(x, 1 - x)$ is generically stable if and only if x maximizes $R_\delta(x)$. Hence there are at most two generically stable conventions and they differ by δ. An argument similar to the proof of Lemma 3 shows that these stable conventions converge to the unique maximum of the strictly quasiconcave function:

$$R(x) = \min_{(a,u) \in T_A} au^+(x)/u(x) \wedge \min_{(b,v) \in T_B} bv^-(1 - x)/v(1 - x). \tag{16}$$

Theorem 4 *Let A and B be two finite populations of agents and let T_A and T_B be the types represented in each class, where at least one agent in each class samples at most half the records. For every precision $\delta > 0$ there exists at least one and at most two generically stable divisions, and as $\delta \to 0$ they converge to the unique division $(x, 1 - x)$ such that x maximizes $R(x)$.*

The solution described in Theorem 4 is a generalization of the Nash solution to heterogeneous populations of bargainers that we shall call the *heterogeneous bargaining solution.* It differs from Harsanyi and Selten's solution (Harsanyi and Selten 1972), which depends on the relative frequency with which the utility functions are represented in each class. Specifically, the Harsanyi–Selten solution maximizes the product

$$\prod_{u_i \in A} u_i^{\lambda_i}(x) \prod_{v_j \in B} v_j^{\mu_j}(1 - x),$$

subject to $0 \leq x \leq 1$, where λ_i is the relative frequency of u_i in population A and μ_j is the relative frequency of v_j in population B. The heterogeneous bargaining solution, by contrast, depends only on the types of bargainers represented in each population, not on their relative frequency.

Example 2 Let population A consist of two types of agents: $(a_1 = 1/4, u_1(x) = x)$ and $(a_2 = 1/3, u_2(x) = x^{1/3})$. Population B consists of two other types of agents: $(b_1 = 1/5, v_1(y) = y)$ and $(b_2 = 1/2, v_2(y) = y^{1/2})$. In population A the criterion $au'(x)/u(x)$ is minimized for every x by the second type: $a_2 u'_2(x)/u_2(x) = 1/(9x)$. In population B the criterion $bv'(y)/v(y)$ is minimized for every y by the first type: $b_1 v'_1(y)/v_1(y) = 1/(5y)$. Hence $R(x) = 1/(9x) \wedge 1/(5(l - x))$. It achieves its unique maximum when $1/(9x) = 1/(5(1 - x))$, that is, when $x = 5/14$. In other words, the stable division is $(5/14, 9/14)$ for any two populations A and B consisting of the types mentioned above, irrespective of the distribution of these types within each population.

7 Mixing Between Classes and the Fifty–Fifty Solution

Under some circumstances it is reasonable to expect that the types represented in each class will be the same, or at least the same with positive probability. For example, suppose that there is some mobility between classes. Every so often a landlord (or the child of a landlord) loses his land and becomes a tenant, and every so often a tenant climbs into the ranks of the landlords. Then at any given time there is a positive probability that the two bargainers could be *any* pair of types drawn from *either* of the two original classes.

This idea can be made more general as follows. Let T be the set of all possible types that either the landlords or tenants could be. Let $\pi(\tau, \tau')$ be the probability that the type-pair $(\tau, \tau') \in T \times T$ is chosen to bargain in any given period. The first entry τ denotes the landlord's type and the second entry τ' denotes the tenant's type. Up to now we have been assuming that T is the union of two classes T_A and T_B and that $\pi(\tau, \tau') > 0$ if and only if $\tau \in T_A$ and $\tau' \in T_B$. The above arguments and theorems generalize, however, to other probability distributions π defined on $T \times T$.

In particular, we say that the distribution π *mixes roles* if every pair in $T \times T$ occurs with positive probability. Mixing is a natural consequence of mobility between classes. It also holds if the classes are rigid but tenants and landlords are drawn from the same "gene pool": every individual has a positive probability of being any type τ, though the probability of being a τ may differ for landlords and tenants. When π mixes roles, it is straightforward to verify that

Theorem 4 still holds with $T_A = T_B = T$ in the expression for $R(x)$. This implies that $R(x)$ is symmetric about one-half.

Corollary 4.1 *Let T be a common set of types and let $\pi(\tau, \tau') > 0$ be the probability that any given landlord and tenant are of types τ and τ', respectively. Then fifty–fifty is the unique generically stable division.*

This result does not say that society ultimately locks into fifty–fifty and stays there forever. Rather, the evolutionary process favors a solution near fifty–fifty when there is mixing between the classes of bargainers. In other words, if memory is large, the noise is small, and there is some mixing between classes, the chances are good that the customary division will be about fifty–fifty.

Acknowledgments

This research was sponsored in part by the Santa Fe Institute. The author is indebted to Brian Arthur, David Canning, Ariel Rubinstein, Sangyoung Sonn, and several anonymous referees for comments on an earlier draft. Special thanks go to Dean Foster for providing suggestions and criticism along the way.

References

Bardhan, P. (1984), *Land. Labor, and Rural Poverty* (Delhi: Oxford University Press).
Binmore, K. G. (1987), “Nash bargaining theory II, and Nash bargaining theory and incomplete information,” in K. G. Binmore and P. Dasgupta, eds., *The Economics of Bargaining* (Oxford: Basil Blackwell).
Canning, D. (1991), “Average Behavior in learning models,” Working paper. Pembroke College, Cambridge.
Carlsson, H. (1991), “A bargaining model where parties make errors,” *Econometrica* 59: 1487–1496.
Harsanyi, J. and Selten, R. (1972), “A generalized Nash solution for two person bargaining games with incomplete information,” *Management Science* 18: 80–106.
Kalai, E. and Smorodinsky, M. (1975), “Other solutions to Nash’s bargaining problem,” *Econometrica* 45: 513–518.
Kandori, M., Mailath, G., and Rob, R. (1993), “Learning, mutation, and long-run equilibria in games,” *Econometrica* 61: 29 56.
Nash, J. (1950), “The bargaining problem,” *Econometrica* 18: 155–162.
Nash, J. (1953), “Two-person cooperative games,” *Econometrica* 21: 128–140.
Raiffa, H. (1953), “Arbitration schemes for generalized two-person games,” in H. Kuhn and A. W. Tucker, eds., *Contributions to the Theory of Games* (Princeton, NJ: Princeton University Press).

Rockafellar, R. T. (1970) *Convex Analysis* (Princeton. NJ: Princeton University Press).
Rubinstein, A. (1982), "Perfect equilibrium in a bargaining model," *Econometrica* 50: 97–109.
Stahl, I. (1972), *Bargaining Theory* (Stockholm: Stockholm School of Economics).
Young, H. P. (1993), "The evolution of conventions," *Econometrica* 61: 57–84.

Preface to Chapter 4

Who Sets the Rules of the Game?

The preceding chapter examined the question of how customary shares arise in a repeated bargaining situation between principals and agents. The protagonists change from one period to the next and the demands made in previous encounters shape the expectations of the bargainers in the current period. A key assumption is that the bargaining process in each period is governed by the Nash demand game: each side demands a share of the pie, they receive their demands if they are compatible, and otherwise they get nothing.

This chapter introduces an alternative model in which not only the shares are being negotiated, but the general terms that will govern the agents' relationship. A specific set of terms will be called a *contract.* Examples include leasing arrangements between tenants and landlords, sales agreements between homeowners and realtors, employment contracts, construction contracts, consulting agreements, and the like.

The terms of a contract are often quite complex and cover various contingencies that might arise, including payment schedules, performance bonuses, notification clauses, and penalties for breach of contract. In any given line of business, however, contractual terms often follow a specific set of conventions or norms. The advantage of such a norm is that accumulated experience with a given contract structure allows loopholes, omissions, and ambiguities to be discovered and corrected; furthermore, legal precedents will accumulate as to what the terms mean and what constitutes a breach of contract. Customary practice gives the parties confidence in how a given contract will actually play out; it also provides a natural focal point for the bargaining process itself.

This chapter shows how customary contracts can evolve through a decentralized, adaptive process in which agents are randomly matched and their expectations are guided by precedent. Consider two populations consisting of potential principals and agents. A principal–agent pair is chosen at random, and each names a contract: they enter into a relationship if and only if they name the *same* contract. Unlike the Nash demand game, there is no partial fall-back in the event of miscoordination. If the parties cannot agree on a specific contract, they will fail to enter into a relationship and as a consequence their payoffs are zero. Once a contract is agreed upon, the terms define the rules that will govern their interaction: the contract defines the "rules of the game." The choice of contract is therefore a meta-game that has the structure of a pure coordination game defined over the set of feasible contracts.

The evolutionary framework is similar to that described in Chapters 2–3. Agents draw a random sample of demands made in prior encounters. With high probability they make an optimal one-shot choice of contract based on their sample information (i.e., they maximize the utility of the proposed contract times the estimated probability it will be accepted). With small probability they deviate from best-reply behavior due to mistakes, experiments, or idiosyncratic utility shocks. The key finding is that, when the random deviations have small probability, such a process leads to a situation in which one contract form is used by most people most of the time. Moreover, this stochastically stable contract is efficient in the sense that no other contract would yield strictly higher expected payoffs for both parties.

In this set-up, unlike the bargaining model in Chapter 3, the stochastically stable contract may differ from the Nash bargaining solution. In particular, if the range of contract outcomes can be represented by a convex set in payoff space (which can be achieved by treating some contracts as lotteries over other contracts), the outcome is the Kalai–Smorodinsky solution of the corresponding bargaining problem. This outcome is efficient and equalizes the ratio of each party's payoff to the maximum payoff it could achieve under some contract. When selection is between just two contracts, this reduces to a very simple criterion: the stochastically stable contract is the one that maximizes the product of the parties' expected payoffs.[1]

This framework can be extended to a model of how institutions—construed as formal and informal rules of the game—evolve through the accumulation of precedent.[2] Conventional game theory either takes the rules of the game as given, or seeks to identify rules that lead to desirable outcomes in equilibrium, as in the mechanism design literature. In the evolutionary framework proposed here, no one "sets the rules"; they emerge as a norm through the accumulation of precedent. The framework shows in particular that *efficient* game forms have an evolutionary advantage over inefficient game forms. Furthermore, within the class of efficient game forms, evolutionary forces tend to favor those that equalize the expected payoffs to the two sides relative to their

[1] As in other evolutionary models, the nature of the random deviations can affect the stochastically stable outcome. For example, if agents only deviate from best reply by demanding a contract that would be more favorable to themselves (if adopted by the other side), then the outcome is the Nash bargaining solution rather than the Kalai–Smorodinsky solution (S. Naidu, S.-H. Hwang, and S. Bowles (2010), "Evolutionary bargaining with intentional idiosyncratic play," *Economics Letters* 109 (1): 31–33.

[2] Douglass North (1991). "Institutions," *Journal of Economic Perspectives* 5 (1): 97–112.

maximum feasible payoffs. One can view this result as saying that equitable institutions and social arrangements are more likely to be sustained over the long run than inequitable arrangements that are systematically biased toward particular groups.

The evolutionary framework also shows, however, that it may take a long time before an equitable outcome is realized. Societies can, and often do, become trapped in inequitable regimes and institutions that discriminate against some groups, such as norms of compensation and promotion that discriminate against women and Blacks, or that put all marital property under the control of the husband, or that leave all inherited property to the eldest son (primogeniture). Once such discriminatory institutions become entrenched they can be very difficult to undo by individual initiative; it make take top-down intervention, coordinated effort, or organized revolt to alter them in a reasonable length of time.[3] The evolutionary framework shows, on the one hand, that discriminatory "rules of the game" may stay in place for long periods of time, and on the other hand that more equitable arrangements will *eventually* emerge and are inherently more stable once they become established. The time it takes for the long-run regime to be realized depends on a variety of factors, including the nature of the game and the topology through which individuals interact. This is the subject of the final two chapters in this volume.

[3] The incorporation of collective action into evolutionary models is discussed by H. P. Young (1998), *Individual Strategy and Social Structure*, Chapter 9, Princeton University Press; J. Newton and S. D. Angus (2015), "Coalitions, tipping points, and the speed of evolution," *Journal of Economic Theory* 157: 172–187, and S. Bowles, J.-K. Choi, S.-H. Hwang, and S. Naidu (2021), "How institutions and cultures change: an evolutionary perspective," in A. Bisin and G. Federico, eds. *The Handbook of Historical Economics*. Elsevier.

4
Conventional Contracts

> Contract as a social phenomenon is the result of a coincidence of free choices.
>
> —Friedrich Kessler

1 Contracting Games

Contracts define the terms on which people enter into relationships with one another. Some contracts are explicit, such as rental agreements between tenants and landlords, or lending agreements between bankers and borrowers. Others are implicit, such as the common understandings that a couple brings to a marriage. Still others occupy a middle ground: employment contracts are usually explicit about some matters, such as the number of hours to be worked, but rather vague about others, such as the relationship between performance and pay. Contracts often have a standard format; their terms follow conventional patterns. Renters and landlords fill in the blanks of standard leases. Building contractors typically negotiate cost-plus or fixed-price contracts. Farmers either share the crop in fixed proportions with their tenants, or rent out the land at a fixed price per acre, and so forth. Not only do the structural terms of contracts follow conventional patterns, frequently their specific terms do too. Cost-plus contracts customarily employ a standard percentage markup (e.g., 20%); lawyers' contingency fees are usually one-third of the jury award; tenants and landlords often divide the harvest fifty–fifty. While opportunity costs undoubtedly constrain the terms that rational agents will accept, there appears to be a strong degree of inertia in these contractual arrangements that results from custom and precedent.

The question that we consider in this chapter is how conventional contracts become conventional, and whether some types of contracts are more likely to become conventional than others. Our point of departure differs from the traditional literature on contracts, which is mainly concerned with the *optimal* terms of a contract given informational conditions, monitoring costs, and

Originally published H. Peyton Young, "Conventional contracts," *The Review of Economic Studies*, Volume 65, Issue 4, October 1998, Pages 773–792. Reproduced with permission.

Spontaneous Order. H. Peyton Young, Oxford University Press. © H. Peyton Young (2024).
DOI: 10.1093/oso/9780198892908.003.0004

other variables that are specific to the contractual situation at hand. Of course, one reason why a contract might be conventional is that it is uniquely optimal, in which case the convention and optimality stories come together. Our primary interest, however, is in cases where there is a *range* of optimal contracts that have different payoff implications for the two sides. Here convention plays a key role by providing a focal point that helps the parties resolve the indeterminacy of the bargaining situation.

At the individual level, we model the contract selection problem as a coordination game. Consider two agents who can choose among m different contracts. In other words they can structure their relationship in one of m different ways over a specified time period. We assume that the terms of the contract are enforceable, and that there is no renegotiation over the period. The game is that each side names a contract. They enter into a relationship if they name the *same* contract; otherwise they are unattached for the period. We make no assumption that the agents select only optimal contracts (this will be derived as a consequence of the model). We do assume however that all contracts are *desirable* (strictly individually rational) in the sense that the expected utility for each side is better than the utility of being unattached.

Altogether there are $m + 1$ states that an individual can occupy in any given period—be a party to one of the m contracts, or be unattached. Let (a_i, b_i) be the expected utility to the row and column player respectively from being in contract i, and let (a_0, b_0) the expected utility from being unattached. There is no loss of generality in assigning a utility of zero to the unattached state for each player, and we shall assume this henceforth. We therefore obtain a pure coordination game in which the diagonal payoffs are strictly positive and the off-diagonal payoffs are zero. We shall call this type of coordination game a *contracting game.*

While contracting games are among the simplest of games in a structural sense, they are among the most difficult to solve in a deductive sense because of the multiplicity of equilibria. Indeed, if none of the equilibrium payoffs Pareto-dominates another, it is not clear that any equilibrium can be eliminated a priori. A canonical example is the Battle of the Sexes game. Moreover, even if an equilibrium is Pareto-dominated, it is nonetheless an equilibrium, so it will still be played by rational players if each expects the other to play it. It is in precisely this type of game that precedent and custom play an important role by conditioning players to expect a given outcome, thus reducing the risk of miscoordination. In this chapter we study the process by which such customs evolve, using the framework developed in Chapter 2 (Young 1993a), which builds on work by Freidlin and Wentzell (1984), Foster and Young (1990), and Kandori, Mailath and Rob (1993).[1]

Given are two populations of players: tenants and landlords, employers and employees, bankers and borrowers, men and women, etc. Players are matched in successive periods with players from the opposite population. The terms that a player demands are governed by his or her expectations about the terms that the other side will demand, and these expectations are derived from observations about the terms that agents in that population actually did demand in previous periods. The evolutionary process is therefore driven by a feedback loop in which precedent shapes present expectations, which determine present actions, which become future precedents. We further assume that the process is buffeted by small stochastic shocks that represent minor disturbances, idiosyncratic aspects of individual behavior, and other random events that are not explicitly modeled.

We shall show that, when the disturbances are small, the dynamics are close to equilibrium most of the time. In fact, they are close to particular equilibria most of the time, and these *stochastically stable* equilibria can be characterized in terms of their welfare properties. First, they are efficient: no other equilibria are strictly better for both parties.[2] Second, they tend to be centrally located within the set of feasible payoffs, not near the extremes in a sense made precise in Theorem 1 below. These results hold for all finite, two-person contracting games. When we put more structure on the payoff set, considerably more can be said. In particular, if the set of payoffs $S = \{(a_i, b_i) : 1 \leq i \leq m\}$ forms a discrete approximation of a convex bargaining set, then the stochastically stable equilibria approximate the Kalai–Smorodinsky bargaining solution. In other words, the expected payoff of each party, divided by its maximum payoff, is essentially the same for both parties.

At a purely formal level these results can be regarded as equilibrium selection theorems for a class of two-person coordination games. Interpreted in the context of contract evolution, the results say that, in the long run, contracts are more likely to be conventional if they offer both sides fairly high expected payoffs given the feasible payoff opportunities.

[1] For related work on convention see Lewis (1969), Ullman-Margalit (1977), Sugden (1986), Ellison (1993), Waneryd (1993, 1994), Binmore (1994), Binmore and Samuelson (1994), Shin and Williamson (1996), Kandori and Rob (1995), Vega-Redondo (1996), and Samuelson (1997).

[2] This generalizes a result of Kandori and Rob (1995), who show that for symmetric contracting games the stable contract is efficient. When the game is symmetric, the Pareto frontier consists of a single point, so the issue of "centrality" is moot.

2 The Evolutionary Selection Process

Let A and B denote the populations of row and column players respectively. Each population consists of k individuals. The process evolves in discrete time intervals $t = 1,2,3,\ldots$. The *state* at the end of period t is a pair of integer-valued vectors $z^t = (x^t, y^t)$ where $x^t{}_i$ is the number of row players who proposed contract i in period t, and $y^t{}_i$ is the number of column players who proposed contract i in period t, $\sum x^t{}_i = \sum y^t{}_i = k$. Notice that the state is described by the distribution of choices in the population, not the choices of specific individuals. This reflects the idea that individuals are constantly entering and exiting the bargaining populations. We can therefore think of each member of the population as a *position* that is occupied by different individuals over time.

A *matching* is a subset of k pairs from $A \times B$ such that every member is contained in exactly one pair. Let M be the set of all matchings. At the beginning of period $t + 1$, everyone in the population is matched with someone from the opposite population by a random draw from M. For simplicity we shall think of the draw as being uniform, but we could equally well assume that the probability of a matching depends on the geographical or social proximity of individuals, which affects the probability that they meet. All that matters for our purposes is that each matching have a positive probability that is bounded away from zero over all time periods.[3]

Once a matching has been chosen, each person in each matched group names a contract—the proposed terms of their relationship. If both name the same contract i, they enter into this contract in period $t + 1$ and receive the expected payoffs a_i, b_i respectively. If they name different contracts (i.e., they cannot agree on terms), they are unattached for the period and their payoffs are zero. Each individual decides what terms to demand by consulting the frequency distribution of demands made by the other side in the previous period. He uses this distribution to predict the likelihood that his current partner will make various demands now. These assumptions obviously sacrifice some degree of realism, but still have considerable justification in a large-population setting.

The myopic best reply of a row player, given the state $z = (x, y)$, is to demand a contract i that maximizes $a_i y_i$, $1 \leq i \leq m$. Similarly, the best reply of the column player is to maximize $b_i x_i$, $1 \leq i \leq m$. If two or more contracts constitute a best reply for some agent, we can suppose that she chooses among them according to a fixed probability distribution that has full support. These

[3] The amount of local interaction does not materially affect the stochastically stable outcome, but it does affect the waiting time until the process reaches that outcome. See Section 10 below.

tie-breaking rules (which may be unique to each agent) are fixed throughout and will not be listed explicitly as parameters of the process.

Individuals can deviate from best reply in two ways: through inertia and random error. Inertia means that the player adopts the same choice that he (or his predecessor) did in the previous period. Error encompasses random shocks to utility, experimentation, and various idiosyncratic aspects of behavior that we do not model explicitly. Let $\nu \in (0,1)$ be the inertia probability and let $\varepsilon \in [0,1)$ be the error probability. In each period, a player sticks to his previous choice with probability ν, chooses a best reply with probability $(1 - \nu)(1 - \varepsilon)$, and chooses a contract at random with probability $(1 - \nu)\varepsilon$.[4] These behavioral rules define a Markov chain $P^{k,\nu,\varepsilon}$ on the finite state space Z consisting of all pairs $z = (x, y) \in R^m \times R^m$, where $\sum x_i = \sum y_i = k$, and x_i, y_i are non-negative integers.

When ε is positive, there is a positive probability of moving from any state z to any other state z' in one period, because everyone could make an error (i.e., choose a contract at random). Hence the process is irreducible. It is a standard result that a finite, stationary, irreducible Markov chain has a unique stationary distribution $\mu^{k,\nu,\varepsilon}(z)$. Over almost all realizations of the process, $\mu^{k,\nu,\varepsilon}(z)$ is the relative frequency with which the process visits state z.

A *convention* is a state in which all players demand the same contract, which we shall call a *conventional contract.* Such a state has the form $z^i = (ke^i, ke^i)$, where e^i is the unit vector that has 1 in position i and 0 elsewhere.

An *absorbing state* is a state z such that, if the process is in state z in some period, it remains in that state forever. When $\varepsilon = 0$, every convention is an absorbing state. Indeed, if i is conventional, the unique best reply for every agent is to choose i, so the convention perpetuates itself. We claim that the conventions are the *only* absorbing states. To see this, suppose that $z = (x, y)$ is absorbing. Let i be a contract such that $x_i > 0$. Obviously i is row's best reply to y, for otherwise x_i could not be the frequency of i in the next period (which it must be because (x, y) is an absorbing state). Moreover, i must be a *strict* best reply to y, because otherwise there is a positive probability that no row player will play i in the next period (recall that the tie-breaking rules have full support), thus violating the hypothesis that (x, y) occurs in the next period with probability one. It follows that $x = ke^i$. Since the support of y consists only of best replies to x, and the best reply to x is uniquely i, it must be that $y = ke^i$. Hence z is a convention.

[4] For notational simplicity, we state results in terms of a uniform inertia rate ν and a uniform error rate ε. However, the results continue to hold even when the inertia probabilities differ among individuals. One can also assume non uniform error probabilities so long as they approach zero at the same rate, that is, so long as the *relative* probability of error between any two individuals is fixed.

When the disturbance term ε is small but positive, it can be shown that the stationary distribution $\mu^{k,\nu,\varepsilon}(.)$ puts almost all of the probability on one or more conventions z^i. To be precise, there is a unique, non-empty set of conventions $Z^* \subseteq \{z^1, z^2, \ldots, z^m\}$ such that[5]

$$\lim_{\varepsilon \to 0} \mu^{k,\nu,\varepsilon}(z) > 0 \quad \text{if and only if } z \in Z^*. \tag{1}$$

Equivalently, Z^* is the minimal set of states such that, given any $p < 1$, $\sum_{z \in Z^*} \mu^{k,\nu,\varepsilon}(z) \geq p$ for all sufficiently small $\varepsilon > 0$. Z^* is called the set of *stochastically stable states* (Foster and Young (1990)). The meaning of stochastic stability is that the process is very likely to be in such a state when the perturbation probability ε is small but non-vanishing. In other words, when conventions are occasionally but persistently challenged by innovations, the stable one(s) are more likely to be observed than the others over the long run.

Of course, since the state space Z is determined by the population size k, it is conceivable that stochastic stability also depends on k. Fortunately this is not an issue when k is large: for generic contracting games there exists a unique contract i such that the corresponding convention z^i is stochastically stable for all sufficiently large k. Even for nongeneric contracting games, there exists a unique *set* of contracts that are stochastically stable for infinitely many values of k. In general, we say that contract i is *stable* if for every $\nu \in (0,1)$ there are infinitely many values of k such that $\lim_{\varepsilon \to 0^+} \mu^{k,\nu,\varepsilon}(z^i) > 0$.

3 Maximin Contracts

A contract is *efficient* if there exists no other contract that yields strictly higher payoffs to both parties. It is *strictly efficient* if there is no other contract that yields at least as high a payoff to one party and a strictly higher payoff to the other. Let $a^+ = \max\{a_i : 1 \leq i \leq m\}$ denote the maximum feasible payoff to the row players under some contract; similarly let $b^+ = \max\{b_i : 1 \leq i \leq m\}$ denote the maximum feasible payoff to the column players. There is no loss of generality in scaling the utility functions so that $a^+ = b^+ = 1$, and we shall assume henceforth that this normalization has been made. Define the *welfare index* of contract i to be the smaller of a_i and b_i,

$$w_i = a_i \wedge b_i, \tag{2}$$

and let

$$w^+ = \max_i w_i. \tag{3}$$

[5] This result follows from Theorem 4 in Chapter 2 (Young 1993a).

A contract with welfare index w^+ is a *maximin contract*. A maximin contract favors the least-favored class in the sense that no other contract improves the position of the least-favored class relative to its most-preferred contract. Figure 4.1 illustrates the welfare indices of eight different contracts. The intersections of the dotted lines with the diagonal determine the welfare indices w_i, and w^+ is the value that lies furthest out along the diagonal.

A maximin contract obviously involves interpersonal utility comparisons. But why is it legitimate to compare utilities in this way? From the individual's perspective it is not legitimate, but we are not assuming that individuals make such comparisons. (How could they if they do not know the others' utility functions?) The claim is that *society* implicitly makes interpersonal comparisons through the feedback effect of precedents on expectations. Moreover, there is a simple intuitive explanation for this phenomenon. By assumption, each individual chooses a strategy based on the strategy's expected payoff times the probability with which she expects the other side to choose it. In a stochastic environment, this means that individuals with different utility functions adjust their choice behavior at different rates. Interpersonal comparisons therefore arise because *differences in von Neumann–Morgenstern utility imply different rates of behavioral adaptation when there is stochastic variability in the environment*. In the remainder of the chapter we examine the specific implications of this idea for the adaptive learning process described above.

4 A Contract Selection Theorem

Let a^- denote the minimum payoff to the row player among all contracts in which the column players get their maximum, that is, $a^- = \min\{a_i : b_i = 1\}$. Similarly let b^- be the minimum payoff to the column player among all contracts in which the row players get their maximum (see Figure 4.1). Let

$$w^- = a^- \vee b^-. \qquad (4)$$

Normally we would expect w^- to be small, because eking out large gains for one class will generally impose a cost on the other class due to substitution possibilities. We claim that the smaller w^- is, the closer is the stochastically stable outcome to the maximin solution. Specifically, define the *distortion parameter* α as follows

$$\alpha = w^- \left(1 - (w^+)^2\right) / (1 + w^-)(w^+ + w^-). \qquad (5)$$

Note that α is small when w^- is small and/or w^+ is close to one.

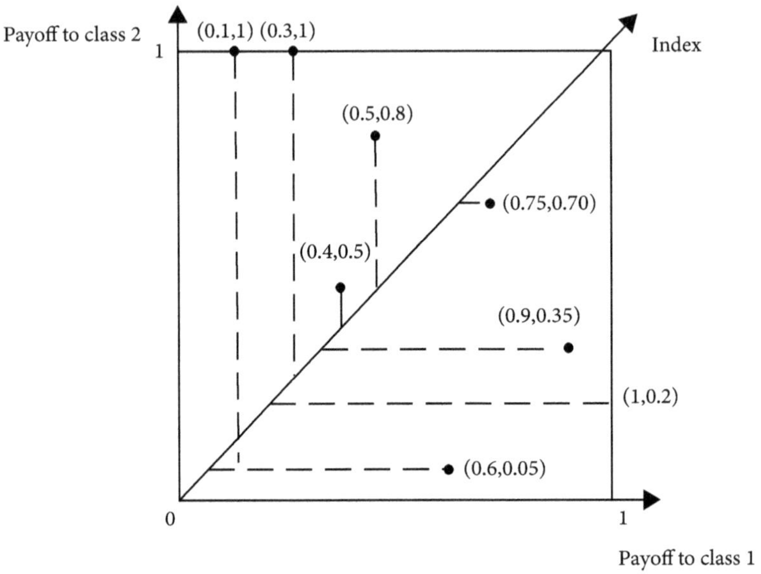

Figure 4.1 Welfare indices for eight hypothetical contracts.

Theorem 1 *Every stable conventional contract is efficient and approximately maximin in the sense that its welfare index is at least* $(w^+ - \alpha)/(1 + \alpha)$.

The usefulness of this bound depends on the problem at hand. In the example of Figure 4.1, $w^+ = 0.700$, $w^- = 0.200$, and $\alpha = 0.067$. Thus the theorem says that the stochastically stable outcome has welfare index at least $(0.700 - 0.067)/1.067 = 0.593$. It is immediately evident from the figure that the only contract that lies above this threshold is the maximin contract. If there were other such contracts the theorem would not eliminate them; in general, however, the addition of more contracts will tend to increase w^+ and to decrease w^-, which decreases α and correspondingly increases the cutting power of the theorem.

While the criterion in Theorem 1 does not provide a sharp lower bound on welfare for most contracting games, it is nevertheless best possible for some games, as we shall show by example in Section 6 below. Moreover the criterion can be further sharpened if we specialize to particular classes of games. In particular the stable contract is the same as the maximin contract in 2×2 games and symmetric contracting games. Moreover, if the payoffs form a discrete approximation of a convex bargaining set, the stochastically stable outcome comes arbitrarily close to the maximin contract, which coincides with the Kalai–Smorodinsky solution, as we show in Section 7.

5 Computation of Stable Equilibria using Rooted Trees

The method for computing the stochastically stable equilibria relies on techniques in stochastic dynamical systems theory (see the Appendix to Chapter 2).[6] Consider m contracts with expected payoffs $\{(a_i, b_i) : 1 \leq i \leq m\}$, and suppose that the process is currently in convention i. (We shall sometimes refer to a convention by its index.) To displace i in favor of some other convention, say j, requires that a certain number of row or column players choose j by mistake. Ignoring integer problems for the moment, we need to find the minimum proportion p^* such that if p^* of the row players switch to j and $1 - p^*$ stay with i, then j is a best reply for the column players. Evidently p^* is the solution of $p^* b_j = (1 - p^*)b_i$, that is, $p^* = b_i/(b_i + b_j)$. Similarly, let q^* be the minimum proportion such that if q^* of the column players choose j and $1 - p^*$ choose i, then j is a best reply for the row players. Clearly, $q^* = a_i/(a_i + a_j)$. Letting k be the population size, it follows that the minimum number of errors necessary to tip the process out of convention i and into the basin of attraction of convention j is

$$r_{ij}^k = \lceil kr_{ij} \rceil, \text{where } r_{ij} = a_i/\left(a_i + a_j\right) \wedge b_i/\left(b_i + b_j\right). \tag{6}$$

(In general, $\lceil x \rceil$ denotes the least integer greater than or equal to x.) The integer $r^k{}_{ij}$ is the *resistance* to transiting from convention i to convention j. When k is large, the numbers $r^k{}_{ij}$ are more or less proportional to the real numbers r_{ij}, which we shall call the *reduced resistances* to transiting between conventions.

Now construct a graph having m vertices, one for each contract. For each ordered pair of vertices (i, j), $i \neq j$, draw a directed edge from i to j and give it weight $r^k{}_{ij}$. An *i-tree* is a subset of $m - 1$ directed edges in the graph such that for every vertex $j \neq i$ there is a unique directed path from j to i. The vertex i is called the *root* of the tree. The *resistance* of the tree is the sum of the resistances on its edges. The *stochastic potential* of contract i is the smallest resistance among all i-trees. It can be shown that *the stable contracts are precisely those with minimum stochastic potential.*[7] Note in particular that the resistances are independent of how we scale each player's utility function; hence stability is scale-independent as well.

We illustrate these ideas with an example involving marriage contracts. Let the two populations consist of men and women. In each period, the eligible

[6] See also Freidlin and Wentzell (1984) andKandori, Mailath, and Rob (1993). Ellison (1996) develops a different computational approach that is convenient for analyzing certain cases and does not rely on the notion of rooted trees.

[7] This is an application ofTheorem 4 in Chapter 2; see the Appendix for details.

men are matched with the eligible women. A matched pair gets married if (and only if) they can agree on control of the marital property. For simplicity assume that there are only three possible contracts: the husband has full control over the property, the wife has full control over the property, or they share control.[8] If they name the same contract they get married, if they name different ones they break up. Let the payoffs be as follows:

Marriage Contracting Game

		Men cede control	Men share control	Men take control
	take control	5,1	0,0	0,0
Women	share control	0,0	3,3	0,0
	cede control	0,0	0,0	1,5

The fact that each outcome has the same total utility is not significant; indeed the utility functions can be scaled in any way that we wish without affecting the resistances between the coordination equilibria.

Figure 4.2 shows the weighted directed graph from which stochastic potential is calculated. From this it follows that sharing is the unique stable contract. Concretely this means that, when the noise is small, most people will be using the same contract most of the time, and this customary contract is more likely to be the sharing contract than either of the others.[9]

Of course we should not take this result too literally, as the model is highly stylized; nevertheless the explanation for the result is robust and makes intuitive sense. Conventions with extreme payoffs are relatively easy to dislodge because some group is highly dissatisfied, which means they do not have much to lose by trying something different. By contrast, conventions whose payoffs are centrally located in the feasible payoff set are relatively hard to dislodge, because it takes a larger accumulation of stochastic shocks to induce any group to change. The net effect, over the long run, is to push society away from the boundaries of the feasible payoff set and toward the middle of the efficiency frontier. (The proof of this result requires a considerably more complex argument, but this is the essence of the matter.)

[8] The division of property rights in marriage has in fact evolved substantially over the last several hundred years. See for example Holcombe (1983).

[9] Joint ownership is stable even if the expected payoffs to sharing are lower than 3, say due to higher transactions costs incurred by joint management of the property. Any payoff greater than $\sqrt{5}$ will work.

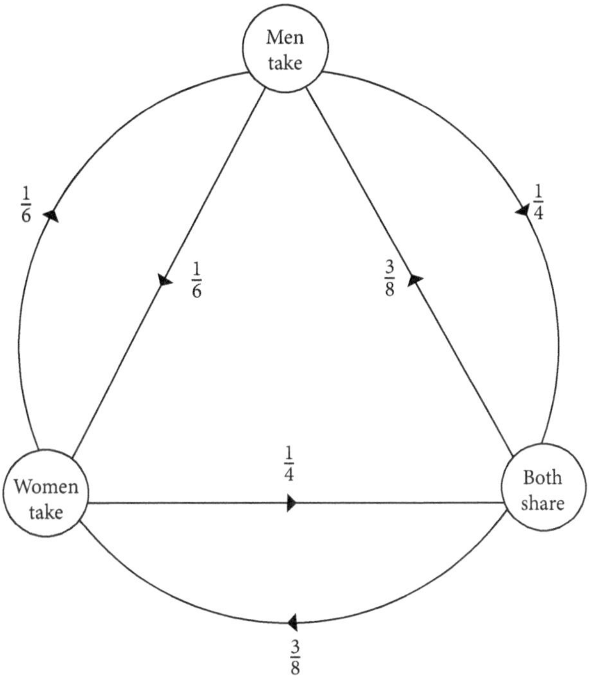

Figure 4.2 Reduced resistances between equilibria in marriage contracting game.

6 Examples Showing Departures from Strict Efficiency and Exact Maximin

We now show why Theorem 1 cannot be substantially strengthened. First we exhibit a situation where the stable contract is efficient but not strictly so. In Figure 4.3, contract 1 weakly Pareto dominates contract 2, but the least-resistant 1-tree (left panel) and the least-resistant 2-tree (right panel) are isomorphic and have the same reduced resistance. Hence both contracts are stable.

The next example shows that the bound in the theorem is tight when there are at least four games. Choose real numbers w^-, w, and w^+ such that

$$0 < w^- < w < w^+ < 1.$$

Consider four contracts with the payoffs shown in Figure 4.4, where w^- and w^+ are as defined in (3) and (4). Figure 4.4(a) shows the least-resistant tree rooted at the maximin outcome (w^+, w^+); Figure 4.4(b) illustrates the least-resistant

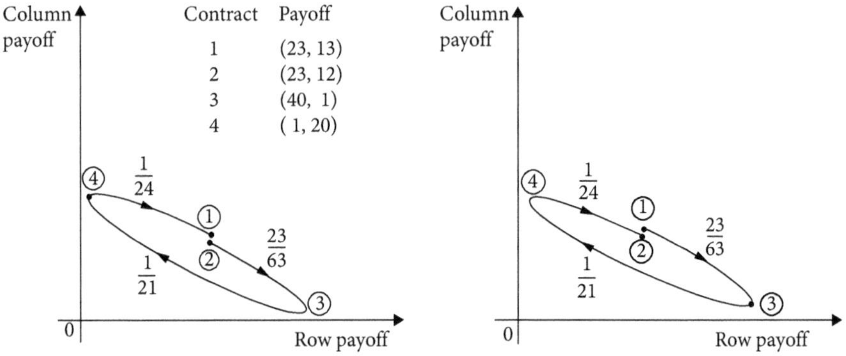

Figure 4.3 Left: Least resistant l-tree, Right: Least resistant 2-tree.

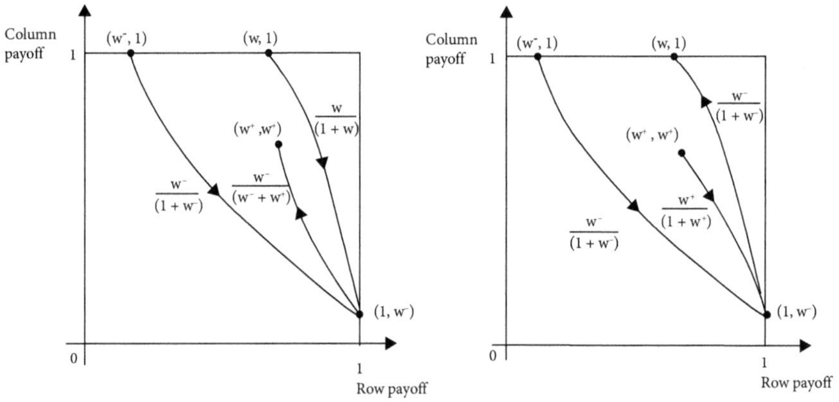

Figure 4.4 Left: Least-resistant tree rooted at vertex with payoffs (w^+, w^+). Right: Least-resistant tree rooted at vertex with payoffs $(w, 1)$.

tree rooted at the outcome $(w, 1)$. The latter has the same or lower stochastic potential than the maximin outcome if

$$\frac{w^-}{1+w^-} + \frac{w^+}{1+w^+} + \frac{w^-}{1+w^-} \leq \frac{w^-}{1+w^-} + \frac{w^-}{w^- + w^+} + \frac{w}{1+w}.$$

This reduces to $w \geq (w^+ - \alpha)/(1 + \alpha)$, where α is defined as in (5). Hence the bound in Theorem 1 is best possible when the game is 4×4. The argument extends easily to larger games.

The example in Figure 4.4 is non-convex in the sense that the maximin payoffs are strictly interior to the convex hull of the feasible payoff set. More generally, say that a discrete set S of payoff vectors is *northeast-convex* if every vector that is efficient with respect to S is also efficient with respect to the convex hull of S. One might imagine that this condition guarantees stability of the maximin outcome, but this is not the case. Consider the four payoff vectors $S = \{(0.1, 1), (1, 0.1), (0.7, 0.98), (0.8, 0.75)\}$. As Figure 4.5 shows, this set

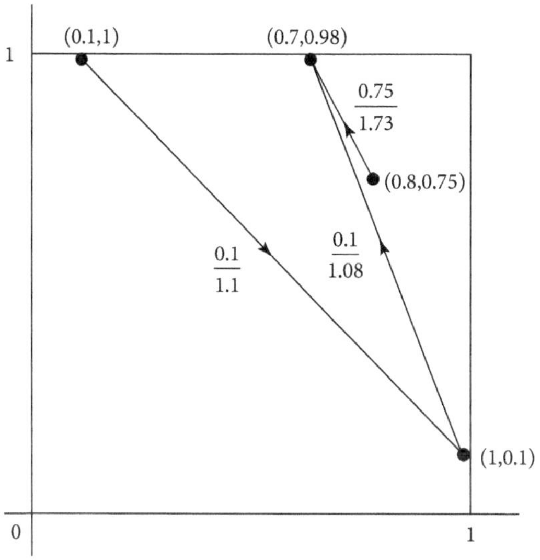

Figure 4.5 The least-resistant rooted tree for a northeast-convex set of outcomes.

is northeast convex; nevertheless (0.7, 0.98) is the unique stable outcome, whereas (0.8, 0.75) is the maximin outcome. The rooted tree of least total reduced resistance is depicted in Figure 4.5.

7 The Kalai–Smorodinsky Solution

When we add more structure to the set of payoffs, we obtain sharper results. Suppose in particular that the set of payoffs closely approximates a bargaining set. By a *bargaining set* we mean a compact, convex, comprehensive, full-dimensional subset C in R^2_+. As before, we can normalize utilities so that

$$\max\{x : (x, y) \in C\} = \max\{y : (x, y) \in C\} = 1.$$

The *Kalai–Smorodinsky solution* (Kalai and Smorodinsky, 1975) is the unique maximal vector of form $(x^*, x^*) \in C$.

Discretize C as follows: for each small $\delta > 0$, let C_δ consist of all payoff vectors $(a, b) \in C$ such that a/δ and b/δ are strictly positive integers. Clearly $C_\delta \to C$ in the Hausdorff metric. Rescaling the utility functions if necessary, define $w^+(\delta)$ and $w^-(\delta)$ as in (3) and (4) respectively. Since C is comprehensive, $w^-(\delta) \to 0$ as $\delta \to 0$. Since C is convex, $w^+(\delta) \to w^+$ as $\delta \to 0$. Defining $\alpha(\delta)$ analogously to (5), it follows that $\alpha(\delta)\to 0$ as $\delta\to 0$. Hence the stable conventions guaranteed by Theorem 1 must be close to the Kalai–Smorodinsky solution when δ is small.[10]

To formulate this result concisely, let us say that a pair $(a, b) \in C_\delta$ is *stable* if the associated convention is stochastically stable with respect to the process $P^{k,\nu,\varepsilon}$ for infinitely many population sizes k.

Theorem 2 *Let C be a bargaining set and let C_δ be a discrete approximation of C with precision δ. As δ goes to zero, the stable payoffs of C_δ converge to the Kalai–Smorodinsky solution for C.*

As was shown in Chapter 3 (Young 1993b), the Nash bargaining solution can be derived using similar methods. Consider a fixed pie that can be divided between two agents. Let σ be the share demanded by the row player and σ' the share demanded by the column player. Suppose that the one-shot game is the Nash demand game: each party demands a strictly positive share, and they get what they ask for if the demands sum to one or less; otherwise they get nothing. Let $u(\sigma)$ be the utility of the row players as a function of the row player's share $\sigma \in [0,1]$. Similarly, let $v(\sigma')$ be the utility of the column players as a function of their share $\sigma' \in [0,1]$. As usual we can normalize utilities so that $u(0) = v(0) = 0$, i.e., the disagreement outcome is that they get nothing.

Let us discretize the state space by only allowing demands (σ, σ') that are positive integer multiples of some small $\delta > 0$, called the *precision* of the process. In a model similar to the one described above, it can be shown that, when δ is sufficiently small, the stable division(s) come arbitrarily close to the Nash bargaining solution, namely, the division $(\sigma^*, 1 - \sigma^*)$ that maximizes $u(\sigma)v(1 - \sigma)$ subject to $0 < \sigma < 1$.

Although the Nash demand game is a coordination game, it is not a contracting game in the sense of this chapter. The reason is that the parties can obtain positive payoffs even if they miscoordinate. For example, if one side demands 1/4 and the other demands 1/3, they get their demands but leave 5/12 of the pie on the table. In a contracting game, by contrast, the parties must coordinate on a *complete contract* to get anything. (In the preceding example, both would have to name the contract (1/4, 1/3) for it to be implemented). We would argue that this assumption is more realistic, especially in situations where there is no natural conception of a "partial" contract for the parties to fall back on when they fail to agree on terms.

[10] This result may also be proved without the help of Theorem 1, by showing directly that the least resistant tree is rooted at a point whose payoff is close to the Kalai–Smorodinsky solution.

8 Special Cases

In this section we mention two other classes of games for which the stable outcome is the same as the maximin outcome. One such class is the set of 2×2 contracting games, where the payoffs take the form

$$\begin{bmatrix} a_1, b_1 & 0, 0 \\ 0, 0 & a_2, b_2 \end{bmatrix}, \quad a_1, b_1, a_2, b_2 > 0.$$

Equilibrium (1,1) is risk dominant if $a_1 b_1 \geq a_2 b_2$, and equilibrium (2,2) is risk dominant if the reverse inequality holds (Harsanyi and Selten 1988). It is known that in any 2×2 coordination game, the stochastically stable equilibria are the same as the risk dominant equilibria.[11] We claim that these are the same as the maximin equilibria. This is clear enough if the risk dominant equilibrium is efficient and the other equilibrium is inefficient. Otherwise we may suppose without loss of generality that $a_1 \leq a_2$ and $b_1 \geq b_2$. The minimum welfare under equilibrium (1,1) is $w_1 = a_1/a_2$, whereas the minimum welfare under equilibrium (2,2) is $w_2 = b_2/b_1$. Thus equilibrium (1,1) is maximin if and only if $w_1 \geq w_2$, which is equivalent to $a_1 b_1 \geq a_2 b_2$, which is precisely the criterion of risk dominance.

We remark that risk dominance and stochastic stability are not equivalent in general contracting games. Consider the payoff matrix

$$\begin{bmatrix} 1, 10 & 0, 0 & 0, 0 & 0, 0 \\ 0, 0 & 4, 10 & 0, 0 & 0, 0 \\ 0, 0 & 0, 0 & 6, 6 & 0, 0 \\ 0, 0 & 0, 0 & 0, 0 & 10, 1 \end{bmatrix}.$$

A straightforward calculation using the method of rooted trees shows that (6,6) has the lowest stochastic potential, though (4,10) is risk dominant.

Another concept of equilibrium selection in the literature is p-dominance (Morris, Rob, and Shin 1995). Equilibrium i is *p-dominant* if i is the unique best reply by either player whenever the other plays i with probability at least p. In a certain sense a p-dominant equilibrium is more robust the smaller p is, because smaller p makes less demands on ex ante coordination. In the above example, equilibrium 3 is uniquely p-dominant for any $p \in (5/8, 5/7)$, and it is also stochastically stable. However, if we change the payoff vector (4,10) to

[11] This was proved (using somewhat different adaptive models) by Kandori, Mailath, and Rob (1993) for symmetric 2×2 coordination games, and by Young (1993a) for asymmetric 2×2 coordination games; see Chapter 2.

(5,10), then equilibrium 2 is stochastically stable, but equilibrium 3 still has the lowest p-dominance. The two concepts are therefore not equivalent.[12]

A second class of games for which the maximin and stable equilibria coincide are symmetric contracting games, that is, the row and column payoffs are identical ($a_i = b_i$ for all i). This situation was first analyzed by Kandori and Rob (1995). Without loss of generality we can index the contracts in decreasing order of welfare: $a_1 \geq a_2 \geq \ldots \geq a_m$. Thus the Pareto frontier consists of a single payoff vector: (a_1, a_1). The assertion is that the stochastically stable contracts are precisely the efficient contracts. Assume for the moment that contract 1 is uniquely efficient. Let us also suppose that k is large enough so that integer rounding does not pose a problem. Choose any $i > 1$. From formula (6) it follows that the edge of least reduced resistance exiting from vertex i is directed toward vertex 1, and its reduced resistance is strictly less than 1/2.

On the other hand, every edge exiting from vertex 1 has reduced resistance strictly greater than 1/2. It follows that, among all rooted trees, the tree rooted at vertex 1 and consisting of the directed edges $T=\{(i,1): i = 2,3,\ldots,m\}$ is the one with least total reduced resistance. Hence for all sufficiently large population sizes k, contract 1 uniquely minimizes stochastic potential, and is therefore the unique stochastically stable outcome. A similar argument shows that when several different contracts are efficient (and hence have the same payoffs), all of them are stable, and for large enough k they are the only stable contracts.

9 Coordination Games with Four or more Players

We remark that the above results do not extend to n-person coordination games when $n \geq 4$.[13] To illustrate the difficulty, suppose that there are exactly two contracts, each of which involves n individuals, and that the payoffs to all agents are equal. Let $a_1 > 0$ be the payoff to each agent under contract 1 and $a_2 > 0$ the payoff to each agent under contract 2. The coordination game is that all players must name the same contract, or they do not enter into a relationship and their payoffs are zero. In particular, we suppose that there is no fallback alternative even when almost all agents (say $n - 1$ out of n) name the same contract; coordination requires complete unanimity. (This is a somewhat extreme assumption of course.)

[12] However, if an equilibrium is 1/2-dominant, then it is the unique efficient outcome, and it is also the unique stable outcome. This case is considered by Ellison (1996) and Maruta (1995).

[13] The author is indebted to an anonymous referee for drawing attention to this point.

Let society be in convention 2. Suppose that, in each of the first $n-1$ populations, a proportion p of the agents choose contract 1 by mistake. Each population contains k individuals, so this involves $pk(n-1)$ errors (ignoring rounding issues). An individual in the remaining (n-th) population will choose contract 1 as a best reply if $a_1 p^{n-1} \geq a_2(1-p)^{n-1}$. This is equivalent to $p \geq a_2^{\theta}/\left(a_2^{\theta}+a_1^{\theta}\right)$, where $\theta = 1/(n-1)$. However, an individual in the n-th population will also choose contract 1 with positive probability if *all* members of exactly *one* of the other populations choose contract 1. (Under these conditions he infers that coordination is impossible, so all contracts have equal expected payoff, and he chooses each with positive probability.) Since each population has k members, the latter event involves k errors. We conclude that the resistance to going from convention 2 to convention 1 is

$$r_{21}^{k} = k \wedge \left[(n-1)ka_2^{\theta}/\left(a_2^{\theta}+a_1^{\theta}\right)\right], \quad \text{where } \theta = 1/(n-1). \tag{7}$$

Similarly,

$$r_{12}^{k} = k \wedge \left[(n-1)ka_1^{\theta}/\left(a_1^{\theta}+a_2^{\theta}\right)\right]. \tag{8}$$

Now suppose that contract 1 is strictly better than contract 2: $a_1 > a_2$. Note that this implies $r^k{}_{12} = k$ whenever $n \geq 3$. If $r^k{}_{21} = r^k{}_{12} = k$, then both contracts 1 and 2 are stochastically stable. For contract 1 to be uniquely stable for all sufficiently large k, we must therefore have $\mathrm{r}^k{}_{21} < k$. This will only be the case if

$$(n-1)a_2^{\theta}/\left(a_2^{\theta}+a_1^{\theta}\right) < 1,$$

which is equivalent to

$$a_1/a_2 > (n-2)^{n-1}.$$

This means that the payoff from contract 1 must be considerably larger than the payoff from contract 2 if the former is to be the unique stable outcome. More generally, when $n \geq 4$ and complete unanimity is required, the equilibrium selection process can be quite insensitive to payoff differences, and equilibria that are both inefficient and not very central to the payoff distribution may survive the evolutionary selection process.

10 Discussion

The model described above, like all models, presents a very simplified picture of reality. Here we shall mention several ways in which it could be made more realistic, and also discuss some of its inherent limitations. One problematic assumption in the model is that individuals react to the *entire distribution* of choices by other agents in the preceding period. This is clearly absurd when the populations are large; people do not have that much information. A more plausible scenario is that agents know, through hearsay and personal experience, what *a few* other people in the other population have been doing. A natural way of incorporating this idea into the model is to suppose that each agent knows a random sample of size s drawn from the other population, where $s < k$. The agent follows the same decision rule as before, based on the information contained in his sample (not the whole population). It is straightforward to show that the long-run behavior of this process is essentially the same as in the full information model: the absorbing states are the conventions, and the resistance to transiting from convention i to convention j is $\lceil sr_{ij} \rceil$ instead of $\lceil kr_{ij} \rceil$. Thus when s is sufficiently large, the two models yield essentially the same long-run outcomes.[14]

A second unrealistic assumption is that changes in behavior arise solely from the cumulative impact of many independent, idiosyncratic choices by individuals. In reality, changes may be correlated for a variety of reasons: perhaps individuals react to a common event—say a technology shock—or they might be influenced by the actions of a highly visible person (a role model). Without trying to minimize the variety and complexity of such influences, we can nevertheless say that correlated changes do not *by themselves* change the substance of the argument, as long as they represent sufficiently small fractions of the total population. To see why, assume that the current norm is z^i. To tip the process into the basin of attraction of some other norm z^j requires that at least $\lceil kr_{ij} \rceil$ people in one of the populations demand j instead of i. When these changes arise from uncorrelated "mutations" or "aberrations" in behavior, each having probability ε, the probability of this event is on the order of $\varepsilon^{\lceil kr_{ij} \rceil}$ Suppose instead that mutations occur in groups. Think of each mutation as being triggered by an event that hits a given population and causes p people within that population to demand a particular contract i (call this an

[14] The reason for requiring the sample size to be large is that otherwise the process may fail to discriminate between equilibria with different payoffs. To take an extreme case: suppose that each person bases his decision on the behavior of exactly one person drawn at random from the other population. To tip the process out of a convention takes only one mutation, which propagates with positive probability through the entire population (assuming the inertia rate is positive). It follows that the resistance to transiting from any convention to any other convention equals 1, which implies that all conventions are stochastically stable. For a given s (or a given k in the full information model) the extent to which the stable outcomes differ from the maximin outcome is given by the same expression as in Theorem 1, except that α is replaced by $\alpha + 4/s$. See the remark at the conclusion of the Appendix.

"*i-event*"). For example, an *i*-event might be a book that extols the virtues of contract i and causes p people to insist on i in their dealings with others.

Let ε^n be the probability that each population receives n such hits within a given period. For each pair of norms z^i and z^j, the number of *individuals* who must change their minds in order to trip the process from z^i to z^j is the same as before, namely, $\lceil kr_{ij} \rceil$. In the correlated model, it therefore takes a succession of at least $\lceil (k/p)r_{ij} \rceil$ j-events to tip the process from z^i to z^j. Assuming that $p \ll kr_{ij}$, this event has probability on the order of $\varepsilon^{\lceil kr_{ij}/p \rceil}$ Therefore the *relative* resistances between norms are essentially the same as in the uncorrelated model. Hence Theorems 1 and 2 continue to hold provided that the correlations involve small enough fractions of the whole population.

The degree of correlation is connected with another important feature of these models, namely, the length of time it takes the process to tip from one regime into another. For purposes of discussion suppose there are just two contracts: let contract 1 have expected payoffs (a_1,b_1) and contract 2 have expected payoffs (a_2,b_2). If $a_1b_1 > a_2b_2$ contract 1 is stable and contract 2 is not. Suppose, however, that society begins in the state where 2 is conventional. How long does it take to tip into the 1-convention? If the populations are of size k, it takes on the order of rk mutations to tip the process into the 1-regime, where $r = a_2/(a_1 + a_2)$. Assuming that mutations are independent, the probability of this event is on the order of ε^{rk}, and the number of periods until it first happens is on the order of $(1/\varepsilon)^{rk}$. This could be a long wait. If instead mutations come in groups of size p, the waiting time is on the order of $(1/\varepsilon)^{rk/p}$. This could still be a long wait, but not nearly as long as in the case of no correlation. Note, moreover, that correlation changes the absolute transit times between conventions, but not the relative times. Hence the long-run probabilities of the various conventions remain the same as before.[15]

The waiting time also depends crucially on the *topology* of the interaction structure. Suppose, for example, that each individual has just a few "neighbors" in each population, and that he reacts only to what his neighbors are doing. If neighborhoods are sufficiently close-knit—which means, roughly speaking, that neighbors of neighbors tend themselves to be neighbors—it only takes a small number of nearby mutants to cause an entire neighborhood to tip. If everyone lives in such a neighborhood, it does not take that long for society at large to tip (Ellison 1993; Young 1998, chapter 6).[16]

[15] Chapter 13 analyzes the expected waiting times to transit between conventions when agents are embedded in a social network.

[16] See Ellison (1996) for a discussion of other structural factors that determine waiting time.

In sum, the inertia of the process depends on a variety of factors, including the size of the error term, the degree of correlation in mutation, the amount of information on which agents base their decisions, and the amount of "neighborliness" in the interaction structure. It also depends, of course, on the frequency with which agents renegotiate contracts. One must therefore be cautious in predicting how long it takes (in real time) for a stable outcome to manifest itself. On the whole, one would expect waiting times to be substantial, and it should come as no surprise when societies operate for long periods of time under regimes that are neither optimal nor stochastically stable. Conventions, once established, can lock people into conventional ways of thinking that are hard to undo.

Nevertheless, conventions can be undone, and over time they are. The substance of the preceding results is that, given sufficient time, there is a selective bias in favor of conventions that are both efficient and egalitarian. The reason for the latter is that conventions whose payoffs lie near the boundary of the feasible payoff set tend to be unstable. In such a convention, some group is dissatisfied, and the more dissatisfied a group is, the more easily it is seduced by alternative arrangements that give its members hope of getting more. Social change, in other words, is driven by those who have the most to gain from change. Over the long run, this tends to favor contractual relationships that are efficient and that offer each side a fairly large share of the potential benefits they can realize from cooperation.

Appendix

Proof of Theorem 1

Let G be a two-person coordination game with strictly positive diagonal payoffs (a_1,b_1), (a_2,b_2),..., (a_m,b_m), and off-diagonal payoffs (0,0). Assume that the payoffs have been normalized so that max a_i = max b_i = 1. Let $P^{k,\nu,\varepsilon}$ be the Markov process on the finite state space Z defined in Section 2. Throughout we shall assume that the inertia rate $\nu \in (0, 1)$ is fixed, so we shall drop it from the notation and write $P^{k,\varepsilon}$ instead of $p^{k,\nu,\varepsilon}$.

Consider the situation in which $\varepsilon = 0$. A *recurrent class* of the process $P^{k,0}$ is a subset of states $Y \subseteq Z$ such that the process never leaves Y once it is in Y, and every state in Y can be reached with positive probability from every other state in Y. Every convention $\{z^i\}$ constitutes a singleton recurrent class (i.e., an absorbing state) of $P^{k,0}$.

We claim that the conventions $\{z^i\}$ are, in fact, the *only* recurrent classes of $P^{k,0}$. To see why, consider any state $z^t = (x^t, y^t)$ at the end of some time period t. There is a positive probability that, in period $t + 1$, all row players choose a best reply to y^t, and that all column players choose a best reply to x^t. Moreover, if there are ties in best reply, the full support of the tie-breaking rule implies that there is a positive probability that everyone in the same population chooses the *same* best reply. Thus at the end of period $t + 1$ there is a positive probability that the process will be in a state of form $z^{t+1} = (ke^i, ke^j)$ for some i and j. At the beginning of period $t + 2$, there is a positive probability that every row player sticks to his previous choice out of inertia, and that every column player chooses a best reply to ke^i. Thus by the end of period $t + 2$ there is a positive probability of reaching the convention $z^i = (ke^i, ke^i)$. We have therefore shown that, from an arbitrary state z^t, there is a positive probability of reaching a convention within two periods. It follows that the conventions are the only recurrent classes of $P^{k,0}$.

Now consider the situation where $\varepsilon > 0$. Let $z = z^t$ be the state at the end of some period t, and let $z' = z^{t+1}$ be the state at the end of the next period. An *error* in the transition $z \to z'$ is a choice by some player in period $t + 1$ that is neither inertial nor a best reply given the state z. For every two states z and z', let $r(z, z')$ be the number of errors in the transition $z \to z'$ (possibly $r(z, z') = 0$). It can be verified that the Markov chain $P^{k,\varepsilon}$ has the following three properties:

$$\begin{aligned}&\text{(i) } P^{k,\varepsilon} \text{ is aperiodic and irreducible whenever } \varepsilon > 0;\\&\text{(ii) } \lim_{\varepsilon \to 0} P^{k,\varepsilon}_{zz'} = P^{k,0}_{zz'} \text{ for all } z, z' \in Z;\\&\text{(iii) } 0 < \lim_{\varepsilon \to 0} \varepsilon^{-r(z,z')} P^{k,\varepsilon}_{zz'} < \infty \text{ for all } z, z' \in Z.\end{aligned} \tag{9}$$

These three properties mean that $P^{k,\varepsilon}$ is a *regular perturbation* of $P^{k,0}$; see Chapter 2 (Young 1993a).It follows from (by now) standard arguments that $P^{k,\varepsilon}$ has a unique stationary distribution $\mu^{k,\varepsilon}$ and that $lim_{\varepsilon \to 0}\mu^{k,\varepsilon}(z) = \mu^{k,0}(z)$ is a stationary distribution of the process $P^{k,0}$.

A state z is *stochastically stable* if $\mu^{k,0}(z) > 0$. Since $\mu^{k,0}(z)$ is a stationary distribution of $P^{k,0}$, and the only recurrent classes of $P^{k,0}$ are the conventions, it follows that the conventions are the *only* states that can be stochastically stable. The proof of Theorem 1 amounts to showing which of the conventions can be stochastically stable using the computational procedure described in Section 5.

Define the *basin of attraction* of a convention z^i to be the set of all states z such that the process $P^{k,0}$ can transit from z to z^i in a finite number of periods (with no errors). As shown in (6), the least number of errors needed to tip the process from z^i into the basin of attraction of z^j is

$$r^k_{ij} = \lceil k r_{ij} \rceil \quad \text{where } r_{ij} = a_i/\left(a_i + a_j\right) \wedge b_i/\left(b_i + b_j\right) \tag{10}$$

Construct a directed graph with m vertices, one for each of the m coordination equilibria. Given equilibria i and j, let the "weight" on the directed edge $i \to j$ *be* r_{ij}. Define the *stochastic potential* of equilibrium i, $\gamma(i)$, to be the least total weight among all i-trees. Similarly, define the k-stochastic potential of i, $\gamma^k(i)$, to be the least total weight among all i-trees when the weights are $r^k{}_{ij}$ instead of r_{ij}. It follows directly from Theorem 4 in Chapter 2 (Young 1993a) that *the stochastically stable conventions of the process $P^{k,\varepsilon}$ correspond to those i that minimize $\gamma^k(i)$.*

Recall that a coordination equilibrium i is *stable* if it is stochastically stable in $P^{k,\varepsilon}$ for infinitely many values of k. On one hand, among all those i that minimize $\gamma(i)$, there is clearly at least one that minimizes $\gamma^k(i)$ for infinitely many values of k. Hence stable equilibria exist. On the other hand, if i does not minimize $\gamma(i)$, then it does not minimize $\gamma^k(i)$ for all sufficiently large k, so it cannot be stable. Thus every stable equilibrium minimizes $\gamma(i)$. To prove Theorem 1, it suffices to show that every equilibrium i that minimizes $\gamma(i)$ is efficient and approximately maximin.

Let j be such an equilibrium. By the preceding, there exists a j-tree T_j such that $r\left(T_j\right) \leq r(T)$ for all rooted trees T, where $r(T)$denotes the sum of the reduced resistances on the edges in T. We need to show that this implies

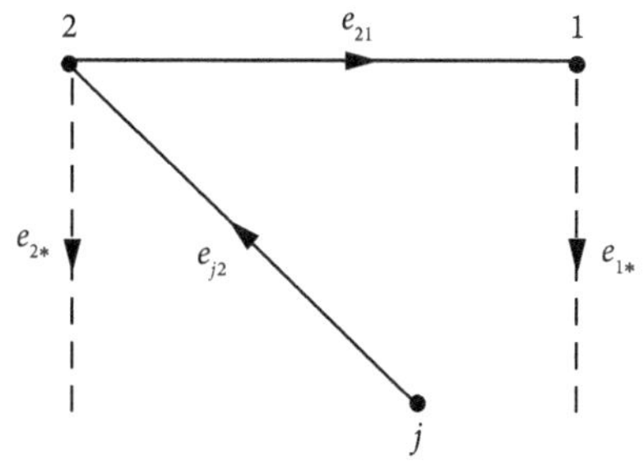

Figure 4.6 Adding and deleting edges in a j-tree.

$$w_j \geq (w^+ - \alpha)/(1 + \alpha). \tag{11}$$

Clearly this follows if $w_j = w^+$. Hence we may assume that $w_j < w^+$, that is, j is not a maximin equilibrium. Without loss of generality let us suppose that 1 is a maximin equilibrium ($w_1 = w^+$). and that in equilibrium j, $w_j = a_j \leq b_j < w^+$.

Let i be an equilibrium with payoffs $a_i = 1, j$ and $b_i = b^-$. *Since* $a_j < w^+ \leq 1$, j and i must be distinct. Assume for the moment that i ≠ l. (Later we shall dispose of the case $i = 1$.) Without loss of generality we can assume that $i = 2$ and $j \geq 3$. Thus $a_2 = 1$ and $b_2 = b^-$.

Fix a least-resistant j-tree T_j. There is a unique edge e_{1*} in T_j exiting from vertex 1, and a unique edge e_{2*} exiting from vertex 2 (see Figure 4.6). Delete the edges e_{1*} and e_{2*} from T_1 and add the edges e_{j2} from j to 2 and e_{21} from 2 to 1. This results in a l-tree, say T_1. (This is true whether or not e_{1*} and e_{2*} point toward the same vertex.) By assumption $r(T_j) \leq r(T_1)$, so

$$r_{1*} + r_{2*} \leq r_{j2} + r_{21}. \tag{12}$$

Let us evaluate the four reduced resistances in this expression. By (10),

$$r_{j2} = a_j/\left(a_j + a_2\right) \wedge b_j/\left(b_j + b_2\right) = a_j/\left(a_j + 1\right) \wedge b_j/\left(b_j + b^-\right).$$

Since $w_j = a_j \leq b_j$ by hypothesis, it follows that

$$r_{j2} = w_j/\left(w_j + 1\right). \tag{13}$$

By (10), the minimum reduced resistance from 1 to any other vertex is

$$\min_i \{a_1/(a_1 + a_i) \wedge b_1/(b_1 + b_i)\}.$$

Since 1 is a maximin equilibrium we have $a_1, b_1 \geq w^+$. Moreover, $a_i, b_i \leq 1$ for all k. Hence

$$\min_i \{a_1/(a_1 + a_i) \wedge b_1/(b_1 + b_i)\} \geq w^+/(w^+ + 1).$$

It follows that

$$r_{1*} \geq w^+/(w^+ + 1). \tag{14}$$

A similar argument shows that

$$r_{2*} \geq w_2/(w_2 + 1). \tag{15}$$

Finally, since $a_1, b_1 \geq w^+$, we have

$$\begin{aligned} r_{21} &= a_2/(a_2 + a_1) \wedge b_2/(b_2 + b_1) \\ &= 1/(1 + a_1) \wedge b^-/(b^- + b_1) \leq 1/(1 + w^+) \wedge b^-/(b^- + w^+). \end{aligned}$$

Since $b^- \leq w^+$, it follows that

$$r_{21} \leq b^-/(b^- + w^+). \tag{16}$$

Combining (12)–(16) yields the inequality

$$w^+/(w^+ + 1) + b^-/(b^- + 1) \leq w_j/(w_j + 1) + b^-/(b^- + w^+). \tag{17}$$

After some algebraic manipulation this leads to the equivalent expression

$$w_j \geq (w^+ - \alpha(b^-))/(1 + \alpha(b^-)), \tag{18}$$

where $\alpha(x) = [x(1 - (w^+)^2)]/[(1+x)(w^+ + x)]$. It is straightforward to show that $\alpha(x)$ is increasing on the domain $D = \{x : 0 \leq x^2 \leq w^+\}$. In particular, $b^-, w^- \in D$, because $0 \leq b^-, w^- \leq w^+ \leq 1$. Since. $b^- \leq w^-$ by definition, $\alpha(b^-) \leq \alpha(w^-)$.

Letting $\alpha = \alpha(w^-)$ we therefore obtain

$$w_j \geq (w^+ - \alpha)/(1 + \alpha),$$

which is the inequality claimed in the theorem.

It remains to dispose of the case $i = 1$, that is, the maximin equilibrium has payoffs $(1, b^-)$. Change the tree T_j by deleting the edge e_{1*}. and adding the edge e_{j1}. The result is a l-tree T_1. Now

$$r_{j1} = a_j/(a_j + a_1) \wedge b_j/(b_j + b_1) = a_j/(a_j + 1) \wedge b_j/(b_j + b^-).$$

Since $w_j = a_j \le b_j$, it follows that

$$r_{j1} = w_j / \left(w_j + 1\right).$$

On the other hand, (10) implies that

$$r_{1^*} \ge w^+ / \left(w^+ + 1\right)$$

Putting these facts together we obtain

$$r_{1^*} \ge w^+ / \left(w^+ + 1\right) > w_j / \left(w_j + 1\right) = r_{j1}. \qquad (19)$$

This implies that $r(T_1) < r(T_j)$, which is a contradiction.

We now establish Pareto optimality. Suppose, by way of contradiction, that i is stable and there is some j such that

$$a_i < a_j \text{ and } b_i < b_j \qquad (20)$$

Since i is stable, we know that $\gamma(i) \le \gamma(j)$. Let T_i be an i-tree having minimum reduced resistance among all rooted trees. If T_i contains an edge from j to i, replace it by the oppositely directed edge e_{ij}. This creates a j-tree whose reduced resistance is strictly less than that of T_i, which would imply $\gamma(\text{i}) > \gamma(j)$. Therefore e_{ij} cannot be an edge in T_i.

Now consider any edge of form e_{hi}, in T_i. (There is at least one such edge if there is more than one coordination equilibrium.) Replace e_{hi} with the edge e_{hj}, and make a similar substitution for all edges that point toward i. In T_i there is also a unique edge exiting *from* j, say $e_{jj'}$, where $j' \neq i$ by the earlier part of the argument. Replace e_{jj} with the edge $e_{jj'}$. It is easy to check that these substitutions yield a j-tree. We claim that it has strictly lower reduced resistance than T_i. Indeed, (10) and (20) imply that for all $h \neq i, j$,

$$r_{hj} < r_{hi} \text{ and } r_{ih} < r_{jh}.$$

Hence the above substitutions result in a j-tree that has strictly lower reduced resistance than T_i, which implies that $\gamma(i) > \gamma(j)$. This contradiction shows that no stochastically stable convention is strictly Pareto dominated, and completes the proof of Theorem 1.

Remark

It is of some interest to ask how closely the stable outcome(s) come to the maximin outcome for *specific* values of k. Clearly, the smaller k is, the less good is the approximation because of the rounding inherent in the definition of reduced resistance (10). The argument establishing (17) can be modified to show that if j is stochastically stable for $P^{k,\varepsilon}$, and if the maximin payoffs are not $(1, b^-)$, then

$$\lceil kw^+/(w^+ + 1)\rceil + \lceil kb^-/(b^- + 1)\rceil \leq \lceil kw_j/(w_j + 1)\rceil + \lceil kb^-/(b^- + w^+)\rceil. \quad (21)$$

This implies

$$kw^+/(w^+ + 1) + kb^-/(b^- + 1) \leq kw_j/(w_j + 1) + 1 + kb^-/(b^- + w^+) + 1, \quad (22)$$

that is,

$$w^+/(w^+ + 1) + b^-/(b^- + 1) \leq w_j/(w_j + 1) + b^-/(b^- + w^+) + 2/k. \quad (23)$$

Following the lines of the preceding argument, we obtain the lower bound

$$w_j \geq \frac{(w^+ - \alpha - (2/k)(1 + w^+))}{(1 + \alpha + (2/k)(1 + w^+))} \geq \frac{w^+ - \alpha - 4/k}{1 + \alpha + 4/k}. \quad (24)$$

Suppose on the other hand that the maximin payoffs are $(1, b^-)$, and that j is stochastically stable but not maximin. Then we obtain a contradiction as in (19) unless $r_{1^*} \leq r_{j1}$, which implies

$$\lceil kw^+/(w^+ + 1)\rceil \leq \lceil kw_j/(w_j + 1)\rceil.$$

Hence

$$kw^+/(w^+ + 1) \leq kw_j/(w_j + 1) + 1,$$

After some algebraic manipulation this leads to

$$w_j \geq \frac{w^+ - (1 + w^+)/k}{1 + (1 + w^+)/k} \geq \frac{w^+ - 2/k}{1 + 2/k}. \quad (25)$$

Thus in any case (24) holds, which estimates the distortion as a function of k.

Acknowledgment

This work was supported by the National Science Foundation, grant SBR-9601743. I am indebted to Kenneth Arrow, Ken Binmore, Samuel Bowles, Joseph Harrington, Mancur Olson, and John Roemer for stimulating suggestions, and to an anonymous referee for pointing out an error in an earlier version, which was entitled "Social Coordination and Social Change."

References

Binmore, K. (1994), *Game Theory and the Social Contract I: Playing Fair.* (Cambridge, MA: MIT).

Binmore, K. and Samuelson, L. (1994), "An economist's perspective on the evolution of norms," *Journal of Institutional and Theoretical Economics* 150, 45–63.

Blume, L. (1993), "The statistical mechanics of strategic interaction," *Games and Economic Behavior* 4, 387–424.

Ellison, G. (1993), "Learning, local interaction, and coordination," *Econometrica* 61, 1047–1071.

Ellison, G. (1996), "Basins of Attraction and Long-Run Equilibria" (Mimeo, Department of Economics, MIT).

Foster, D. and Young, H. P. (1990), "Stochastic evolutionary game dynamics," *Theoretical Population Biology* 38, 219–232.

Freidlin, M. and Wentzell, A. (1984), *Random Perturbations of Dynamical Systems* (Berlin: Springer-Verlag).

Harsanyi, J. C. and Selten, R. A. (1988). *General Theory of Equilibrium Selection in Games.* (Cambridge, MA: MIT)

Holcombe, L. (1983), *Wives and Property* (Oxford: Martin Robertson).

Kalai, E. and Smorodinsky, M. (1975), "Other solutions to Nash's bargaining problem," *Econometrica* 43, 513–518

Kandori, M. and Rob, R. (1995), "Evolution of equilibria in the long run: a general theory and applications," *Journal of Economic Theory* 65, 29–56.

Kandori, M., Mailath, G., and Rob, R. (1993), "Learning, mutation, and long-run equilibria in games," *Econometrica* 61, 29–56.

Kessler, F. (1986), "Contracts as a principle of order," in F. Kessler, G. Gilmore, and A. T. Kronman (eds.), *Contracts* (3rd ed.).

Lewis, D. (1969), *Convention: A Philosophical Study* (Cambridge, MA: Harvard University Press).

Maruta, T. (1995), "On the Relationship Between Risk-Dominance and Stochastic Stability" (Mimeo, MEDS Department, Kellogg Graduate School of Management, Northwestern University).

Morris, S., Rob, R., and Shin, H. S. (1995), "*p*-dominance and belief potential," *Econometrica* 63, 145–167.

Samuelson, L. (1997), *Evolutionary Games and Equilibrium Selection* (Cambridge, MA: MIT).

Shin, H. S. and Williamson, T. (1996), "How much common belief is necessary for a convention?," *Games and Economic Behavior* 12, 252–268.

Sugden, R. (1986), *The Evolution of Rights, Cooperation, and Welfare* (New York: Basil Blackwell).

Ullman-Margalit, E. (1977), *The Emergence of Norms* (Oxford: Oxford University Press).

Vega-Redondo, F. (1996), *Evolution, Games, and Economic Behavior* (Oxford: Oxford University Press).

Warneryd, K. (1993), "Anarchy, uncertainty, and the emergence of property rights," *Economics and Politics* 5: 1–14.

Warneryd, K. (1994), "Transaction cost, institutions and evolution," *Journal of Economic Behavior and Organization* 25, 219–239.

Young, H. P. (1993a), "The evolution of conventions," *Econometrica* 61, 57–84.

Young, H. P. (1998), *Individual Strategy and Social Structure: An Evolutionary Theory of Institutions* (Princeton, NJ: Princeton University Press).

Preface to Chapter 5

The Role of Custom in Setting Commissions, Fees, and Shares

The evolutionary bargaining model described in Chapter 3 predicts that bargaining outcomes will often gravitate toward conventions or customs that mask idiosyncratic differences in individuals' utility functions. The model is highly simplified, and omits the role of exogenous factors that in practice constrain the shares that bargainers will agree to. This chapter presents an alternative framework that incorporates such constraints and that can be brought to bear on empirical data.

In general, a *share contract* or *performance contract* is an arrangement in which a principal promises to give an agent a specified proportion of the value created by their relationship. Such arrangements are often expressed in terms of percentage commissions, fees, shares, or bonuses. They strike a balance between risk-sharing and incentive provision when effort is difficult to monitor and there are potential shocks beyond the parties' control that affect the value of the output.

If such contracts are competitively negotiated, one would expect the size of the share to vary in accordance with the mean and variance of the returns, the risk aversion of the parties, the agent's ability, and other relevant factors. In practice, however, shares often cluster around "usual and customary" levels even when there is substantial heterogeneity among principal–agent pairs.

Real estate commissions are a case in point. In the US the standard (modal) commission for selling a house is 6 percent, which is paid by the seller. The rate is a "norm" in the sense that it is customary and expected in most locales. A homeowner who approaches an agent about selling her house will typically be quoted 6 percent as the "going rate" irrespective of the value of the house. This particular percentage dates to the 1940s, when it was fixed by the National Association of Realtors. Despite numerous lawsuits and a 1950 Supreme Court decision banning the practice as anti-competitive, local real estate boards continued to "suggest" 6 percent without fixing it outright.[1] Thus

[1] United States v. Real Estate Boards 339 US 485 (1950).

it exhibits the phenomenon of path dependency: once a contractual norm is in place it becomes difficult to dislodge because it anchors expectations.

Another contemporary example of a contractual norm can be seen in the hedge fund industry. Until very recently the standard was to charge "two and twenty": 2 percent per year in management fees, and 20 percent of the profit earned over and above some established benchmark rate. Where did this arrangement come from? It turns out that it too is strongly embedded in precedent. Alfred Winslow Jones, who originated the concept of hedge funds in the 1940s, charged a 20 percent performance fee and claimed as justification that this was the rate charged by Phoenician merchants in ancient times.[2] In spite of the proliferation of hedge funds and the huge growth in assets under management, this particular commission structure has remained the industry norm until very recently; indeed a 20 percent performance bonus remains the most common rate in spite of the wide variation in fund returns.

This chapter examines how evolutionary models can help us understand the high degree of uniformity observed in such settings. The model can be used to evaluate the dynamics of adjustment in the intermediate as well as the long run, the former being particularly relevant to empirical cases. We take agricultural share contracts in Illinois (a leading agricultural state in the US) as the motivating example, because the data on such contracts are exceptionally rich. The US Department of Agriculture has for many years collected detailed survey data on the terms of such contracts, and how they are related to the underlying circumstances of the agents that should matter in an economic sense. These include the size and soil characteristics of different farms, the variability in their yields, and side payments for various inputs. One can therefore test the excess uniformity hypothesis against other potential explanations, such as assortative matching between tenants and landlords or side payments that offset the apparent uniformity of the cropsharing contract.

The framework in this chapter differs in several respects from the approach taken in Chapters 3- 4. First, the basic form of the contract is taken as given; the only parameter to be decided is the division itself. Second, the amount to be divided is highly uncertain because it depends on exogenous factors such as market prices and the weather. Third, agents' outside options vary between locales, because they depend on the local supply of labor and wages in alternative forms of employment such as factory work. Contract terms respond to these fundamentals to some extent, but they are also sensitive to prevailing

[2] Sebastian Mallaby (2010), *More Money than God: Hedge Funds and the Making of a New Elite*, Penguin, 30.

customs in each local area. The force of custom is treated as a parameter of the model that can, in principle, be estimated from detailed event data.

The analysis shows how equilibrium configurations emerge from out-of-equilibrium conditions, and how they depend on a *combination* of economic fundamentals and the force of customary practice. On the one hand, if there is no spillover effect from custom, the evolutionary model predicts that every farm will have a contract that depends on the particular characteristics of the tenant and the landlord, including their risk preferences, outside options, and discount rates, as well as the inherent quality of the land (just as in a standard principal-agent model) On the other hand, if there is a sufficiently strong spillover effect from custom, the evolutionary model predicts that contract terms will be more or less uniform irrespective of economic fundamentals. For intermediate degrees of spillover, the model predicts that regional norms are the most likely outcome. In other words, equilibrium configurations will be "patchy," exhibiting a high degree of conformity within particular regions (reflecting average conditions within each region), and there will be abrupt changes in contract terms between regions—a kind of "quantum" effect. These predictions help explain the otherwise puzzling regional patterns that we observe in the empirical distribution of cropsharing contracts in Illinois agriculture.

From a modeling perspective, this set-up differs from the evolutionary bargaining model described in Chapter 3. Here we are positing that one side (the landlord) sets the terms, which the other side (the tenant) can either accept or refuse. This simplifying assumption allows the evolutionary process to be modeled as a potential game, which is especially convenient for analyzing its dynamical behavior. An interesting variant of the model would be to treat the two sides (landlords and tenants) symmetrically, say by supposing that they make simultaneous demands that constitute perturbed optimal responses to the demands of others in their neighborhood. We conjecture that under moderate degrees of social spillover this variation of the process will also lead to the establishment of regional norms, but this remains an open question for future research.

5

Competition and Custom in Economic Contracts

A Case Study of Illinois Agriculture

With Mary A. Burke

> [T]he constraining force of custom and public opinion ... resembled the force which holds rain-drops on the lower edges of a window frame: the repose is complete till the window is violently shaken, and then they fall together.
>
> —Alfred Marshall

Economists have long been puzzled by the extent to which local custom, rather than competition, shapes the terms of certain kinds of contracts. A well-known example is cropsharing contracts, whereby a landlord leases his farm to a tenant laborer in return for a fixed share of the crops. The high degree of uniformity in the terms of such contracts has attracted the attention (though by no means the approval) of almost all writers on the subject, both ancient and modern. In speaking of the system then prevalent in Italy and France, for example, John Stuart Mill remarks:

> This proportion ... is usually (as is implied by the words *metayer, mezzaiuolo*, and *medietarius*) one-half. There are places, however, such as the rich volcanic soil of the province of Naples, where the landlord takes two-thirds. ... Whether the proportion is two-thirds or one-half, it is a fixed proportion, not variable from farm to farm, or from tenant to tenant (Mill 1848 p. 303).

Originally published Young, H., Peyton, and Mary A. Burke. 2001. "Competition and Custom in Economic Contracts: A Case Study of Illinois Agriculture." *American Economic Review*, 93 (3): 559–573. Reproduced with permission.

Spontaneous Order. H. Peyton Young with Mary A. Burke, Oxford University Press. © H. Peyton Young (2024).
DOI: 10.1093/oso/9780198892908.003.0005

Similarly, in a more recent study of contract forms in West Bengal, Ashok Rudra writes:

> the proportion 50:50 for paddy shows a great resilience in that it is known to have existed for a long time in this state ... irrespective of soil conditions, improved or backward methods of cultivation, and other factors which could be expected to affect the profitability of farm business. (Rudra 1975 p. A58)

For those who find these observations disturbing or implausible, there are two comforting explanations. One is that cropsharing is largely a feature of pre-economic, custom-bound cultures; in modern societies, contracts are surely structured more rationally and are governed by competitive forces. Alfred Marshall seems to have been of this view, asserting that the sway of custom is a feature of "primitive times and backward countries," and that the "English" system of straight land rent contributed to its superior position in agriculture (Marshall 1920 chapter X). Other commentators went even further, claiming that "wherever this system [of cropsharing] prevails, it may be taken for granted that a useless and miserable population is found" (Arthur Young 1792).

A second possibility is that people pay their respects to custom without really following it. In other words, while the nominal terms of contracts may appear to be uniform, this is largely an illusion: the side terms of the contract and the actual performance of the parties yield factor returns that are in line with competitive forces (Steven N. S. Cheung 1969). This argument is not supported by the evidence either, as we show in a companion paper (Burke and Young 2000). Moreover, even if the data did support it, one would still have to explain why people bother to maintain the illusion of "fairness" unless they actually do care about it, a point that seems to have been overlooked by exponents of this view.

In this chapter we shall argue that custom is a real force in setting contract terms, even in modern economies (see Akerlof 1980, 1997). The evidence that we marshal in support of this claim is based on a statewide survey of contract terms that prevailed on nearly 1,000 farms in the state of Illinois in 1995. These data provide information on the contractual division of both outputs and inputs by each party as a function of the size of the farm and the quality of its soil. Conventional theories fail to explain certain key features of these data, particularly the observed spatial pattern of contract forms. There are regional "patches" where contractual terms are nearly uniform, separated by boundaries where contractual norms jump substantially from one set of terms to another. These regional customs are roughly related to average differences in economic fundamentals, but they mask the considerable amount of heterogeneity that exists within each region. We introduce a dynamic model of contract formation that explicitly takes the role of custom into account, and show how it resolves these and other apparent puzzles in the data.

1 Focal Shares

The data come from a sample survey conducted by the Illinois Cooperative Agricultural Extension Service in 1995. Tenants were asked to report in detail the terms of the contract they operated under, including the shares of each output and each input, cash payments (if any), and responsibilities for providing farm machinery. The stated purpose of the survey was to provide information that would enable the agency to give better advice to local farmers; individual respondents could not be identified from the data and there is no reason to think they were motivated to misreport.

Almost all of the reported contracts fall into one of two categories. *Cropsharing contracts* specify fixed shares of outputs and inputs for each party, and generally involve no cash payments. *Land rent contracts* specify a fixed rental rate per acre; the other inputs are usually provided by the tenant. Many other contracts exist in theory, including pure wage contracts and hybrid forms of rent, wage, and share contracts (Stiglitz 1974, 1989), but it appears that they are seldom used in practice. Of the 1,704 responses in the 1995 survey, cropsharing contracts were the most frequent (55 percent), land rent contracts the next most frequent (41 percent), and all other contract forms (mostly livestock and pasture leases) constituted less than 4 percent. Here we shall restrict ourselves to an analysis of the cropsharing contracts, though we conjecture that the force of custom may operate on land rent contracts as well.

In principle, the terms of cropsharing contracts can be extremely varied and flexible. For example, one could specify different shares for each of the major outputs (corn, soybeans, and wheat), as well as for each of the major inputs (seed, fertilizer, and machinery). But in practice very few of the possible contracts are used. Indeed, almost all of the Illinois contracts apply the same share to all of the outputs, and these shares are almost invariably expressed in denominators of two, three, or five. Figure 5.1 shows the frequency distribution of shares of corn output, which is virtually the same as the frequency distribution of shares for soybeans and wheat. Fifty–fifty division is by far and away the most common contract, just as Mill, Marshall, and others noted for the European agriculture of their day.

A similar pattern holds in developing countries, such as the village economies of India and Africa (Bardhan 1976, 1984; Bardhan and Rudra 1980,

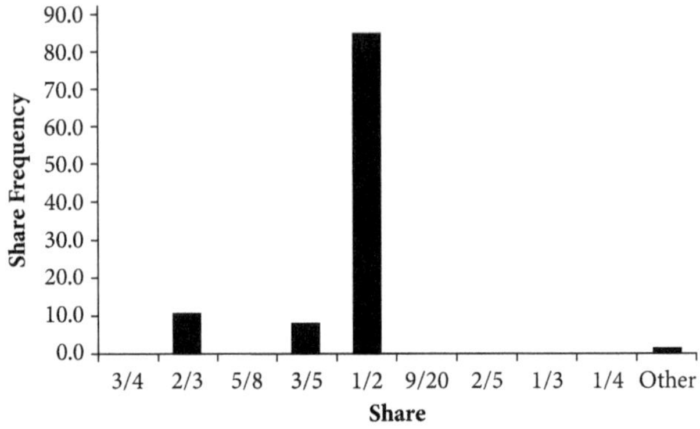

Figure 5.1 Crop share frequencies in Illinois: Tenant's share of the corn crop (Frequencies in percent).
Source: Illinois Cooperative Agricultural Extension Service *Farm Leasing Survey*, 1995.

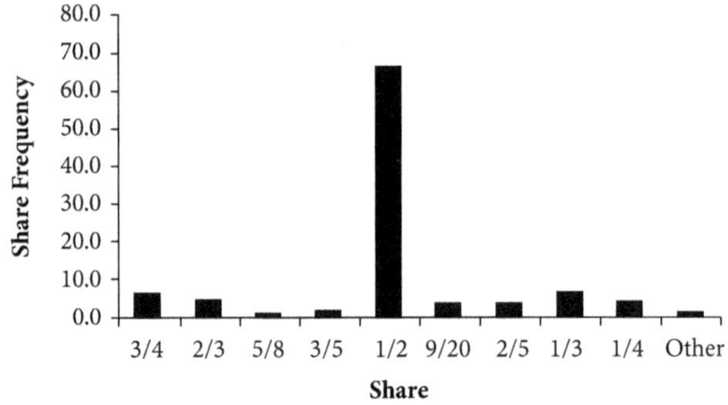

Figure 5.2 Crop share frequencies in West Bengal, 1975–1976: Tenant's share of paddy (Frequencies in percent).
Source: Pranab Bardhan 1984 Table 9.2.

1981, 1986a, 1986b; Robertson 1987). To illustrate, consider Figure 5.2, which shows the distribution of shares among paddy farmers in West Bengal.[1]

A common explanation of this phenomenon is that equal division expresses a relationship of *fairness* between owner and tenant, and fairness is of fundamental importance in traditional cultures where market psychology is weak. If we take this explanation at face value, however, then Illinois

[1] We should say that Bardhan's remarkable study inspired us to undertake this project. For other treatments of cropsharing contracts see Johnson, 1950; Stiglitz, 1974; Reid, Jr., 1975, 1976; Bell and Zusman 1976; Roumasset and James 1979; Bliss and N. H. Stern 1982; Braverman and Stiglitz 1982; Murrell 1983; Binswanger and Rosenzweig 1984; Allen and Lueck 1993; Hayami and Otsuka 1993.

agriculturalists would appear to be even more traditional, and less market oriented, than those in India. This is scarcely credible: contemporary Illinois agriculture is a sophisticated, capital-intensive, and highly decentralized business. Landowners typically own plots of several hundred acres, and tenants typically farm even larger amounts (sometimes over 1,000 acres) by contracting with several different owners. These tenant laborers have substantial capital investments in equipment, and they tend to be well informed about recent developments in seeds, fertilizers, and cultivation techniques. Markets in both crops and in labor are dynamic due to low transportation costs. In short, there is every reason to think that landlords and tenants would be highly sensitive to market forces, and that tradition would play a secondary role in determining the returns to labor and land.

The data in Figures 5.1 and 5.2 suggest, however, that market forces might not be enough to overcome the focal power of particular shares. This interpretation is reinforced by observing that the frequency of contract forms is a decreasing function of the size of the denominator. In Figure 5.2, for example, 66.9 percent of the contracts have denominator 2, 11.1 percent have denominator 3, 10.5 percent have denominator 4, and 5.6 percent have denominator 5. (An exception is the denominator 20, which is more prevalent than the denominator 8.) A similar pattern holds for Figure 5.1. In West Bengal, this monotonicity produces a trimodal distribution with peaks at ⅓, ½, and ⅔. Similarly, Illinois has a bimodal distribution with over 90 percent of the mass concentrated on ⅔ and ½.

It might be, of course, that these "easy" fractions are just a way of rounding off the economically justifiable shares, that is, the shares that yield competitive returns to labor and land. If this were the case, however, then we would have to accept that: (i) a share of about fifty–fifty represents the competitive returns to labor and land on most farms and for different crops in both Illinois and India;[2] (ii) land quality is not unimodally distributed, but has peaks that fall at or near numbers that justify ⅔, ½, or ⅓. Both of these explanations seem rather implausible. We hypothesize that a different mechanism is at work that has to do with the psychology of bargaining. "Easy" fractions—those with small denominators—are inherently more focal than "hard" fractions. When a principal and an agent bargain over how to split a pie, they are drawn to these

[2] Not to mention Africa; see Robertson (1987).

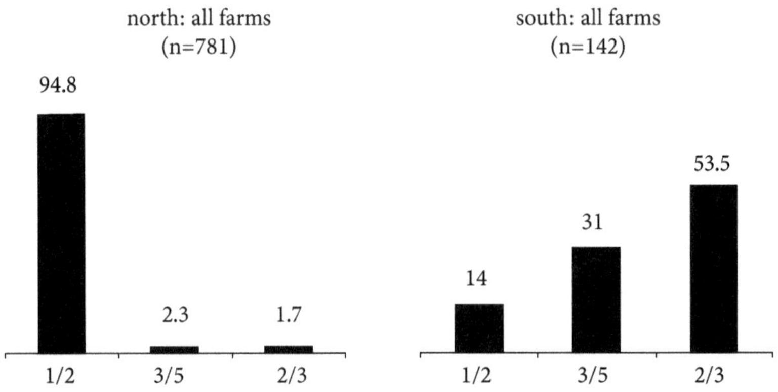

Figure 5.3 Distribution of share contracts by region: Illinois, 1995.

natural focal points, thereby avoiding protracted and confrontational haggling that might damage the subsequent relationship.[3]

If we accept that focal points matter, then we are left with the question of how much they matter. An extreme position would be that cropsharing contracts are constructed around focal points that completely ignore economic fundamentals, such as the quality of the land, the quality of the labor, or the wage to labor in alternative employments. This hypothesis is not supported by the data either. To see why, let us disaggregate the Illinois contract data by region. The state consists of two parts that are noticeably different in their soil characteristics and geology. In the north, the land is flat and the soils are, on average, highly productive. In the south, the land is hillier, the topsoil is not as thick, and on average it is less productive. This north–south division is quite sharply defined, especially in the middle and eastern part of the state, where it corresponds to the southern boundary of the last major glaciation (Mausel et al. 1975). When we compare contract frequencies in the northern and southern parts of the state, substantial differences appear (see Figures 5.3 and 5.4).

In the north, contract terms are concentrated on ½ – ½, whereas in the south the predominant contracts are ⅔ – ⅓ and 3/5 – 2/5. (Here as elsewhere we write the tenant's share first, and the landlord's second.) Moreover there is a simple explanation for this shift: the land in the south is, on average, inherently less productive than the land in the north. Hence the share for the tenant must be higher in the south if net returns to the tenants in the two regions are to be comparable.

[3] The importance of focal points in bargaining was first pointed out by Schelling (1960). See Roth (1985) and Binmore et al. (1993) for empirical studies of this phenomenon. To our knowledge the proposition that focalness decreases as the size of the denominator increases has not been tested in laboratory settings.

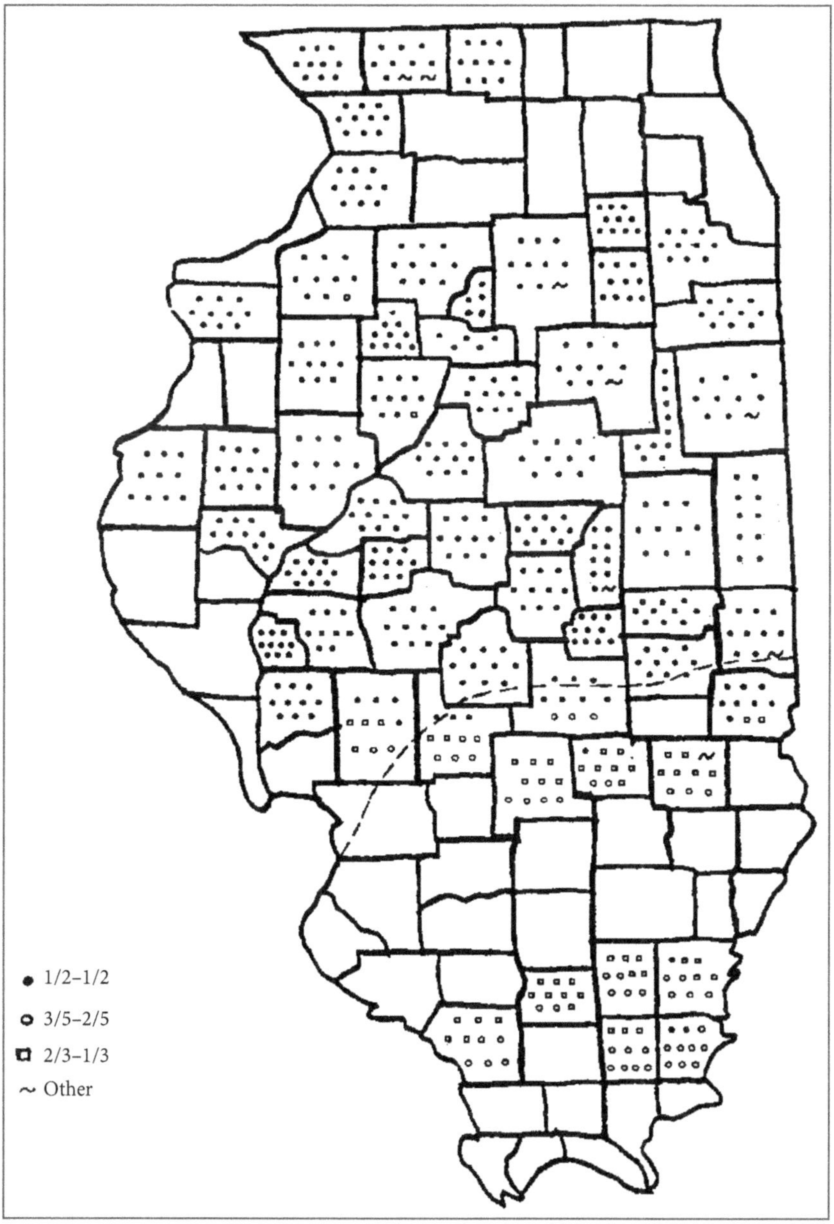

Figure 5.4 Distribution of share contracts by county: Illinois, 1995.
Note: Blank counties had too few data to be statistically meaningful.

Of course, this is very far from saying that the observed shift from ½ – ½ to ⅔ – ⅓ is exactly the right amount to compensate for the difference in average land quality. Indeed it would be quite a coincidence if ½ – ⅓ = ⅙ represents just the right additional compensation necessary to equilibrate the tenant-labor market between north and south. A more plausible explanation is that thirds and fifths have the most focal power when market forces push the parties away from equal division.[4] While the quantum jump required to move from halves to fifths or thirds may bring things closer to market equilibrium, it will at best be a rough approximation.

2 Competition vs. Custom

How does one compute the shares that would, in principle, exactly equilibrate factor markets? The answer depends on such factors as variability in returns, degree of risk aversion, and monitoring costs. It also depends importantly on the quality of the inputs and, in particular, on the productivity of the land. Fortunately the survey instrument includes a measure of each farm's inherent productivity, called the "soil quality index," which estimates the expected yield per acre holding labor input constant. This measure is constructed as follows. Soils are classified into some 435 distinct types based on chemical composition, texture, color, and moisture content (Fehrenbacher et al. 1978).[5] For each type of soil, average yields are computed on test plots under a standardized measure of labor input called "standard management practice" (Odell and Oschwald 1970). Based on these test plot results, the index of a soil is calibrated so that expected yield, holding labor input constant, is a linearly increasing function of the index. The soil index, s_i, of a specific farm i, is the average of the soil indices on that farm, weighted according to the number of acres of each soil type. This index, which is reported for each farm in our data set, amounts to an exogenously estimated predictor of the average yield of basic crops grown on that farm under standard management practice.[6]

A standard model of contracting is to suppose that landlords set contract terms in order to maximize returns subject to tenants earning their reservation wage (Johnson 1950; Cheung 1969; Roumasset and James 1979; Hayami and Otsuka 1993). In this literature it is assumed that landlords can costlessly enforce the terms of the contract, including the level of labor input. When factor markets are in equilibrium, contract shares are determined by the local

[4] Compare Mill's remark about the customary share on rich soils around Naples (see the opening page of this chapter).

[5] In some areas contests are held in which people try to identify the soil type by inspecting a sample, much as one would identify the origin and vintage of a wine by tasting a glass.

[6] The following analogy may be helpful: one could rate the reliability of different makes of automobile by test driving them under "standard conditions" using subjects who follow "standard driving practice." One could then compute the frequency (or cost) of repairs needed over, say, the first 20,000 miles driven. From this one could construct a reliability index for each type of automobile that is analogous to the productivity index of a soil.

reservation wage and the soil quality. The specific predictions depend on the assumed form of the production function. Under most specifications the tenant's share either stays constant or decreases as soil quality increases.[7] Since the average soil quality in the northern part of Illinois is higher than the average soil quality in the south, it comes as no surprise that the shares in the north are, on average, lower.

While this confirms the predictions of theory in the broadest sense, however, a closer look at the data shows that little else about the neoclassical predictions are confirmed. The difficulty is that soil quality is highly variable within each of these two regions, and in fact even within quite localized areas. Within a given county, for example, it is not at all unusual for the highest-rated farms to be nearly twice as productive (in terms of expected yield per acre) as the lowest-rated farms. Standard theory would predict that the shares should reflect these local differences, especially since labor (and other factor inputs) can be assumed to be fully mobile locally.

The data tell a different story, however. Figure 5.5 shows the distribution of shares in a representative northern county (upper panel) and a representative southern county (lower panel). In the north, fifty–fifty is used on almost all farms despite the fact that there is a wide range of soil qualities on which it is used. In the south, the ⅔ – ⅓ contract is the most frequent and there is greater variation; nevertheless, only three contracts (½ – ½, 3/5 – 2/5, ⅔ – ⅓) are used with any appreciable frequency. Notice also that the same contract is used on soils of substantially different qualities, while substantially different contracts are used on soils of the same quality. Indeed most of the contracts in the south give the tenant more—in fact substantially more—than contracts in the north *holding soil quality fixed.*

How can these facts be reconciled with standard theory? One possibility is that labor is not very mobile between the regions, and reservation wages differ between the two because of alternative employment opportunities. In Burke and Young (2000) we examine this hypothesis in more detail and show that, if anything, it makes the puzzle greater, since the wage differential between the two regions goes in the wrong direction: average wages in the south are lower, even though laborers on southern farms generally get a larger fraction of the crop than do laborers on northern farms of similar quality.

A second possibility is that the adherence to customary shares is more apparent than real, that is, other terms of the contract vary enough to pull the

[7] For a Cobb–Douglas production function and risk-neutral agents the equilibrium share is constant. However, this specification implies that labor input per acre should increase substantially with soil quality, a prediction that is not borne out by the data (Burke and Young 2000).

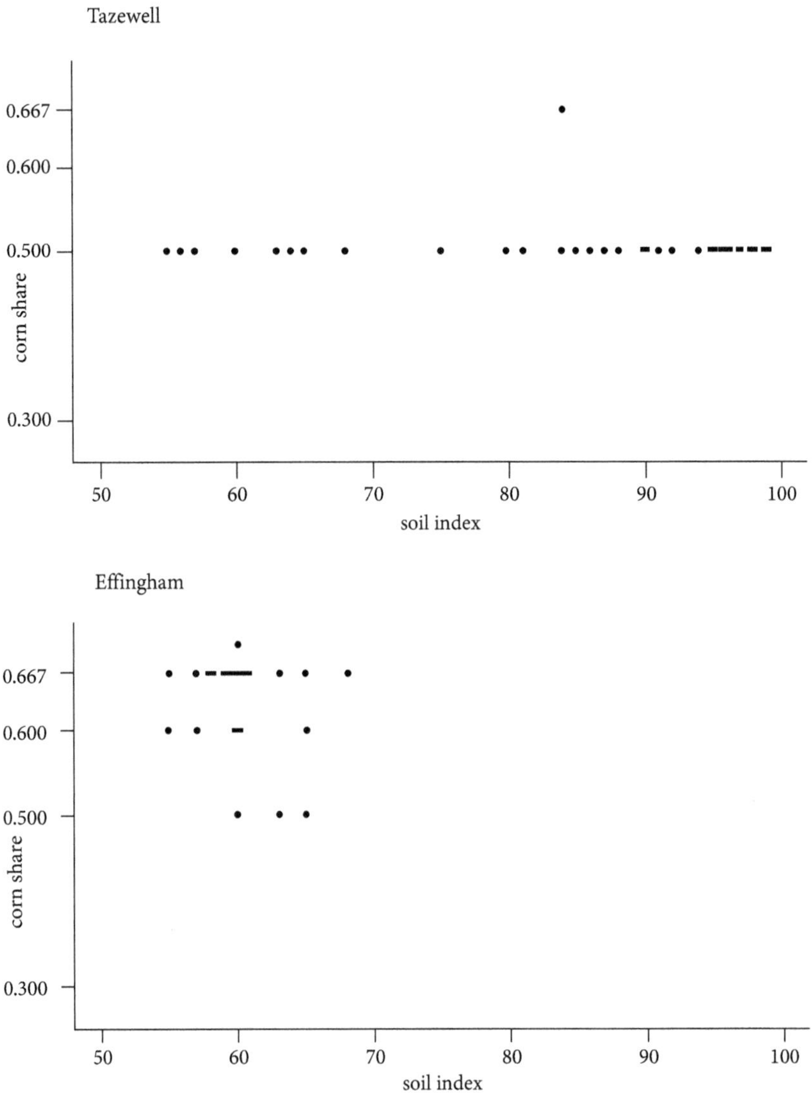

Figure 5.5 Distribution of shares: Tazewell County (north), Effingham County (south).

outcomes back into line with neoclassical predictions. This explanation does not hold up, however, when we examine the actual variation in the other terms. We have already noted that almost all of the contracts apply the *same* share to all of the crops (e.g., corn, wheat, and soybeans), so there is little variation there. Next, let us consider the division of responsibilities for providing key inputs, namely, seed, fertilizer, and machinery. In virtually all cropsharing contracts, the tenant is solely responsible for providing mobile farm equipment (tractors, ploughs, harvesters, etc.), so there is almost no variation in this dimension. Moreover, in the vast majority of contracts, fertilizer and seed are either shared fifty–fifty or are provided solely by the tenant (see Tables 5.1 and 5.2). Taken together, these constitute the bulk of the economically relevant parameters in the contract.[8]

A third possibility is that high-quality tenants migrate to farms of high productivity, that is, the market equilibrates through assortative matching rather than through variation in contract terms. We point out that this would still pose a challenge to conventional theory, since it is not clear why the contract terms should fail to adjust unless some notion of custom is at work. Leaving this issue aside, one can test the assortative matching hypothesis by examining the rate at which total output per acre rises as soil quality increases. The hypothesis is supported if the rate of increase is higher than it would be using the benchmark inputs under which the soil index was originally calibrated. The data do not support this hypothesis either: tenants on high-quality land succeed in capturing a substantial portion of the land rent without increasing labor productivity, and this is directly attributable to the failure of the shares to adjust to local differences in soil quality (Burke and Young 2000).

Table 5.1 Tenant's corn output share (rows) vs. bulk fertilizer share (columns)

	0.25	0.34	0.4	0.5	0.53	0.587	0.6	0.666	1
0.200	0	0	0	1	0	0	0	0	0
0.250	0	0	0	1	0	0	0	0	0
0.300	0	0	0	1	0	0	0	0	0
0.400	0	0	0	1	0	0	0	0	0
0.450	0	0	0	1	0	0	0	0	0
0.500	1	1	1	717	1	1	11	0	24
0.530	0	0	0	1	0	0	0	0	0
0.587	0	0	0	1	0	0	0	0	0
0.600	0	0	0	16	0	0	19	9	18
0.666	0	0	0	11	0	0	23	25	29
0.667	0	0	0	1	0	0	0	0	0
0.700	0	0	0	0	0	0	1	0	0
0.750	0	0	0	0	0	0	1	0	0
1.00	0	0	0	2	0	0	0	0	0

[8] Of course, it is possible that adjustments occur in aspects of the contract that are not observable. It can be shown, however, that the observable differences in contract terms are reflected in the actual net returns to labor, that is, they flow through to the tenants' bottom line (Burke and Young 2000).

Table 5.2 Tenant's corn output share (rows) vs. corn seed share (columns)

	0.25	0.4	0.5	0.53	0.587	0.6	0.666	1
0.200	0	0	0	0	0	0	0	1
0.250	1	0	0	0	0	0	0	0
0.300	0	0	0	0	0	0	0	0
0.400	0	1	0	0	0	0	0	0
0.450	0	0	1	0	0	0	0	0
0.500	0	0	748	0	0	0	0	7
0.530	0	0	0	1	0	0	0	0
0.587	0	0	0	0	1	0	0	0
0.600	0	0	1	0	0	2	0	59
0.666	0	0	0	0	0	0	4	85
0.700	0	0	0	0	0	0	0	1
0.750	0	0	0	0	0	0	0	1
1.00	0	0	0	0	0	0	0	2

3 A Dynamic Model of Contractual Custom

Let us recapitulate the empirical facts that need to be explained. First, standard theory predicts that contract terms should vary continuously as a function of economic fundamentals. The most important of these are: the reservation wage for labor, the quality of the land, the quality of the labor, and the degree of risk aversion of the contracting parties. While the reservation wage may be uniform in a given locality, the other three parameters are almost certainly not uniform. Thus we would expect that contracts are locally heterogeneous. As we have seen, however, the data are not consistent with this view. Rather, the terms are concentrated on a very small number of combinations that have a priori focal significance.

The tendency of contract terms to cluster on a few discrete values we shall call the *quantum effect*. The lack of heterogeneity we shall call the *local conformity effect* (Young 1998a p. 148). Notice that these are not the same thing: even though the quantum effect limits the number of distinct contracts that might be observed, contract terms could still vary substantially from one farm to the next, resulting in a great deal of local diversity. This is not confirmed by the data even though local differences in fundamentals might call for it.

We now propose an extension of the neoclassical framework that captures these effects. As in the standard model, we maintain the assumption that landlords propose contracts, which tenants accept or reject depending on whether they exceed or fall short of the reservation wage. In the standard model, the landlord proposes a contract that maximizes his expected returns subject to being acceptable to the tenant. In our setup, the landlord is influenced by expected returns plus two other factors: (i) the *exogenous* salience of a contract, e.g., whether it is based on easy fractions and round numbers; and (ii) the *endogenous* salience of a contract, as measured by the frequency with which it is used by others.

The process of contract choice works as follows. From time to time landlords recontract with their tenants. The landlord must offer the tenant a contract whose utility is at least equal to the utility of the reservation wage. Among all such *acceptable* contracts, the landlord offers one that maximizes his expected utility with high probability. In other words, the landlord follows a stochastic choice rule rather than strictly optimizing, a point of view that seems to fit well with empirical evidence (see, for example, Suppes and Atkinson 1960; Roth and Erev 1995). This adjustment process yields a stochastic dynamical system in which the distribution of contract choices evolves over time. The object of the analysis is to ascertain which distributions are most likely to emerge, starting from arbitrary initial conditions.

To be more specific, assume that agents live at the vertices of a graph, which represent distinct locations in some social or geographical space. Let V denote the set of n vertices, and let $i \in V$ denote a particular vertex. If the choice of agent j has an influence on agent i, we connect i and j by a directed edge (i, j); the amount of influence is given by a weight $w_{ij} \geq 0$. If influence is determined by geographical proximity, then we would have $w_{ij} = w_{ji}$ for all i and j. This *symmetry* assumption will be maintained in what follows.

Let X denote the finite set of contracts that people are willing to consider. At each time t, the contract in force at location i is denoted by $x^t_i \in X$, and the vector $\mathbf{x}^t \in X^n$ denotes the *state* of the process. At random times governed by a Poisson random variable ω_i, the landowner and tenant at location i renegotiate their contract. Assume that the random variables ω_i are independent and identically distributed, and that time is scaled so that, on average, there is one renegotiation per time period in each location. (Allowing differences in the rate of renegotiation at different locations does not change the results in any important way.)

Suppose that the current state is $\mathbf{x}$ and the contract at location i is about to be renegotiated. The landowner offers the tenant a contract, and the tenant accepts if and only if the contract's expected utility to the tenant is above some reservation utility level u^0_i. The value of u^0_i may depend on the particular location i, e.g., on local wages in alternative employment. We shall assume that the subset X_i of *acceptable* contracts that meet this reservation level is nonempty for each i. The landlord is assumed to choose a contract $x_i \in X_i$ that maximizes his utility.[9]

We treat the choice of contract at a given location as a state-dependent random variable. Let $P(x_i|\mathrm{x}_{-i})$ denote the conditional probability that contract x_i is adopted at location i when x_{-i} is the state of the other locations. We posit that $P(x_i|\mathbf{x}_{-i})$ is monotone increasing in the three factors mentioned earlier: the expected returns from the contract, the focalness of the contract, and its frequency of use in the "neighborhood." Denote the expected monetary returns to the landlord from contract x_i by $v_i(x_i)$. The focalness of the contract will be represented by a real-valued function $f(x_i)$, whose operational meaning will become clear in a moment. The local frequency of use is given by the function

$$w_i(x_i, \mathbf{x}_{-i}) = \sum_{j \neq i} w_{ij}\delta(x_i, x_j),$$

where $\delta(x_i, x_j) = 1$ if $x_i = x_j$, and $\delta(x_i, x_j) = 0$ otherwise.

A convenient functional form is to suppose that the log probability of choosing a given contract is a linear function of these three factors. In other words, for some $\beta \geq 0$, and every x_i and $y_i \in X_i$

$$\log P(x_i \mid \mathbf{x}_{-i}) - \log P(y_i \mid \mathbf{x}_{-i}) = \beta\left[v_i(x_i) - v_i(y_i) + w_i(x_i, \mathbf{x}_{-i}) - w_i(y_i, \mathbf{x}_{-i}) + f(x_i) - f(y_i)\right]. \tag{1}$$

Equivalently,

$$P(x_i|\mathbf{x}_{-i}) = \frac{\exp\beta\left[v_i(x_i) + w_i(x_i, \mathbf{x}_{-i}) + f(x_i)\right]}{\sum_j \exp\beta\left[v_j(x_j) + w_j(x_j, \mathbf{x}_{-j}) + f(x_j)\right]}. \tag{2}$$

This is known as a *loglinear response function*, and β is the *response parameter* (Blume 1995; Brock and Durlauf 1995; McKelvey and Palfrey 1995; Durlauf 1997).

This model can be interpreted as a perturbed optimal response process in the following sense. Suppose that we represent the utility of agent i in state $\mathbf{x}$ by

[9] This take-it-or-leave-it model of the bargaining process is fairly standard in the principal–agent literature. Models of custom that give a symmetric role to the contracting parties are discussed in Chapter 2 (Young 1993a), Chapter 3 (Young 1993b), Chapter 4 (Young 1998a) and Young (1998b).

$$U_i(\mathbf{x}) = v_i(x_i) + w_i(\mathbf{x}) + f(x_i) + \varepsilon_i(\mathbf{x}), \tag{3}$$

where $\varepsilon_i(\mathbf{x})$ is an unobserved utility shock. Suppose that these utility shocks are independent and identically distributed according to the extreme value distribution $P(\varepsilon \leq z) = e^{-e^{-\beta z}}$. If i always chooses a contract $x_i \in X_i$ that maximizes $U_i(\cdot)$ conditional on the others' choices being $\mathbf{x}_{-i}$, then from the observer's standpoint it appears that i is following the stochastic adoption process given by (2). Such models are standard in the discrete choice literature (Daniel McFadden 1974). For purposes of empirical estimation, however, it is not the utility function but the choice probabilities that matter. In principle, these could be estimated from a sufficiently rich set of event histories that give the temporal sequence of contract adoptions at different locations.

The functional form (2) is a convenient way of expressing the idea that the probability of a contract at a given location is an increasing function of three factors: return to land, focalness, and local frequency of use. The formula states, for example, that if two contracts x and y provide the same expected returns at a given location, and if they are used with the same frequency, but x is more inherently focal than y (say because x uses fractions with denominator 2 while y uses fractions with denominator 119), then x is more likely to be adopted than y. Alternatively, if x and y have the same expected returns, and both are based on very easy fractions, but x is much more commonly used than y in the neighborhood, then x is more likely to be adopted than y, and so forth.

Unfortunately, the Illinois data set does not allow us to estimate the choice probabilities directly, because the location of farms is known only down to the county level. (Furthermore, even if the precise locations were known, the sample is not large enough to include many instances of farms that are physically close to one another.) Thus we are proposing the relationship (2) as one that seems reasonable in principle, rather than one that we can directly test with our data set.

There is considerable experimental evidence, however, that round numbers and endogeneous notions of fairness matter in similar bargaining situations. An experimental setup that is closely related to the present one is the ultimatum game. In this game, the principal offers a fraction of the pie to an agent, which the latter can accept or reject (Guth et al. 1982). A standard finding is that agents often refuse offers that are perceived to be "stingy," even though this behavior would seem to be self-defeating. However, two other features of players' behavior are noteworthy for our purposes. First, the principals very often make offers that are in round numbers. Second, they adjust their offers to the prevailing conception of what constitutes a fair or acceptable offer, and this conception varies substantially from one culture to the next.

For example, Roth et al. (1991) conducted experiments in four different countries—Japan, Israel, the United States, and Yugoslavia—using the same experimental protocols. The experimenters corrected for differences in local currency by having the subjects express their offers in tokens, each token being worth the equivalent of about one US cent. In almost every case, over half of the offers were expressed in multiples of 50 tokens (Roth et al., 1991 Figures 5.3–5.4).[10] Furthermore, Roth et al. found that, while the modal offers differed among countries, the probability of an offer being accepted depended not on its absolute size, but on the extent to which it deviated from the modal offer. In other words, an agent's willingness to accept an offer appears to be governed by norms and expectations that differ from one culture to another. Both of these findings lend support (albeit indirect) to the model that we have proposed above.

The probabilistic choice model (2), together with the Poisson updating processes, defines a time-homogeneous Markov process on a finite state space, namely, the set of all states $\mathbf{x}$ that specify the contract in use at each location. This process is ergodic and has a unique invariant distribution μ. For each state $\mathbf{x}$, $\mu(\mathbf{x})$ represents the long-run relative frequency with which state $\mathbf{x}$ is visited in almost all realizations of the process.

In the present case the invariant distribution takes a particularly tractable form. Define the *potential* of state $\mathbf{x}$ to be

$$\rho(\mathbf{x}) = \sum_{i=1}^{n} v_i(x_i) + \sum_{i=1}^{n} f(x_i) + (1/2)\sum_{i=1}^{n} w_i(\mathbf{x}) \qquad (4)$$

Potential equals the total utility from those factors that involve no social interactions, plus one-half of the utility from those that do involve social interactions. It follows from standard arguments (see, e.g., Liggett 1985; Blume 1995; Young 1998a, chapter 6) that the long-run distribution of the process has the following simple form, known as a Gibbs representation:

[10] The focalness of round numbers has been documented in many other contexts; see, for example, Albers and Albers (1983).

$$\mu(\mathbf{x}) = e^{\beta p(\mathbf{x})} / \sum_{\mathbf{y}} e^{\beta p(\mathbf{y})}. \tag{5}$$

It follows that, when β is large, the probability is close to one that the process will be in a state that maximizes potential. We summarize this in the following result.

Theorem *Starting from an arbitrary initial state, the long-run probability of being in any given state* $\mathbf{x}$ *is proportional to* $e^{\beta p(\mathbf{x})}$. *When* β *is large but finite, the probability is close to one that the process will be in a state that globally maximizes potential. These are known as the stochastically stable states of the process.*

While the exponential form of the long-run distribution is especially convenient to work with, and depends on the loglinear representation in (2), it needs to be emphasized that the general approach can be applied without this assumption. Suppose that the renegotiation process at each location can be represented as a stochastic choice over contracts, whose distribution is perhaps state dependent but not time dependent. Suppose further that, at each location, every contract in the feasible set is chosen with a probability that is bounded away from zero over all possible states. Together with the Poisson updating processes, this yields an irreducible, finite Markov chain that has a unique invariant distribution. When the variance of the individual choice probabilities is sufficiently small, the process will typically occupy a few states with high probability, and these stochastically stable states can be identified with considerable precision using the method of large deviations (see Freidlin and Wentzell 1984; Kandori et al. 1993) and Chapter 2 (Young 1993a).

4 Application of the Model to Cropsharing

We now show how this approach can help to explain some of the apparently anomalous features of the Illinois data. For purposes of illustration, let us assume that the only contractual variable is the tenant's share x of total output. Let us also suppose that labor input is fully monitored, and that the contract specifies that it must correspond to standard management practice. At each farm i, we shall represent the productivity by a real number s_i, which gives the expected output per acre, measured in dollars, under standard management practice. (For example, $s_i = 80$ means that total net income on farm i is, on average, \$80 per acre.) The tenant's expected income on farm i is thus the contract share x_i times the index s_i times the size of the farm. For mathematical convenience, assume that all farms have the same size, which we may as well suppose to be unity. (This does not affect the analysis in any important way.) The tenant accepts a contract x_i only if his expected return $x_i s_i$ is at least m_i, where m_i is the reservation or market wage at location i.[11] The expected return to the landlord from such a contract is $v_i(x_i) = (1 - x_i)s_i$.

Let us assume that all contracts are equally focal from an a priori standpoint, say $f(x_i) = 0$ for all i and all x_i. To model the impact of custom, suppose that each of agent i's neighbors exerts the same degree of social influence. In other words, there is a number $\gamma \geq 0$ such that $w_{ij} = \gamma$ if farm i is adjacent to farm j, and $w_{ij} = 0$ otherwise. Let $n_i(\mathbf{x})$ be the number of i's neighbors using the same contract as i does in state $\mathbf{x}$. Then the potential function takes the form

$$\rho(\mathbf{x}) = \sum_{i=1}^{n} (1 - x_i)s_i + (\gamma/2) \sum_{i=1}^{n} n_i(\mathbf{x}). \tag{6}$$

The term $\sum (1 - x_i)s_i$ is just the total return to land, which we shall denote by $r(\mathbf{x})$. The term $(1/2) \sum n_i(\mathrm{x})$ is just the total number of edges (neighbor-pairs) that are coordinated on the same contract in state $\mathbf{x}$, which we shall denote by $c(\mathbf{x})$. This is the *degree of local conformity* in the system, and γ is the *conformity parameter*. We thus obtain the particularly transparent formulation

$$\rho(\mathbf{x}) = r(\mathbf{x}) + \gamma c(\mathbf{x}) \tag{7}$$

and

$$\mu(\mathbf{x}) \propto e^{\beta[r(\mathbf{x}) + \gamma c(\mathbf{x})]}.$$

In other words, *the log probability of each state* $\mathbf{x}$ *is a linear function of the total rent to land plus the degree of local conformity.* Given specific values of the conformity parameter γ and the response parameter β, we can compute the relative probability of various states of the process, and from this deduce the likelihood of different geographic distributions of contracts. In fact, one can say a fair amount about the qualitative behavior of the process even when one does not know specific values of the parameters.

[11] If tenants are not risk-neutral, the lower bound m_i depends on the tenant's degree of risk aversion, the variability of the returns, and the alternative wage.

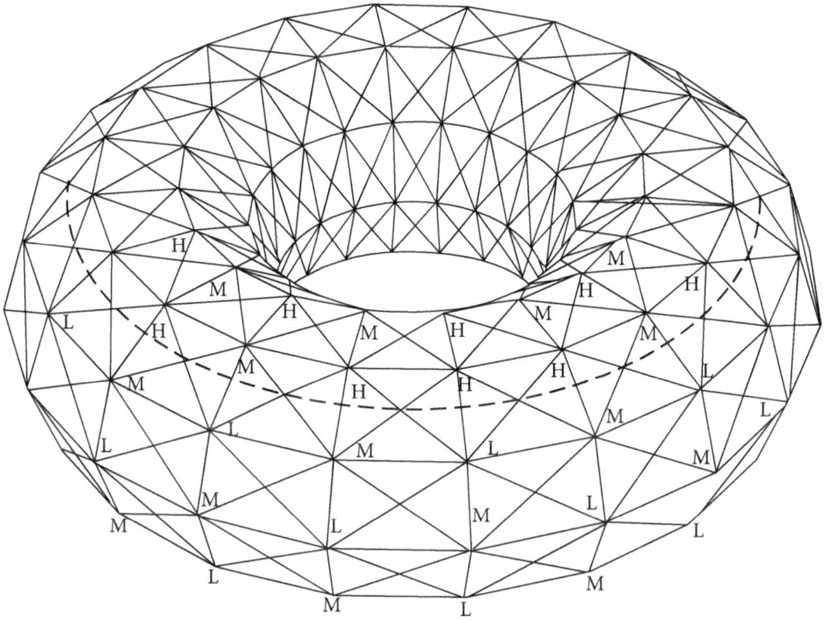

Figure 5.6 The hypothetical state of Torusota.
Note: Each vertex represents a farm, and soil qualities are High, Medium, or Low

5 An Illustrative Example

We illustrate with a concrete example that is meant to capture some of the key features of the Illinois case. Consider the hypothetical state of Torusota shown in Figure 5.6. In the northern part of the state—above the dashed line—soils are evenly divided between High (H) and Medium (M) quality soils, whereas in the southern part they are evenly divided between Medium and Low (L) quality soils. As in Illinois the soil types are interspersed, but average soil quality is higher in the north than it is in the south.

Let n be the number of farms. Each farm has exactly eight neighbors, so there are $4n$ edges altogether. As in the preceding section, contracts are one dimensional and specify the share of output for the tenant. Suppose there are nine salient contracts: $x = 10\%, 20\%, \ldots, 90\%$. (Contracts in which the tenant receives 0% or 100% are not considered.) For the sake of concreteness, assume that High soils have index 85, Medium soils have index 70, and Low soils have index 60. Let the reservation wage be 32.

We wish to determine the states with maximum potential. The answer depends on the size of γ, that is, on the trade-off rate between economic and social payoff. Consider first the case where $\gamma = 0$, that is, there are no conformity effects. Maximizing potential is then equivalent to maximizing the total rent to land, subject always to the constraint that labor earns at least its reservation wage on each class of soil. The contracts with this property are 40 percent on High soil, 50 percent on Medium soil, and 60 percent on Low soil. The returns to labor under this arrangement are: 34 on H, 35 on M, and 36 on L. Notice that labor actually earns a small premium over the reservation wage ($w = 32$) on each class of soil. This *quantum premium* is attributable to the discrete nature of the contracts: no landlord can impose a less generous contract (rounded to the nearest 10 percent) without losing his tenant. Except for the quantum premium, this outcome is the same as would be predicted by a standard market-clearing model, in which labor is paid its reservation wage and all the rent goes to land. We shall call this the *competitive equilibrium* state $\mathbf{w}$.

Notice that, in contrast to conventional equilibrium models, our framework actually gives an account of how the state $\mathbf{w}$ comes about. Suppose that the process begins in some initial state $\mathbf{x}^0$ at time zero. As landlords and tenants renegotiate their contracts, the process gravitates toward the equilibrium state $\mathbf{w}$ and eventually reaches it with probability one. Moreover, if β is not too small, the process stays close to $\mathbf{w}$ much of the time, though it will rarely be *exactly* in equilibrium.

These points may be illustrated by simulating the adjustment process. We chose $n = 400$, that is, 200 farms in the north and 200 in the south, and a moderate level of noise ($\beta = 0.20$). Starting from a random initial seed, we then simulated the process for three levels of conformity: $\gamma = 0$, 3, and 8. For each of these values the process settles into a characteristic pattern after the initial bias has worn off. Figure 5.7 shows a typical distribution of contract shares after 1,000 periods have elapsed. When $\gamma = 0$ (bottom panel), the contracts are matched quite closely with land quality, and the state is close to the competitive equilibrium. When the level of conformity is somewhat higher (middle panel), the dominant contract in the north is 50 percent, in the south it is 60 percent, and there are pockets here and there of other contracts. (This looks quite similar to the Illinois case.) Somewhat surprisingly, however, a further increase in the conformity level (top panel) does not cause the two regional customs to merge into a single global custom; it merely leads to greater uniformity in each of the two regions.

To understand why this is so, let us suppose for the moment that everyone is using the same contract x. Since everyone must be earning their

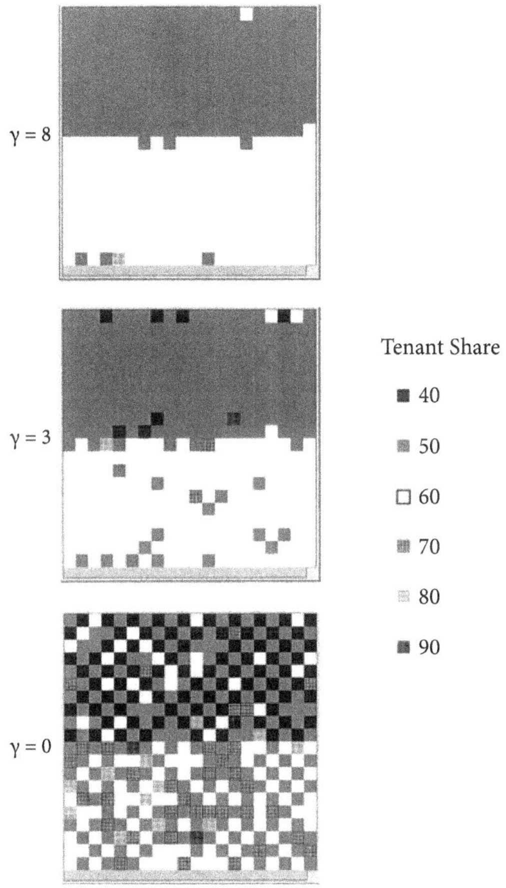

Figure 5.7 Simulated outcomes of the process for $n = 400$, $\beta = 0.20$ (Tenant shares in percent).

reservation wage, x must be at least 60 percent. (Otherwise southern tenants on low-quality soil would earn less than $w = 32$.) Moreover, among all such global customs, $x = 60\%$ maximizes the total rent to land. Hence the 60-percent custom, which we shall denote by **y**, maximizes potential among all global customs.

But it does not maximize potential among all states. To see this, let **z** be the state in which everyone in the north uses the 50-percent contract, while everyone in the south uses the 60-percent contract. From the standpoint of potential this is almost as good as **y**, because the only negative social externalities are suffered by those who live near the north–south boundary. Let us assume that the number of such agents is on the order of $\sqrt{n}$, where n is the total number of farms. Thus the *proportion* of farms near the boundary can be made as small as we like by choosing n large enough. But **z** offers a higher land rent to all the northern farms as compared to **y**. Assume that there are $n/2$ farms in the north, split equally between High and Medium soils, and that there are $n/2$ farms in the south, split equally between Medium and Low soils. Then the total income difference between **z** and **y** is $7n/4$ on the Medium soil farms in the north, and $8.5n/4$ on the High soil farms in the north, for a total gain of $31n/8$. It follows that, if γ is large enough, then for all sufficiently large n, the regionalized custom **z** has higher potential than the global custom **y**.[12]

While the details are particular to this example, the logic is quite general. Consider any distribution of soil qualities that is heterogeneous locally, but exhibits substantial shifts in average quality between geographic regions. For intermediate values of conformity γ, it is reasonable to expect that potential will be maximized by a distribution of contracts that is uniform locally, but diverse globally—in other words the distribution is characterized by *regional customs.*

Such a state will typically have higher potential than the competitive equilibrium, because the latter involves substantial losses in social utility when land quality is heterogeneous. It will typically have higher potential than a global custom, because it allows landlords to capture more rent at relatively little loss in social utility, provided that the boundaries between the regions are not too long (i.e., there are relatively few farms near the boundaries).

In effect, these regional customs form a compromise between completely uniform contracts on the one hand, and fully differentiated contracts on the other. Given the nature of the model, we should not expect perfect uniformity within any given region, nor should we expect *sharp* changes in custom at the boundary. The model suggests instead that there will be occasional departures from custom within regions (due to idiosyncratic influences), and considerable variation near the boundaries. This seems to be a reasonably accurate characterization of the distribution of contracts in the state of Illinois.

Acknowledgments

This research was supported by National Science Foundation Grant No. SES-9818975. The authors thank the referees for suggesting improvements on an earlier draft.

[12] A more detailed calculation shows that **z** uniquely maximizes potential among all states whenever γ is sufficiently large and n is sufficiently large relative to γ.

References

Akerlof, George A. (1980), "A theory of social custom, of which unemployment may be one consequence," *Quarterly Journal of Economics* 94 (4): 749–775.

Akerlof, George A. (1997), "Social distance and social decisions," *Econometrica* 65 (5): 1005–1027.

Albers, Wulf and Albers, Gisela (1983),"On the prominence structure of the decimal system," in R. W. Scholz, ed., *Decision Making under Uncertainty* (Amsterdam: Elsevier, pp. 271–287).

Allen, Douglas W. and Lueck, Dean (1993), "Transaction costs and the design of cropshare contracts," *Rand Journal of Economics* 24 (1): 78–100.

Bardhan, Pranab K. (1976), "Variations in extent and forms of agricultural tenancy—I: Analysis of Indian data across regions and over time," *Economic and Political Weekly* September 11: 1505–1546.

Bardhan, Pranab K. (1984), *Land, Labor, and Rural Poverty: Essays in Development Economics* (New York: Columbia University Press).

Bardhan, Pranab and Rudra, Ashok (1980), "Terms and conditions of sharecropping contracts: an analysis of village survey data in India," *Journal of Development Studies* 16 (3): 287–302.

Bardhan, Pranab and Rudra, Ashok (1981), "Terms and conditions of labour contracts in agriculture: results of a survey in West Bengal, 1979," *Oxford Bulletin of Economics and Statistics* 43(1): 89–111.

Bardhan, Pranab and Rudra, Ashok (1986a), "Labour mobility and the boundaries of the village moral economy," *Journal of Peasant Studies* 13: 90–115.

Bardhan, Pranab and Rudra, Ashok (1986b), *Agrarian Relations in West Bengal: Results of Two Surveys* (Bombay: Somaiya).

Bell, Clive and Zusman, Pinhas (1976), "A bargaining theoretic approach to cropsharing contracts," *American EconomicReview* 66 (4): 578–588.

Binmore, Ken, Swierzbinski, J., Hsu, S., and Proulx, C. (1993), "Focal points and bargaining," *International Journal of Game Theory* 22 (4): 381–409.

Binswanger, Hans P. and Rosenzweig, Mark R., eds. (1984), *Contractual Arrangements, Employment, and Wages in Rural Labor Markets in Asia* (New Haven, CT: Yale University Press).

Bliss, Christopher J. and Stern, N. H. (1982), *Palanpur: The Economy of an Indian Village* (Oxford: Clarendon Press).

Blume, Lawrence E. (1995), "The statistical mechanics of best-response strategy revision," *Games and Economic Behavior* 11 (2): 111–145.

Braverman, Avishay and Stiglitz, Joseph E. (1982), "Sharecropping and the interlinking of agrarian markets," *American Economic Review* 72 (4): 695–715.

Brock, William A. and Durlauf, Steven N. (1995), "Discrete Choice with Social Interactions I: Theory." National Bureau of Economic Research (Cambridge, MA) Working Paper No. 5291 October.

Burke, Mary A. and Young, H. Peyton (2000), "Terms of agricultural contracts: theory and evidence," Working Paper No. 16, Center on Social and Economic Dynamics, the Brookings Institution and Johns Hopkins University, [www.brookings.edu/es/dynamics].

Cheung, Steven N. S. (1969), *The Theory of Share Tenancy* (Chicago: University of Chicago Press).

Durlauf, Steven N. (1997), "Statistical mechanical approaches to socioeconomic behavior," in W. Brian Arthur, Steven N. Durlauf, and David Lane, eds., *The Economy as a Complex Evolving System*, Vol. 2 (Redwood City, CA: Addison-Wesley, pp. 81–104).

Fehrenbacher, J. B., Pope, R. A., Jansen, I. J., Alexander, J. D., and Ray, B. W. (1978), *Soil Productivity in Illinois.* Circular 1156, College of Agriculture, Cooperative Extension Service, University of Illinois at Urbana-Champaign.

Freidlin, Mark and Wentzell, Alexander (1984), *Random Perturbations of Dynamical Systems* (Berlin: Springer-Verlag).

Guth, Werner, Schmittberger, Rolf, and Schwarze, Bernd (1982), "An experimental analysis of ultimatum bargaining," *Journal of Economic Behavior and Organization* 3 (4): 367–388.

Hayami, Yujiro and Otsuka, Keijiro (1993), *The Economics of Contract Choice: An Agrarian Perspective* (Oxford: Clarendon Press).

Illinois Cooperative Extension Service (1995), *1995 Cooperative Extension Service Farm Leasing Survey.* Department of Agricultural and Consumer Economics, Cooperative Extension Service, University of Illinois at Urbana-Champaign,.

Johnson, D. Gale (1950), "Resource allocation under share contracts," *Journal of Political Economy* 58 (2): 111–123.

Kandori, Michihiro, Mailath, George J., and Rob, Rafael (1993), "Learning, mutation, and long run equilibria in games," *Econometrica* 61 (1): 29–56.

Liggett, Thomas M. (1985), *Interacting Particle Systems* (New York: Springer-Verlag).

McFadden, Daniel (1974), "Conditional logit analysis of qualitative choice behavior," in Paul Zarembka, ed., *Frontiers in Econometrics* (New York: Academic Press, pp. 105–142).

Mckelvey, Richard D. and Palfrey, Thomas R. (1995), "Quantal response equilibria for normal form games," *Games and Economic Behavior* 10 (1): 6–38.

Marshall, Alfred (1920), *Principles of Economics, 8th edition* (London: Macmillan).

Mausel, P. W., Runge, E. C. A., and Carmer, S. G. (1975), *Soil Productivity Indexes for Illinois Counties and Soil Associations.* Bulletin 752, College of Agriculture, Agricultural Experiment Station, University of Illinois at Urbana-Champaign.

Mill, John Stuart (1929), *Principles of Political Economy* (1848) (London: Longmans Green).

Murrell, Peter (1983), "The economics of sharing: a transactions cost analysis of contractual choice in farming," *Bell Journal of Economics* 14 (1): 283–293.

Odell, R. T. and Oschwald, W. R. (1970), *Productivity of Illinois Soils.* Circular 1016, College of Agriculture, Cooperative Extension Service, University of Illinois at Urbana-Champaign,.

Reid, Joseph D., Jr. (1975), "Sharecropping in history and theory," *Agricultural History* 49 (2): 426–440.

Reid, Joseph D., Jr. (1976), "Sharecropping and agricultural uncertainty," *Economic Development and Cultural Change* 24 (3): 549–576.

Robertson, A. F. (1987), *The Dynamics of Productive Relationships: African Share Contracts in Comparative Perspective* (Cambridge: Cambridge University Press).

Roth, Alvin E. (1985), "Toward a Focal-point theory of bargaining," in Alvin E. Roth, ed., *Game-Theoretic Models of Bargaining* (New York: Cambridge University Press).

Roth, Alvin E. and Erev, Ido (1995), "Learning in extensive-form games: experimental data and simple dynamic models in the intermediate term," *Games and Economic Behavior* 8 (1): 164–212.

Roth, Alvin E., Prasnikar, Vesna, Okuno-Fujiwara, Masahiro, and Zamir, Shmuel (1991), "Bargaining and market behavior in Jerusalem, Ljubljana, Pittsburgh, and Tokyo: An experimental study," *American Economic Review* 81 (5): 1068–1095.

Roumasset, James and James, William (1979), "Explaining variations in share contracts: land quality, population pressure and technological change," *Australian Journal of Agricultural Economics* 23 (2): 116–127.
Rudra, Ashok (1975), "Sharecropping arrangements in West Bengal," *Economic and Political Weekly* 10: A58–A63.
Schelling, Thomas C. (1960), *The Strategy of Conflict* (Cambridge, MA: Harvard University Press).
Stiglitz, Joseph E. (1974), "Incentives and risk sharing in sharecropping," Review of Economic Studies 41 (2): 219–255.
Stiglitz, Joseph E. (1989), "Rational peasants, efficient institutions, and a theory of rural organization: methodological remarks for development economics," in Pranab Bardhan, ed., *The Economic Theory of Agrarian Institutions* (Oxford: Oxford University Press, pp. 18–29).
Suppes, Patrick and Atkinson, Richard (1960), *Markov Learning Models for Multi-person Interactions* (Stanford, CA: Stanford University Press).
Young, Arthur (1792), *Travels during the Years 1787, 1788, and 1789* (Bury St. Edmonds: J. Rackham).
Young, H. Peyton (1993a), "The evolution of conventions," *Econometrica* 61 (1): 57–84.
Young, H. Peyton (1993b), "An evolutionary model of bargaining," *Journal of Economic Theory* 59 (1): 145–168.
Young, H. Peyton (1998a), "Conventional contracts," *Review of Economic Studies* 65 (4): 773–792.
Young, H. Peyton (1998b), *Individual Strategy and Social Structure: An Evolutionary Theory of Institutions* (Princeton, NJ: Princeton University Press).

PART II
LEARNING

Preface to Chapters 6–8

Learning to Play Without Knowing the Game

The next three chapters contrast different ways of modelling subjects' learning behavior. Do high rationality rules always lead to equilibrium? How much knowledge do subjects need for equilibration to occur? Are there low rationality adaptive rules that lead to equilibrium even when subjects have little or no knowledge of others' behaviors? The chapters in this section provide an overview of results for different classes of learning rules, as well as evidence about how agents actually do adapt in experimental settings.

Prior chapters focused on situations where agents are myopic best responders. Faced with a decision, they collect sample information about the actions taken by others in recent periods, they summarize this information as an empirical probability distribution, and choose a best reply with high probability. In this way, past behavior shapes agents' expectations about others' behavior going forward. Variants of this set-up, such as fictitious play, are well-established in the game theory literature, but they represent a significant departure from higher forms of rationality in several respects. For one thing, the agents are assumed to be myopic: they fail to take into account future developments beyond the next period. Furthermore, this type of learning behavior fails to account for the decision rules being used by the other agents. In particular, if others are updating using a similar precedents-based procedure, might it not be better to respond, not to the precedents per se, but to the others' responses to the precedents? Of course, this leads to a new problem: why should one not assume that the other agents are anticipating in the same fashion?

In principle, Bayesian updating provides one way to close the loop, but it presumes an implausible degree of rationality and computational ability on the part of the agents. The fundamental problem is that interactive learning is inherently much more complex than learning about the value of an exogenous parameter or state of the world, which is the typical situation that Bayesian learning is meant to address. *In a system of interacting agents, the act of learning changes the thing to be learned.* If agent A is trying to learn about agent B, A's behavior will naturally depend on what she has learned so far, and also on what she hopes to learn in the future. But A's behavior can be observed by B, hence B's behavior may change as a result of A's attempts to learn it. The same holds for B's attempts to learn about A.

This feedback loop is a central and inescapable feature of interactive learning situations. Consequently, methods which work for single-agent learning problems may fail in multi-agent settings. As Chapter 6 shows, this turns out to be the case: there is an *uncertainty principle* for strategic interactions which states that if there is enough ex ante uncertainty about the other players' payoffs (and hence their potential behavior), then in some games it is impossible for perfectly rational players to learn to predict one another's behavior, even over an infinite number of repetitions of the game.[1] This impossibility result rests on very demanding (albeit conventional) assumptions about agents' rationality, and what it means to learn to predict an opponent's behavior.

There are several ways of circumventing the difficulties that beset Bayesian rational learning in a multi-agent setting. One approach is to assume that players' repeated game strategies can be restricted in some way. For example, we might assume that the opponents' repeated-game strategies are countable in number. Suppose further that given any prior distribution over these strategies it is optimal to choose a strategy in the same restricted class. Then with probability one players' behavior will eventually come close to Nash equilibrium behavior.[2] Note, however, that this approach makes an implicit assumption that agents know a great about their opponents' possible behaviors, which in practice may be unrealistic.

An alternative approach is a learning procedure based on statistical hypothesis testing. At each point in time, the players entertain hypotheses about the strategies their opponents are using, and compare the actual behavior of their opponents to the predicted ones. A player rejects his current hypothesis if the opponents' observed behaviors are sufficiently unlikely (given the hypothesis), and then chooses a new hypothesis at random. Subject to certain regularity conditions, this learning process eventually leads to approximate equilibrium behavior in the following sense: a large proportion of the time the players' hypothesis-testing strategies constitute an ε-equilibrium of the repeated game

[1] For other results in a similar spirit see K. Binmore (1987), "Modelling rational players: Part I," *Economics and Philosophy* 3: 179–214, J. S. Jordan (1993), "Three problems in learning mixed-strategy Nash equilibria," *Games and Economic Behavior* 5: 368–386, and J. H. Nachbar (1997), "Prediction, optimization, and learning in repeated games," *Econometrica* 65: 275–309.

[2] E. Kalai and E. Lehrer (1993), "Rational learning leads to Nash equilibrium," *Econometrica* 61: 1019–1045.

from that time on; in fact the strategies are close to being subgame perfect for long stretches of time. Furthermore, the players are eventually "good predictors" in the sense that, a large fraction of the time, they predict the probabilistic behavior of their opponent very closely. In this sense the players can learn to predict more accurately than their Bayesian rational counterparts in some classes of games.[3]

One limitation of all of these "traditional" learning methods—including Bayesian updating, hypothesis testing, and best response to precedent—is that they depend on the observability of others' actions. In games with many dispersed players, this assumption is not particularly plausible. It turns out, however, that there are simple and intuitive learning rules, incorporating some degree of experimentation, that require no information about the payoffs and behaviors of the other players, and that lead to equilibrium behavior in a wide class of strategic situations. Chapter 7 describes one such rule called *trial-and-error learning.* The key idea is that an agent's pattern of received payoffs triggers different emotional states or 'moods,' that give rise to different types of search. If payoffs are steady or increasing, then with small but positive probability agents search for something better. If, however, they experience decreasing payoffs without experimenting, they become "discontent," and initiate a wide-area search for a new strategy, which is retained with a probability that is increasing in its realized payoff.[4] Chapter 7 shows that, any finite game with generic payoffs and at least one pure Nash equilibrium, this process eventually approximates equilibrium behavior in the sense that a pure Nash equilibrium is played a high proportion of the time. (Note, however, that this is not the same thing as *convergence* to equilibrium, because every so often behaviors move far away from equilibrium due to random experimentation.)

Are the basic features of trial-and-error learning observed in practice? Chapter 8 explores this question in an experimental setting where subjects engage in a voluntary contributions game. Contributing incurs a cost, and the net returns from contributing increase with the number of others who contribute. In one version of this set-up, the returns from contributing are set

[3] For details of this approach see D. P. Foster and H. P. Young (2003), "Learning, hypothesis testing, and Nash equilibrium," *Games and Economic Behavior* 45: 73–96.

[4] For other learning rules with somewhat similar features see M. Nowak et al. (1993), "A strategy of win-stay, lose-shift that outperforms tit-for-tat in the prisoner's dilemma game," *Nature* 364 (6432), 56–58; U. Motro and A. Shmida, (1995), "Near–far search: an evolutionary stable foraging strategy," *J. Theoretical Biology* 173: 15–22; A. E. Eiben and C. A. Schippers (1998), "On evolutionary exploration and exploitation," *Fundam. Inf.* 35 (1): 35–50; B. Skyrms (2012), "Learning to signal with probe and adjust," *Episteme* 9 (2): 139–150.

high enough that contributing is a dominant strategy for every player; in an alternative version the returns are set low enough to make *not contributing* a dominant strategy.

Each version of the game was repeated for twenty periods under different informational assumptions. In the low information treatment, players were not told the structure of the game, and they were unable to observe what other players were doing or what their payoffs were. In effect each player was isolated in a "black box" and knew nothing about their interactions with others. In the "standard" information treatment, players were fully informed about the game structure and provided with summary statistics about the behavior of the other players in prior periods. In the enhanced treatment, they were also informed about the behavior and payoffs of other players at an individual level.

The chapter investigates how players adapt their behavior under these different informational conditions, and focuses in particular on how their search behavior is affected by the pattern of their realized payoffs. A key finding is that search behavior is different when payoffs are declining, as opposed to steady or increasing. In the former case search behavior is more frequent, and wider ranging, than in the latter case. This is consistent with the asymmetric search modes posited by trial-and-error learning. Moreover, these features are present when subjects have no information about the game, and also when they know the structure of the game as well as the actions and payoffs of the other players. Thus it appears that asymmetric search behaviors are triggered even when subjects have enough information to employ a belief-based model. Another interesting finding is that convergence to approximate equilibrium behavior occurs at about the same rate irrespective of the information treatment. In particular, it seems that subjects are able to feel their way toward equilibrium fairly quickly even when they have no information about others' behavior. These results were obtained for voluntary contribution games; whether they hold for more general classes of games is an open question for future research.

6
On the Impossibility of Predicting the Behavior of Rational Agents

With Dean P. Foster

Economists often assume that people are *rational*: they maximize their expected payoffs given their beliefs about future states of the world. This hypothesis plays a crucial role in game theory, where each player is assumed to choose an optimal strategy given his belief about the strategies of his opponents. In this setting, a belief amounts to a forecast or prediction of the opponents' future behavior, that is, of the probability with which the opponents will take various actions. The prediction is *good* if the forecasted probabilities are close to the actual probabilities. Together prediction and rationality justify the central solution concept of the theory. Namely, if each player correctly predicts the opponents' strategies and if each chooses an optimal strategy given his prediction, then the strategies form a Nash equilibrium of the repeated game. But under what circumstances will rational players actually learn to predict the behavior of others starting from out-of-equilibrium conditions?

In this chapter we show that there are very simple games of incomplete information such that players almost never learn to predict their opponents' behavior even approximately, and they almost never come close to playing a Nash equilibrium. This impossibility result and its proof builds on the existing literature on learning in repeated games (Jordan 1991, 1993, 1995; Kalai and Lehrer 1993; Lehrer and Smorodinsky 1997; Nachbar 1997, 1999, 2001); for other critiques of Bayesian learning in economic environments see (Binmore 1987, 1990, 1993; Blume and Easley 1998). The present chapter demonstrates the incompatibility between rationality and prediction without placing any restrictions on the players' prior beliefs, their learning rules, or the degree to which they are forward-looking.

Originally published as "On the impossibility of predicting the behavior of rational agents," *PNAS*, 98 (22) 12848–12853 https://www.pnas.org/doi/10.1073/pnas.211534898. Reproduced with permission.

Spontaneous Order. H. Peyton Young with Dean P. Foster, Oxford University Press. © H. Peyton Young (2024).
DOI: 10.1093/oso/9780198892908.003.0006

An Example. We begin by illustrating the problem in a concrete case. Consider two individuals, A and B, who are playing the game of matching pennies. Simultaneously each turns a penny face up or face down. If the pennies match (both are heads or both are tails), then B buys a prize for A; if they do not match, A buys a prize for B. Assume first that the prize is one dollar and that the utility of both players is linear in money. Then the game has a unique Nash equilibrium in which each player randomizes by choosing heads (H) and tails (T) with equal probability. If both adopt this strategy, then each is optimizing given the strategy of the other. Moreover, although neither can predict the *realized action* of the opponent in any given period, each can predict his strategy, namely, the *probabilities* with which the actions will be taken. In this case no tension exists between rationality and prediction because the game has a unique equilibrium, and the players know what it is.

Now change the situation by assuming that if both players choose H, then B buys an ice cream cone for A, whereas if both choose T then B buys A a milk shake. Similarly, if A chooses H and B chooses T then A buys B a coke, whereas if the opposite occurs then A buys B a bag of chips. Assume that the game is played once each day, the players' tastes do not change from one day to the next, and they have a fixed positive utility for each of the prizes and also for money. Unlike the previous situation, this is a game of incomplete information in which neither player knows the other's payoffs.

		B's action		B's action	
		H	T	H	T
A's action	H	eat cone	buy coke	buy cone	drink coke
	T	buy chips	drink shake	eat chips	buy shake
		Outcomes for A		**Outcomes for B**	

For expositional simplicity assume first that the players are myopic, that is, they do not worry about the effect of their actions on the future course of the game. Imagine that the following sequence of actions has occurred over the first ten periods.

Period	1	2	3	4	5	6	7	8	9	10	11
A:	H	T	T	H	H	H	T	H	T	H	?
B:	T	H	T	H	T	H	T	H	T	H	?

The immediate problem for each player is to predict the *intention* of the opponent in period 11 and to choose an optimal response. The opponent's intention might be to play H for sure, T for sure, or to randomize with probability p for H and $1 - p$ for T. If the opponent's intention is to randomize, then obviously one cannot predict his realized action, but it does not seem too much to ask that one predict the approximate probability with which he intends to play each action. We claim, however, that this is essentially impossible.

To see why, let's put ourselves in A's shoes. The behavior of B suggests an alternating pattern, perhaps leading us to predict that B will play T next period. Because we are rational, we will (given our prediction) play T for sure next period. But if B is a good predictor, then she must be able to predict that with high probability we are in fact going to play T next period. This prediction by B leads her to play H next period, thus falsifying our original prediction that she is about to play T.

The point is that if either side makes a prediction that leads them to play H or T for sure, the other side must predict that they are going to do so with high probability, which means that they too will choose H or T for sure. But there is no pair of predictions such that both are approximately correct and the optimal responses are H or T for sure. It follows that, for both players to be good predictors of the opponent's next-period behavior, at least one of them must be intending to play a mixed strategy next period, and the other must predict this.

Suppose, for example, that player B intends to play a mixed strategy in period 11. Because B is rational, she only plays a mixed strategy if she is *exactly indifferent* between playing H and T given her predictions about A. (If there is a slight difference in payoff between the two actions, strict rationality requires that the one with higher payoff be chosen exclusively.) Now B's predictions about A's behavior in the 11th period are based on the observed history of play in the first 10 periods. Let us say that the particular history given above leads B to predict that A will play H with a probability of 0.127. Because B intends to play mixed, it must be the case that B's expected utility from playing H or T is identical given B's utility function u_{B} for the various outcomes. In other words, it must be that

$$.127u_{\mathrm{B}}(\text{buy cone}) + .873u_{\mathrm{B}}(\text{eat chips}) = .127u_{\mathrm{B}}(\text{drink coke}) + .873u_{\mathrm{B}}(\text{buy shake}).$$

But there is no reason to think that B's utilities actually do satisfy this equation exactly. More precisely, let us suppose that B's utility for each outcome could be any real number within a certain interval, and that B's actual utility (B's *type*) is the result of a random draw from among these possible values. (The draw occurs once and for all before the game begins.) Following Jordan (1993), we claim that the probability is zero that the above equation will be satisfied. The reason is that there is only a finite number of distinct predictions that B could make at this point in time, because B's prediction can only be based on A's (and B's) previous observed behavior together with B's initial beliefs. Because this argument holds for every period, the probability is zero that B will ever be indifferent. From this and the preceding argument it follows that in any given period, one or both players must be making a bad prediction. Moreover, they cannot be playing a Nash equilibrium in any given period (or even close to a Nash equilibrium), because this would require them to play mixed strategies, which means that both must be indifferent.

Jordan (1993) was the first to employ this kind of argument to show that myopic players effectively cannot learn mixed equilibria no matter what their beliefs are. Moreover, as we have just seen, the same argument shows that at least one of them cannot learn to predict the behavior of the other. The limitation of Jordan's result is that it assumes players are completely myopic. Forward-looking behavior allows for a richer repertoire of learning strategies and more time to detect complex patterns in the behavior of one's opponent. Nevertheless, the incompatibility between rationality and prediction continues to hold even in this case, as we shall show below.

A second closely related body of work is provided by Nachbar (1997, 1999, 2001). He was the first to argue that there is a fundamental tension between prediction and rationality in the context of Bayesian learning even when players are forward-looking. Nachbar's critique was prompted by an earlier paper by Kalai and Lehrer (1993), which laid out conditions under which Bayesian rational players would in fact be able to learn to predict the behavior of their opponents. Suppose that each player begins the game with a prior belief over the possible repeated game strategies that his opponents might use. Kalai and Lehrer show that, if these prior beliefs contain a "grain of truth," that is, they put positive probability (however small) on the *actual* repeated game strategies of the opponents, then players learn to predict with probability one.

As Nachbar points out, however, the grain-of-truth condition may be very difficult to satisfy in practice. To illustrate, consider the preceding example and suppose that the players are perfectly myopic. Then the unique equilibrium of the repeated game is for A to play H with some fixed probability p^* each period and for B to play H with some fixed probability q^* each period. These values are not known to the players because p^* depends on B's payoffs, whereas q^* depends on A's payoffs. Can they be learned through Bayesian updating of a diffuse prior? Suppose that each player begins with a belief that the other is playing an i.i.d. strategy with an unknown parameter (the probability of playing H), where the beliefs have full support on the interval [0, 1]. In any given period, the players almost surely will have updated beliefs that lead them to play H or T with a probability of 1 in that period, because the expected payoffs from H and T are not exactly equal. However, their updated beliefs lead them to predict that their opponent is almost surely going to play a *mixed* strategy next period. Thus their predictions almost certainly are not close to their actual strategies. Furthermore, as the game proceeds, rationality causes them to play H for sure in some periods and T for sure in others. Hence their actual strategies are not i.i.d. and not in the support of their beliefs. More generally, Nachbar (1997, 1999, 2001) argues that in games such as this it is difficult to identify any plausible family of beliefs such that the players' best response strategies are in the support of their beliefs [another paper in the same general spirit is provided by Miller and Sanchirico (1997)].

In this chapter we are agnostic about whether or not the players are Bayesian and what the structure of their priors might be. Instead we show that no matter how players use the information revealed by repeated play, they will fail to learn to predict the opponents' behavior in some kinds of games.

Before turning to a precise statement of our result, we should point out that it is *prediction by the players* that is problematical; to an *observer* the average behavior of the players may exhibit empirical regularities. For example, it could be that the cumulative frequency distribution of play approaches a Nash equilibrium of the game. In fact this will be the case for fictitious play in which each player uses the empirical distribution of the opponent's play up through a given period to predict his next-period behavior and then chooses a best response given that prediction. In games such as matching pennies, this simple learning rule induces long-run average behavior that converges to the mixed Nash equilibrium of the game (Miyasawa 1961; Monderer and Shapley 1996). There are other models in which a player's average behavior mimics Nash equilibrium from the observer's standpoint (Harsanyi 1973; Fudenberg and Kreps 1993); in fact Nash himself proposed such an interpretation (Nash 1950). But this does not imply that *individual* players ever play Nash equilibrium strategies or that they learn to predict.

The Learning Model We now describe our impossibility result in detail. Consider an n-person game G with finite action space $X = \Pi X_i$ and utility functions $u_i: X \to R$. We shall assume that the payoffs take the form $u_i(x) = u_i^0(x) + \omega_i(x)$, where the $u_i^0(x)$ are payoffs in a benchmark game G^0, and the $\omega_i(x)$ are i.i.d. random variables drawn from a continuous density $\nu(\omega)$, the support of which is the interval $I_\lambda = [-\lambda/2, \lambda/2]$. The parameter $\lambda > 0$ is the range of uncertainty in the payoffs. We shall assume that the distribution of payoffs is common knowledge, but the realized payoff $u_i(x)$ is known only to player i. Errors are drawn once only before play begins, and the resulting one-shot game (called the ν-perturbation of G^0) is played infinitely often.

Each player takes an action once in each time period $t = 1, 2, 3, \ldots$ The *outcome* in period t is an n-tuple of actions $x^t \in X$, where x_i^t is the action taken by i in period t. *A state* of the process at time t is a history of play up to t, that is, a sequence of outcomes $h^t = (x^1, x^2, \ldots x^t)$. Let h^0 represent the null history, H^t the set of all length-t histories, and $H = \cup_t H^t$ the set of all finite histories, i.e., the set of all states. A realization of the process will be denoted by h, and the set of realizations (i.e., the set of infinite histories) will be denoted by H^∞. Histories are observed publicly, that is, there is perfect monitoring.

The discounted payoff to player i from a realization $h = (x^1, x^2, \ldots x^t, \ldots)$ is

$$U_i(h) = (1 - \delta_i) \sum_{t=1}^{\infty} \delta_i^{t-1} u_i(x^t)$$

where δ_i is i's discount factor, $0 \leq \delta_i < 1$ (if $\delta_i = 0$, $U_i(h) = u_i(x^1)$). Let Δ_i denote the set of probability distributions over X_i. Let $\Delta = \prod_j \Delta_j$ denote the product set of mixtures, and let $\Delta_{-i} = \prod_{j \neq i} \Delta_j$ be the product set of mixtures by i's opponents. A *behavioral strategy* for player i specifies a conditional probability distribution over i's actions in each period conditional on the state in the previous period. Thus we can represent i's strategy by a function $q_i^t = g_i(h^{t-1}) \in \Delta_i$, where $q_i^t(x_i)$ is the probability that i plays x_i in period t given that h^{t-1} is the state in period $t - 1$. This is of course a function of i's realized utility function u_i, but we shall not write this dependence explicitly.

A *prior belief* of player i is a probability distribution over all possible combinations of the opponent's strategies. We can decompose any such belief into one-step-ahead forecasts of the opponent's behavior conditional on each possible state. Thus, if h^{t-1} is the state at time $t - 1$, i's forecast about the behavior of her opponents in period t can be represented by a probability distribution $p_{-i}^t = f_i(h^{t-1}) \in \Delta_{-i}$, where $p^t_{-i}(x_{-i})$ is the probability that i assigns to the others playing the combination x_{-i} in period t. The function f_i: $H \rightarrow \Delta_{-i}$ will be called i's *forecasting function*. Given any vector of forecasting functions $f = (f_1, f_2, \ldots f_n)$, one for each player, there exists a set of prior beliefs such that the f_i describe the one-step-ahead forecasts of players with these beliefs [see Kalai and Lehrer (1993)].

Consider the situation just after the players have been informed privately of their realized payoff functions u_i. Because of the independence of the draws among players, no one knows anything he did not already know about the others' payoffs, and this fact is common knowledge. This has an implication for the forecasting functions. Namely, at the beginning of each period t, i knows that j's information consists solely of the publicly observed history h^{t-1} and j's own payoff function u_j. Player j's behavior cannot be conditioned on information that j does not have (namely u_{-j}), and player i's forecast of j's behavior cannot be conditioned on information that i does not have (namely, u_{-i}). Thus i's forecast $\left[f_i\left(h^{t-1}\right)\right]_j$ about j's behavior in each period t does not depend on the realization of the values u_k for *every* k, including $k = i, j$. It follows that the functions f_i do not depend on the realized payoff functions $u_i(\cdot)$, although they may depend on v. Another way of saying this is that the beliefs must be consistent with the players' a priori knowledge of the information structure.

Following Jordan (1993), we shall say that a *learning process* is a pair $(f, g) = (f_1, \ldots, f_n, g_1, \ldots, g_n)$, where $f_i: H \to \Delta_{-i}$ and $g_i: H \to \Delta_i$ for each player i. Given a realization of the process h, we shall denote player i's forecast in period t by $p^t_{-i}(h) = f_i(h^{t-1})$, and i's behavioral strategy in period t by $q^t_i(h) = g_i\left(h^{t-1}\right)$.

The pair (f_i, g_i) induces a probability measure on the set of all realizations H^∞. Similarly, for every state h^{t-1}, f_i and g_i induce a conditional probability distribution on all continuations of h^{t-1}. Denote this conditional distribution by $\mu_i(f_i, g_i|h^{t-1})$. We say that individual i is *rational* if, for every h^{t-1}, i's conditional strategy $g_i(\cdot|h^{t-1})$ optimizes i's expected utility from time t on, given i's conditional forecast $f_i(\cdot|h^{t-1})$. (This is also known as *sequential rationality*.) Specifically, for every alternative choice of strategy $g_i'(\cdot|h^{t-1})$,

$$\int U_i(h)d\mu_i\left(f_i, g_i \mid h^{t-1}\right) \geq \int U_i(h)d\mu_i\left(f_i, g_i' \mid h^{t-1}\right).$$

Prediction Intuitively, player i learns to predict the behavior of his opponent(s) if i's forecast of their next-period behavior comes closer and closer to their actual next-period strategies. This idea may be formalized as follows. Consider a learning process (f, g), and let $\mu(g)$ denote the probability measure induced on H^∞ by the strategies $g = (g_1, g_2, \ldots, g_n)$. We say that player i *learns to predict* if the mean square error of i's next-period predictions goes to zero over almost all histories of play. In other words, for $\mu(g)$-almost all realizations h

$$\lim_{T\to\infty} \sum_{t=1}^{T} |p^t_{-i}(h) - q^t_{-i}(h)|^2/T = 0 \tag{1}$$

Similarly, we shall say that player i *never learns to predict* if the subset of histories for which Eq. 1 holds has μ-measure zero. Note that this condition permits players to make bad forecasts from time to time, provided they do not occur too often.

An Impossibility Theorem We now demonstrate a class of repeated games such that, with probability one, some player never learns to predict his opponent's behavior, and this holds for all prior beliefs. Because our result holds for all beliefs, it must hold for beliefs that are in some sense best possible. A reasonable candidate for "best possible beliefs" are rational expectations beliefs. These have the property that, at every point in time, each player's prediction of his opponent's future behavior is conditioned correctly on the posterior distribution of payoff types revealed by play so far. Jordan (1991, 1993) shows that these posterior distributions converge to the set of Nash equilibria of the game (see also Nyarko 1998). However, this does not imply that the posteriors lead to predictions that are close to being correct for a given opponent. Our result shows, in fact, that these rational expectations predictions are not close to being correct for almost all opponents.

This still leaves open the possibility that for some combinations of beliefs the players' strategies converge to Nash equilibrium even though their predictions do not. In a repeated game convergence to equilibrium can be given a variety of interpretations; we shall show that the process fails to converge to equilibrium in almost any reasonable sense. Let Q^N be the set of all one-period strategy tuples $q \in \Delta$ such that q occurs in *some* time period in *some* Nash equilibrium of the repeated game. For every $q \in \Delta$, let $d(q, Q^N)$ be the minimum Euclidean distance between q and the compact set Q^N. Given a learning process (f, g) and a specific history h, if the behavioral strategies come close to Nash equilibrium on h then at a minimum we would expect the following condition to hold,

$$\lim_{T\to\infty}\left[\sum_{t=1}^{T} d\left(q^t(h), Q^N\right)^2\right]/T = 0. \qquad (2)$$

This implies that, for every $\varepsilon > 0$, play is within ε of some Nash equilibrium at each point in time except possibly for a sparse set of times. We shall show that in some games, the process *fails to come close* to Nash in the sense that Eq. 2 fails to hold for *almost all* histories h.

Theorem *Let v be a continuous density on $[-\lambda/2, \lambda/2]$, and let G be a v-perturbation of a finite, zero-sum, two-person game G^0, all of whose Nash equilibria have full support. Assume that the players are perfectly rational, have arbitrary discount factors less than unity, and that each updates his predictions of the opponent's future behavior by a learning rule that is based solely on observable actions. If λ is sufficiently small, then for v-almost all* *payoff realizations, the probability is 1 that someone never learns to predict and that play fails to come close to Nash.*

We remark that the set of games for which this impossibility result holds is actually much larger than the one stated in the theorem. Consider, for example, any two-person game G with strategy space $Y_1 \times Y_2$ such that $|Y_i| \geq 2$; all Nash equilibria have full support on $Y_1 \times Y_2$, and every action not in Y_i is dominated strictly by some action in Y_i. Then the theorem holds for perturbed versions of this game. Next let us extend G to an n-person game G^* by adjoining $n-2$ players as follows: each new player has a strictly dominant action, and G^* is the two-person subgame that results when they play these actions. It follows that for any finite action space $X = \Pi X_i$, there exists an n-person game G^* on X such that when the payoffs of G^* are perturbed by small i.i.d. random errors, good prediction fails to occur with probability one.

Now consider *any* n-person game G on the finite strategy space $X = \Pi X_i$. Suppose that we perturb the payoffs of G by i.i.d. random errors drawn from a normal distribution or in fact any distribution with a continuous density, the support of which is the whole real line. With positive probability the payoffs of the realized game will be close to the game G^* constructed above. Thus as a corollary we obtain the following.

Corollary *Let G be any finite n-person game, the payoffs of which are perturbed once by i.i.d. normally distributed random errors. Assume that the players are perfectly rational, have arbitrary discount factors less than unity, and that each updates his predictions of the opponents' future behavior by a learning rule that is based solely on observable actions. For almost all payoff realizations, there is a positive probability that someone never learns to predict and that play fails to come close to Nash.*

Proof of the Theorem Because the proof is somewhat involved, we shall explain first why the argument given in the introduction for myopic players does not extend easily to the general case. One difficulty is that patient players might interact through conditional strategies that involve no randomization, and these might be predictable at least some of the time. Eliminating this case requires a delicate probabilistic argument. The second difficulty is that even when players randomize and are therefore indifferent among alternative strategies, this does not imply that the stage-game payoffs are solutions of a linear equation. Rather, they are the roots of a nonlinear function, and we must show that the roots of this function constitute a set of measure zero.

To increase the transparency of the proof, we shall give it for the game of matching pennies. It generalizes readily to any finite zero-sum, two-person game, the stage-game Nash equilibria of which are all strictly interior in the space of mixed strategies. Fix a continuous density ν, the support of which is $[-\lambda/2, \lambda/2]$. To be concrete, we may think of ν as the uniform distribution. The perturbed game has the payoff matrix

	1		2	
1	$1+\omega_{11}$,	$-1+\omega'_{11}$	$-1+\omega_{12}$,	$1+\omega'_{12}$
2	$-1+\omega_{21}$,	$1+\omega'_{21}$	$1+\omega_{22}$,	$-1+\omega'_{22}$,

where ω_{ij}, ω'_{ij} are i.i.d. random variables distributed according to ν.

Fix two rational players, 1 and 2, with discount factors $0 \le \delta_1 \le \delta_2 < 1$. Let their beliefs be f_1, f_2, and let their strategies be $g_1(\cdot|A)$, $g_2(\cdot|B)$, where A and B are the realized values of the players' payoff matrices. The functions f_1, f_2, g_1, g_2 will be fixed throughout the proof. All probability statements will be conditional on them without writing this dependence explicitly. Let $H(A, B)$ be the set of all histories h such that good prediction (Eq. 1) holds when the realized payoffs are (A, B). Let P be the set of pairs (A, B) such that good prediction holds with positive probability, that is, $\mu[H(A, B)] > 0$. First we shall show that $\nu(p) = 0$, that is, there are almost no payoff realizations (A, B) such that both players learn to predict with positive probability. In the second part of the proof we shall show that for almost all (A, B) the process fails to come close to Nash.

Lemma 1 *For every positive integer m, every $0 < \varepsilon' \le \varepsilon < 1$, and every $(A, B) \in P$, there exists a time T, possibly depending on m, ε, ε', A, B, such that with μ-probability at least $1 - \varepsilon'$, each player forecasts the other's next-period strategy within ε in each of the periods $T + 1, \ldots, T + m$.*

Proof Let (A, B) $\in$ P and suppose there were no such time T. Then for every time T the μ-probability would be greater than $\varepsilon' > 0$ that at least one player misforecasts the opponent's behavior by more than ε in one or more of the periods $T + 1, \ldots, T + m$. This would imply that *Eq. 1* is violated for almost all histories, that is, $\mu[H(A, B)] = 0$, which contradicts our assumption that (A, B) $\in$ P.

Lemma 2. *For each (A, B) ∈ P there exists a time T and a history h^T, possibly depending on A, B, such that conditional on h^T each player's expected future payoffs, discounted to T + 1, are bounded above by cλ for some positive number c that depends only on the discount rates.*

Proof Given a small λ > 0, choose $m \geq 1$ such that $\delta_2^m \leq \lambda$ and $0 < \varepsilon' \leq \varepsilon \leq \lambda / m4^m e^m$. As guaranteed by Lemma 1, let h^T be a history such that the μ-probability is at least $1 - \varepsilon'$ that each player forecasts the other's next-period strategy within ε in each of the periods $T + 1, \ldots, T + m$. Let α^*_{T+1} and β^*_{T+1} be the payoffs that players 1 and 2 *expect* to get from period $T + 1$ on, discounted to period $T + 1$. We shall exhibit a positive constant c, depending only on the discount factors, such that $\alpha^*_{T+1}, \beta^*_{T+1} \leq c\lambda$. Note first that each player has the option of playing 50–50 in each period from $T + 1$ on, which has an expected discounted payoff at least $-\lambda/2$. Because each player's strategy is optimal, it follows that $\alpha^*_{T+1}, \beta^*_{T+1} \geq -\lambda/2$.

For each j, $1 \leq j \leq m$, let α_{T+j} be player 1's expected *undiscounted* payoff in period $T + j$ *as forecast* by player 1 at the end of period T. Define β_{T+j} similarly for player 2. Let H_{j,h^T} be the set of all continuations of h^T to time $T + j$. Let $\phi_1(h^{T+j})$ denote player 1's probability assessment of $h^{T+j} \in H_{j,h^T}$ and similarly define $\phi_2(h^{T+j})$ for player 2. The true probability is $\mu_0(h^{T+j})$, where μ_0 is μ conditional on h^T. The set of continuations on which someone makes a bad forecast have μ_0− probability at most ε′. On the remaining *good* continuations, each player errs by at most ε in forecasting his opponent's stage-game behavior in each of j stages. Hence for every good continuation h^{T+j}, $| \phi_i(h^{T+j}) - \mu_0(h^{T+j}) | \leq (1 + \varepsilon)^j - 1 \leq (j\varepsilon)e^{j\varepsilon}$.

Each player's forecasted payoff in period $T + j$ cannot differ from the actual payoff in period $T + j$ by more than 2 + λ no matter how bad the forecast is. There are 4^j continuations to period $T + j$ including good and bad. Over all of the good ones, player 1's forecasted expected payoff differs from his actual expected payoff by at most $4^j(j\varepsilon)e^{j\varepsilon}(2 + \lambda)$. Over all of the bad ones the two differ by at most $\varepsilon'(2 + \lambda)$. Thus the difference between 1's forecasted expected payoff, α_{T+j}, and his actual expected payoff, $\bar{\alpha}_{T+j}$, is at most $(\varepsilon' + 4^j(j\varepsilon)e^{j\varepsilon})(2 + \lambda)$. By assumption, $\varepsilon' \leq \varepsilon \leq \lambda / m\, 4^m e^m$ and $j \leq m$, so $\varepsilon' + 4^j(j\varepsilon)e^{j\varepsilon} \leq \varepsilon + 4^m(m\varepsilon)e^{m\varepsilon} \leq 2\lambda$. Thus $|\alpha_{T+j} - \bar{\alpha}_{T+j}| \leq 2\lambda(2 + \lambda) \leq 6\lambda$. Similarly $|\beta_{T+j} - \bar{\beta}_{T+j}| \leq 6\lambda$. The actual payoffs satisfy $|\bar{\alpha}_{T+j} + \bar{\beta}_{T+j}| \leq \lambda$, from which we conclude that $|\alpha_{T+j} + \beta_{T+j}| \leq 13\lambda$ for $1 \leq j \leq m$.

For each j, $1 \leq j \leq m$, let $\alpha^*_{T+j} = (1 - \delta_1)(\alpha_{T+j} + \delta_1\alpha_{T+j+1} + \delta_1^2\alpha_{T+j+2} + \ldots)$ be player 1's expected payoff from period $T + j$ on, discounted to period $T + j$, as forecast at the end of period T. Similarly define $\beta^*_{T+j} = (1 - \delta_2)(\beta_{T+j} + \delta_2\beta_{T+j+1} + \delta_2^2\beta_{T+j+2} + \ldots;)$. We claim that $\alpha^*_{T+j}, \beta^*_{T+j} \geq -\lambda/2$ for every j. If not, some player could switch his strategy to a 50–50 random mixture from period $T + j$ on, thus increasing his expected payoff from that time on, which would contradict sequential rationality.

Beyond period $T + m$, the forecasts may no longer be good within ε. However, neither player expects to get more than $1 + \lambda/2$ in any period, thus the sum of expected payoffs beyond period $T + m$, discounted to period $T + 1$, cannot be more than $(1 - \delta_2)\ \delta_2^m\ (1 + \lambda/2)$. By choice of m, $\delta_2^m \leq \lambda$, thus the previous expression is at most 2λ when λ is small. Putting this fact together with $| a_{T+j} + \beta_{T+j}| \leq 13\lambda$, it follows that

$$\begin{aligned}\beta^*_{T+1} &\leq (1 - \delta_2)\sum_{j=1}^{m}\delta_2^{j-1}\beta_{T+j} + 2\lambda \\ &\leq (1 - \delta_2)\sum_{j=1}^{m}\delta_2^{j-1}\left(13\lambda - \alpha_{T+j}\right) + 2\lambda \\ &\leq 15\lambda - (1 - \delta_2)\sum_{j=1}^{m}\delta_2^{j-1}\alpha_{T+j}.\end{aligned} \tag{3}$$

The term $\sum_{j=1}^{m}\delta_2^{j-s1}\alpha_{T+j}$ is similar in form to α^*_{T+1} except that the wrong discount factor is being used, and the sum is truncated. Nevertheless, we claim that if α^*_{T+1} is small, then so is the term in question. To see this, consider the identity $\alpha^*_{T+j} = \delta_1\alpha^*_{T+j+1} + (1 - \delta_1)\,\alpha_{T+j}$ which holds for all j. From this we obtain

$$\sum_{j=1}^{m}\delta_2^{j-1}\alpha^*_{T+j} = \delta_1\sum_{j=1}^{m}\delta_2^{j-1}\alpha^*_{T+j+1} + (1 - \delta_1)\sum_{j=1}^{m}\delta_2^{j-1}\alpha_{T+j}$$

and after rearranging terms,

$$\sum_{j=1}^{m}\delta_2^{j-1}\alpha_{T+j} = \left[\alpha^*_{T+1} + (\delta_2 - \delta_1)\sum_{j=1}^{m-1}\delta_2^{j-1}\alpha^*_{T+j+1} - \delta_1\delta_2^{m-1}\alpha^*_{T+m+1}\right] / (1 - \delta_1). \tag{4}$$

All of the terms $\alpha^*_{T+2}, \ldots \alpha^*_{T+m}$ are at least $-\lambda/2$, the term α^*_{T+m+1} is at most $1 + \lambda/2$, and $\delta_1\delta_2^{m-1} \leq \delta_2^m \leq \lambda$. Thus, the right-hand side of Eq. 4 is bounded below by $\alpha^*_{T+1}/(1 - \delta_1) - c'\lambda$, where $c' > 0$ depends only on the discount factors. The left-hand side of Eq. 4 is the summation on the right-hand side of Eq. 3. Substituting this expression into Eq. 3, we see that

$\beta^*_{T+1} + [(1-\delta_2)/(1-\delta_1)]\,\alpha^*_{T+1} \leq 15\lambda + (1-\delta_2)\,c'\lambda$. Because $\alpha^*_{T+1}, \beta^*_{T+1} \geq -\lambda/2$, we conclude that both α^*_{T+1} and β^*_{T+1} are bounded above by $c\lambda$ for some c that depends only on the discount factors δ_1 and δ_2. This concludes the proof of Lemma 2.

Lemma 3 *For every positive integer m and all sufficiently small $\lambda > 0$, if $(A, B) \in P$, then there exists a history h^T such that, conditional on h^T at time T, both players randomize in each of the periods $T + 1, \ldots, T + m$.*

Proof As in the proof of Lemma 2, choose $m \geq 1$ such that $\delta_2^m \leq \lambda$ and let $0 < \varepsilon \leq \lambda/m4^m e^m$. Assume in addition that $\varepsilon' = \varepsilon^{4m}$. Now apply Lemma 1 with $2m$ instead of m: there is a time T such that, with a probability of at least $1 - \varepsilon'$, the next-period forecasts are within ε of being correct for the periods $T + 1, \ldots, T + 2m$.

For each h^{T+j}, $0 \leq j \leq 2m-1$, say that h^{T+j} is *good* if both players' next-period forecasts are within ε of being correct; otherwise h^{T+j} is *bad*. Say that h^{T+j} is γ-*good* if it is good and, conditional on h^{T+j} occurring in period $T + j$, the probability is at most γ that someone makes a bad next-period forecast in any continuation of h^{T+j} through period $T + 2m - 1$.

By choice of T there is at least one state, h^T, that has positive probability under the strategies and is ε'-good. Lemma 2 implies that the expected discounted payoffs from $T + 1$ on are bounded above by $c\lambda$. We claim this implies that both players randomize in period $T + 1$, and in fact each of them chooses each action with probability at least ε. Suppose, to the contrary, that some player (say player 1) chooses action 1 with probability less than ε. Because h^T is good, player 2 forecasts that 1 will play action 2 with probability at least $1 - 2\varepsilon$. But then player 2 could obtain a higher expected payoff by mismatching (playing action 1) in period $T + 1$ and randomizing fifty–fifty in every period thereafter. (The expected payoff from this strategy is at least $[(1 - \delta_2)(1 - 4\varepsilon) - \lambda/2]$, which is greater than $c\lambda$ for all sufficiently small λ and $\varepsilon \leq \lambda$.) This contradiction shows that player 1 chooses each action in period $T + 1$ with probability at least ε, and the same holds for player 2.

It follows that each of the four possible continuations of h^T to period $T + 1$ has probability at least ε^2. Because $\varepsilon^2 > \varepsilon'$ and h^T is ε'-good, none of these four continuations can be bad, and in fact each of them must be at least

$(\varepsilon'/\varepsilon^2)$-good. Now apply Lemma 2 again (redefining ε' to be $\varepsilon^{4m}/\varepsilon^2$) and conclude that, for every continuation of h^T to some h^{T+2}, the conditional expected payoffs from period $T+2$ forward are bounded above by $c\lambda$. As before, we conclude that both players randomize in period $T+2$, each putting at least ε on each action. Continuing in this manner, we deduce that both players randomize in every continuation of h^T to period $T+m$. This concludes the proof of Lemma 3.

The gist of the proof thus far is that, if the payoff realizations (A, B) lead to good predictions with μ-positive probability, then for every sufficiently large positive integer m, there exists a state h^T that induces randomization by both players in each of the next m periods. We now show that this implies that the payoffs are zeroes of a function, the set of zeroes of which has ν-measure zero. This will show that good prediction occurs with ν-measure zero.

Let h^T be any state, and let m be a positive integer. Suppose that player 1 plays action 1 in each of the periods $T+1$ to $T+m$, after which he plays an optimal strategy given his beliefs. We can write his expected utility, discounted to time $T+1$, as a function of his payoff matrix A as follows:

$$U_1(A) = \theta_1 a_{11} + \left(1 - \delta_1^m - \theta_1\right) a_{12} + \delta_1^m R_1(A)$$

Here θ_1 comes from player 2's randomization between actions 1 and 2, and the remainder term $R_1(A)$ is convex and bounded. In fact, $|R_1(A)| \leq (1-\delta_1{}^m)(|a_{11}| + |a_{12}| + |a_{21}| + |a_{22}|)$. Similarly define $U_2(A)$ to be player 1's expected utility from playing action 2 for m periods and an optimal strategy thereafter. This can be written analogously to $U_1(A)$ with a remainder function $R_2(A)$ that satisfies the same bound as $R_1(A)$. All of these functions depend of course on h^T.

It will be convenient to consider a one-dimensional subspace of the payoff matrices A. Namely, for every four real numbers w, x, y, and z, let $\psi_{x,y,z}(w)$ be the 2×2 matrix with entries $a_{11} = w + x$, $a_{12} = w - x$, $a_{21} = y$, and $a_{22} = z$. Given x, y, and z, define the following function: $F_{x,y,z}(w) = U_1(\psi_{x,y,z}(w)) - U_2(\psi_{x,y,z}(w))$. Abbreviating $\psi_{x,y,z}(w)$ by $\psi(w)$, we can write this in the form

$$F_{x,y,z}(w) = K_{x,y,z} + \left(1 - \delta_1^m\right) w + \delta_1^m \{R_1(\psi(w)) - R_2(\psi(w))\},$$

where $K_{x,y,z}$ is a linear function of x, y, z and does not depend on w. The functions $R_i(\psi(w))$ are convex and bounded by the same bound as before. By choosing m to be sufficiently large, we can ensure that $F_{x,y,z}(w)$ is strictly monotone increasing in w. It follows that for any triple x, y, z, there is at most one value of w such that $F_{x,y,z}(w) = 0$. Because x, y, z are drawn from the continuous density ν, we have $P[\{w: F_{x,y,z}(w) = 0|x, y, z)\}] = 0$. By the smoothing theorem (i.e., the law of iterated expectations), it follows that $P[\{(w, x, y, z) \in \mathbf{R}^4: F_{x,y,z}(w) = 0\}] = 0$.

To state this in terms of the matrix A, let $G(A) = F_{x,y,z}((a_{11} + a_{12})/2)$ where $x = (a_{11} - a_{12})/2$, $y = a_{21}$, and $z = a_{22}$. The preceding implies that $P[\{A: G(A) = 0\}) = 0$. Recalling that F (and thus G) are conditional on a particular history h^T, we can write this as $P[\{A: G(A) = 0\}|h^T] = 0$. Hence, $\Sigma_{h^T} P[\{A: G(A) = 0\}| h^T] P(h^T) = 0$. In other words, player 1 is only indifferent between actions 1 and 2 on a set of payoff matrices A having ν-measure zero.

Suppose now that (A, B) is a pair for which good prediction holds. Let h^T be a history as guaranteed by Lemma 3, where m is sufficiently large that F is strictly monotone increasing in w. By Lemma 3, player 1 randomizes in each of the periods $T + 1, \ldots, T + m$. Hence he is indifferent between playing action 1 or action 2 in each of these periods, an event that has ν-measure zero. We conclude that there are ν-almost no payoff realizations (A, B) such that both players learn to predict with positive probability. This establishes the first claim of the theorem. We also note for future reference that we have actually established the following fact.

Lemma 4 *If m is large enough and λ is small enough, then the ν-probability is zero that there exists a state h^T such that, conditional on h^T at time T, both players randomize in each of the periods $T + 1, \ldots, T + m$.*

It remains to be shown that, for ν-almost all (A, B), play fails to come close to the set of Nash equilibria in the sense that condition 2 fails to hold for almost all histories h. The first step is to show that all Nash equilibria of the repeated game are mixed sufficiently in each time period provided that λ is sufficiently small.

Lemma 5 *There exists $\varepsilon > 0$ and $\lambda' > 0$ such that whenever $0 < \lambda \le \lambda'$, every Nash equilibrium of the repeated game puts probability at least 2ε on each action in every time period.*

The proof is similar to that of Lemma 2; in outline it runs as follows. In equilibrium, each player's expected discounted payoff must be at least $-\lambda/2$, because at least this much is guaranteed by randomizing fifty–fifty in every period. Because the actual payoffs in each period sum to λ or less, each player's expected discounted payoff can be bounded from above by $k\lambda$, where k is a positive constant. If some player were to play an action with less than probability

2ε in some period t, the opponent can take a pure action with expected payoff at least $(1 - 4\varepsilon - \lambda/2)$ in period t and get at least $-\lambda/2$ in every period thereafter. When ε and λ are sufficiently small, the expected discounted payoff from such a deviation exceeds $k\lambda$, a contradiction.

Fix $\lambda \in (0, \lambda']$. For each pair of payoff matrices (A, B), let $N(A, B)$ be the set of all histories h such that condition 2 holds, i.e., such that play comes close to Nash in a weak sense. We are going to show that there are μ-almost no such histories for ν-almost all (A, B). This is a consequence of the following.

Lemma 6 *Let (A, B) be a pair of payoff matrices such that condition 2 holds with μ-positive probability. Then for every positive integer m and every sufficiently small λ, there exists a state h^T such that, conditional on h^T, each player randomizes in each of the periods $T + 1, \ldots, T + m$.*

By choosing m large enough, it follows from Lemma 4 that there are ν-almost no payoff realizations (A, B) with this property. In other words, for ν-almost all payoff realizations play fails to come close to Nash. Thus, once we establish Lemma 6, we will have completed the proof of the theorem.

Proof of Lemma 6 Fix a pair (A, B) such that condition 2 holds with μ-positive probability. Choose ε and λ such that every element of Q^N puts probability at least 2ε on each action in each time period as guaranteed by Lemma 5. Let m be a positive integer, and let $\varepsilon' = \varepsilon^{2m}$. Let $q^{t+1}(h^t)$ denote the strategies in period $t + 1$ given the history h^t to period t. There exists a time T such that with μ-probability at least $1 - \varepsilon'$, $d(q^{t+1}(h^t), Q^N) \leq \varepsilon$ for every h^t in the interval $T \leq t \leq T + m - 1$. (If this were not so, condition 2 would hold with μ-probability zero, contrary to our assumption.)

Say that a history h^{t+1} is *good* if $d(q^{t+1}(h^t), Q^N) \leq \varepsilon$. It is *very good* if it is good and all of its successors for the next m periods are good. If a history is good then each action is played in the next period with probability at least ε. Hence every continuation of a good history occurs with probability at least ε^2. If no history at time T is very good, then the μ-probability of a bad history occurring in the interval $T, T + 1, \ldots, T + m - 1$ is at least $\varepsilon^{2m-2} > \varepsilon'$, contrary to our assumption. Hence there exists h^T such that $d(q^{t+1}(h^t), Q^N) \leq \varepsilon$ for every continuation of h^T in the interval $T \leq t \leq T + m - 1$, and hence both players randomize for m periods in succession. By Lemma 4 this happens with ν-probability zero. This concludes the proof of Lemma 6, and thereby the proof of the theorem.

Acknowledgment

We thank the editor, referees, and members of the Santa Fe Institute and the Brookings-Johns Hopkins Center on Social and Economic Dynamics for constructive comments on an earlier draft. This research was supported in part by National Science Foundation Grant SBR 9601743.

Reference

Binmore, K. (1987), "Modelling rational players: part I." *Economics and Philosophy* 3: 179–214.

Binmore, K. (1990), *Essays on Foundations of Game Theory* (Oxford: Basil Blackwell).

Binmore, K. (1993), "Studies in Logic and the Foundations of Game Theory," in D. Prawitz, B. Skyrms, and D.Westerstahl, eds., *Proceedings of the Ninth International Congress of Logic, Methodology, and the Philosophy of Science* (Dordrecht: North Holland, vol. 9, pp. 927–946).

Blume, L. and Easley, D. (1998), "Rational expectations and rational learning," in M. Majumdar, ed., *Organizations with Incomplete Information: Essays in Economic Analysis* (Cambridge: Cambridge University Press, pp. 61–109).

Fudenberg, D. and Kreps, D. (1993), "Learning mixed equilibria." *Games and Economic Behavior* 5: 320–367.

Harsanyi, J. (1973), "Games with randomly disturbed payoffs: a new rationale for mixed-strategy equilibria." *International Journal of Game Theory* 2: 1–23.

Jordan, J. S. (1991), "Bayesian learning in normal form games." *Games and Economic Behavior* 3: 60–91.

Jordan, J. S. (1993), "Three problems in learning mixed-strategy equilibria." *Games and Economic Behavior* 5: 368–386.

Jordan, J. S. (1995), "Bayesian learning in repeated games." *Games and Economic Behavior* 9: 8–20.

Kalai, E. and Lehrer, E. (1993), "Rational learning leads to Nash equilibrium." *Econometrica* 61: 1019–1045.

Lehrer, E. and Smorodinsky, R. (1997), "Repeated large games with incomplete information." *Games and Economic Behavior* 18: 116–134.

Miller, R. and Sanchirico, C. (1997), "Almost everybody disagrees almost all the time: the genericity of weakly merging nowhere." Columbia University Discussion Paper, No. 9697–25.

Miyasawa, K. (1961), *On the Convergence of the Learning Process in a 2 × 2 Non-Zero-Sum Two-Person Game*, Economic Research Program, Research Memorandum no. 33, Princeton, NJ: Princeton University).

Monderer, D. and Shapley, L. (1996), "Fictitious play property for games with identical interests." *Journal of Economic Theory* 68: 258–265.

Nachbar, J. H. (1997), "Prediction, optimization, and learning in repeated games." *Econometrica* 65: 275–309.

Nachbar, J. H. (1999), "Rational Bayesian learning in repeated games." Dept. Economics, Working Paper, Washington Univ., St. Louis.

Nachbar, J. H. (2001), "Bayesian learning in repeated games of incomplete information." *Social Choice Welfare* 18: 303–326.

Nash, J. (1950), "Non-cooperative games." Ph.D. Dissertation (Princeton: Princeton University).

Nyarko, Y. (1998), "Bayesian learning and convergence to Nash equilibria." *Economic Theory* 11: 643–655.

7
Learning by Trial and Error

1 Introduction

Consider a situation in which people interact, but they do not know how their interactions affect their payoffs. In other words, they are engaged in a game, but they do not know what the game is or who the other players are. For example, commuters in a city can choose which routes to take to work. Their choices affect congestion on the roads, which determines the payoffs of other commuters. But no single commuter can be expected to know the others' commuting strategies or how their strategies influence his own commuting time. Similarly, in a market with many competing firms, no single firm is likely to know precisely what the other firms' marketing and pricing strategies are, or how these strategies affect its own profits (even though this assumption is routinely invoked in textbook models of competition). Likewise, traders in a financial market are typically unable to observe the strategies of the other traders, and probably do not even know the full set of players participating in the market.

In situations like these, one would like to have a learning procedure that does not require any information about the opponents' actions or payoffs. Such a rule is said to be *completely uncoupled*.[1] This chapter introduces a simple, completely uncoupled learning rule such that, when used by all players in a game, period-by-period play comes close to pure Nash equilibrium play a high proportion of the time, provided that the game has such an equilibrium and the payoffs are generic.

This rule, called *interactive trial and error learning*, has two key ingredients: (i) players occasionally experiment with alternative strategies, keeping the new strategy if and only if it leads to a strict increase in payoff; (ii) if someone experiences a payoff *decrease* due to a strategy change by *someone else*, he starts

Originally published H. Peyton Young, 2009. "Learning by trial and error," *Games and Economic Behavior*, Elsevier, 65 (2): 626–643, March. Reproduced with permission.

[1] Foster and Young (2006) use the term *radically uncoupled*. A learning rule is *uncoupled* if it does not require information about the opponents' payoffs, though it may depend on their actions (Hart and Mas-Colell 2003).

Spontaneous Order. H. Peyton Young, Oxford University Press. © H. Peyton Young (2024).
DOI: 10.1093/oso/9780198892908.003.0007

a random search for a new strategy, eventually settling on one with a probability that increases monotonically with its realized payoff. A novel feature of the process is that different search procedures are triggered by different psychological states or *moods*, where mood changes are induced by the relationship between a player's realized payoffs and his current payoff expectations.

2 Related Literature

Before defining this procedure in detail, I shall briefly outline its relationship with other learning rules in the literature. In the approach discussed in Chapter 9 of this volume (Marden et al. 2009), each player experiments with a small probability in each period, and adopts the experimental action if and only if it results in a higher payoff. It can be shown that in any potential game—in fact in any weakly acyclic game[2]—this rule implements Nash equilibrium behavior in the sense that a pure Nash equilibrium will be played a high proportion of the time provided that the experimentation probability is sufficiently small. The principal difference between this approach and interactive trial and error learning is that the latter has an additional search phase that is triggered by decreases in payoff caused by someone else. This feature guarantees that in any finite game with generic payoffs and at least one pure Nash equilibrium, such an equilibrium will be played a high proportion of the time provided that the experimentation probability is sufficiently small.

Another closely related learning rule is *regret testing* (Foster and Young 2006). In this procedure a player computes his average per period payoff over a long sequence of plays, and compares this with the average payoff he receives from occasional (random) experiments with alternative strategies. If one of these alternative strategies generates a *sufficiently larger average payoff* than the average payoff from his current strategy, he chooses a new strategy *at random.* (In other words, the strategy with the higher payoff is not necessarily chosen; it merely signals to the player that he is not playing the right strategy.) Foster and Young show that, for all finite two-person games, this rule comes close to Nash equilibrium behavior a large proportion of the time. Subsequently, Germano and Lugosi (2007) showed that a slight variant of the procedure comes close to Nash equilibrium behavior in any finite n-person

[2] A game is *weakly acyclic* in better replies if for every joint action-tuple there exists a sequence of better replies—one player moving at a time—that ends at a pure Nash equilibrium (see Chapter 2). Potential games and congestion games are special cases.

game with generic payoffs. Interactive trial and error learning differs from regret testing in that search is more directed and the rule requires no statistical estimation; however, it only leads to equilibrium behavior in games that have pure equilibria.

A third learning rule that bears some resemblance to the present proposal is due to Karandikar et al. (1998). In this procedure each player has an endogenously generated aspiration level that is based on a smoothed average of his prior payoffs. He changes strategy (with positive probability) when his current payoff falls below his current aspirations. This rule is simple, intuitive, and completely uncoupled. Unlike interactive trial and error learning there is no experimentation per se; rather, the aspiration levels are subjected to small random perturbations. These trembles occasionally cause the players to switch strategies even though the resulting payoffs are *lower* than before. The overall effect is that play transits among strategy-tuples in a way that depends on the rate at which aspirations are updated and also on the probability distribution of the trembles. Unlike the present method, however, this procedure does not necessarily lead to Nash equilibrium behavior even in 2×2 games.

Another, more distantly related, family of learning rules are those based on regret minimization. In general, a player has *ex post regret* if he could have realized a higher average payoff by playing some strategy s' in all those periods when he in fact played s. There exist quite simple learning procedures that minimize ex post regret (Foster and Vohra 1999; Hart and Mas-Colell 2000, 2001); moreover they can be cast in a form that is completely uncoupled (Hart and Mas-Colell 2000). However, unlike interactive trial and error learning, there is no guarantee that behaviors come close to Nash equilibrium most of the time. What can be shown is that regret minimizing rules cause the empirical frequency distribution of play to converge to the set of correlated equilibria (Hart and Mas-Colell 2000; Foster and Vohra 1999). This set includes the Nash equilibria as extreme points but is frequently much larger.

There are a number of learning rules that have a similar *stochastic structure* to the present proposal, in the sense that small trembles in behavior (or perceptions of others' behavior) cause play to shift among alternative strategy combinations. When these trembles are small the probability is high that play is concentrated on particular strategy combinations. The rule of Karandikar et al. has this structure, as do the model-based learning rules proposed by Jehiel (1998) and Foster and Young (2003). A key difference between the latter two approaches and the present one is that model-based learning requires *observability* of the opponents' play, whereas interactive trial and learning does not.

Before examining the properties of interactive trial and error learning in detail, a remark is in order about the sense in which it "implements" equilibrium behavior. We have repeatedly said that interactive trial and error learning cause behaviors to come close to Nash equilibrium a high proportion of the time. Why not just say that behaviors *converge* to Nash equilibrium? Because typically they do *not* converge. In fact, there are very severe limits to what can be achieved if one insists on convergence to Nash equilibrium. To be specific, suppose that a learning rule has the following properties: (i) it is uncoupled, (ii) each player's choice of action depends solely on the frequency distribution of past play (as in fictitious play), and (iii) each player's choice of action, conditional on the state, is deterministic. Hart and Mas-Colell (2003) show that for a large class of games, no such rule causes the players' period-by-period behavior to *converge* to Nash equilibrium.

Matters are not quite so bad when stochastic choice is allowed. In this case there exist simple, uncoupled rules that *converge almost surely* to pure Nash equilibrium behavior in games that have such an equilibrium (Hart and Mas-Colell 2006; Babichenko, 2007).[3] The approach taken in the present chapter shows that one can achieve something similar *even when learning is completely uncoupled*, provided that convergence is weakened to "close most of the time."

3 Interactive Trial and Error Learning

Interactive trial and error learning is a modification of ordinary trial and error learning that takes into account the interactive nature of the learning environment. In ordinary trial and error learning, agents occasionally try out new things and accept them if and only if they lead to higher payoffs. (This is the procedure described in Chapter 9.) In an interactive situation, however, "errors" can arise in two different ways: by trying something that turns out to be no better than what one was doing, or by continuing to do something that turns out to be worse than it was before. The former are *active errors* whereas the latter are *passive errors*. We posit that the learning process is conditioned on whether an agent experiences an active or a passive error. Specifically, we propose that these situations trigger different psychological states or *moods*.

[3] Hart and Mas-Colell (2006) show that the following rule suffices: if everyone played the same action over the last two periods, and if player *i*'s action is a best response to the others' actions, *i* plays that action again; otherwise *i* chooses an action uniformly at random.

In the rule proposed here, an agent can have four different moods: content, discontent, watchful, and hopeful. When an agent is *content*, he occasionally experiments with new strategies, and switches if the new one is better than the old. When *discontent* he tries out new strategies frequently and at random, eventually becoming content with a probability that depends on how well his current strategy is doing. These are the main states, and reflect the idea that search can be of two kinds: careful and directed (when content), or thrashing around (when discontent).

The other two states are transitional, and are triggered by changes in the behavior of *other* agents. Specifically, if an agent is currently content and does not experiment in a given period but his payoff changes anyway (because someone else changed strategy), then he becomes *hopeful* if his payoff went up and *watchful* if it went down. If he is hopeful and his payoff stays up for one more period, he becomes content again with a higher expectation about what his payoff should be. If he is watchful and his payoff stays down for one more period, he becomes discontent, but does not immediately change his payoff expectations. (The assumption of a one-period waiting time is purely for convenience; it can be any fixed number of periods.)

The proposition that an agent's behavior may be conditional on his emotional state has been examined in a number of experimental papers (Capra 2004; Smith and Dickhaut 2005; Kirchsteiger et al. 2006). Here I employ the term "mood" in a more abstract sense: it is simply a state variable that determines how an agent responds to recent payoff history given the agent's current expectations.[4] The names of these states are meant to be suggestive but should not be taken too literally. In particular, I make no claim that people's search behavior actually does change in the manner prescribed by the rule (though it is certainly possible that different payoff histories induce different types of search).

Why does this process lead to equilibrium? The intuitive idea is that active search leads the players toward progressively higher payoffs and higher aspiration levels until one of two things happens: (i) an equilibrium is reached, or (ii) someone's aspirations are disappointed before an equilibrium is reached. In the latter case the disappointed player starts searching at random, which causes the other players to become disappointed with positive probability, which leads to a full-scale random search by everyone. This phase concludes

[4] See Compte and Postelwaite (2007) for another setting in which psychological states affect learning behavior.

when everyone calms down and they start building a new monotone-payoff path. It can be shown that, when the probability of calming down is sufficiently large relative to the probability of experimentation, the process is in a pure Nash equilibrium state much more often than in a disequilibrium state (assuming the game has a pure Nash equilibrium). Of course, there are many variants of the method proposed here that have similar properties, but it would take us too far afield to attempt to formulate the most general such method.

Let us now consider the model in greater detail. Let G be an n-person game with players $i = 1, 2, \ldots, n$, finite joint action space $A = \prod A_i$, and utility functions $u_i: A \to R$. A *state* of player i at a given point in time is a triple $z_i = (m_i, \bar{a}_i, \bar{u}_i)$, where m_i is i's current mood, $\bar{a}_i$ is i's current benchmark action, and $\bar{u}_i$ is i's current benchmark payoff. The four possible moods are content (c), discontent (d), hopeful (h), and watchful (w). A *state* z of the process specifies a state z_i for each player. We shall write this in the form $z = (m, \bar{a}, \bar{u})$, where each of the three components is an n-vector describing the players' moods, action benchmarks, and payoff benchmarks respectively. Let Z be the finite set of states corresponding to a given game G on A.

Given any state $z \in Z$, a joint action-tuple $a \in A$ is realized next period according to a conditional probability distribution $\psi(a|\, z)$. It will be useful to study the structure of these transitions without estimating the transition probabilities precisely (that will come later). One way to do this is by an automaton diagram showing the various transitions among states, but this turns out to be somewhat cumbersome. Instead we shall first describe the nature of the shifts in qualitative terms, then give them in detail. Let us focus on a particular player who is currently content. With small probability he can move (in two periods) to a content state with a higher benchmark payoff, assuming such an improvement is possible. Alternatively, he could move (in two periods) to a discontent state, which happens when someone else changes strategy and makes his payoff go down. From a discontent state he eventually moves to a new content state and corresponding new benchmark payoffs; the key feature here is that each of the new content states has positive probability of occurring.

Next we give the transitions in detail. Fix a particular player i. There are four cases to consider, depending on the player's current mood.

Content: $z_i = (c, \bar{a}_i, \bar{u}_i)$. Player i experiments next period with probability ε, and does not experiment with probability $1 - \varepsilon$. Denote i's choice of action next period by a_i. (Obviously a_i differs from $\bar{a}_i$ only if i experimented.). The possible transitions are:

$$z_i = (c, \bar{a}_i, \bar{u}_i) \longrightarrow \begin{array}{lll} (c, \bar{a}_i, \bar{u}_i) & a_i \neq \bar{a}_i,\ u_i(a) \leq \bar{u}_i & (1) \\ (c, a_i, u_i(a)) & a_i \neq \bar{a}_i,\ u_i(a) > \bar{u}_i & (2) \\ (w, \bar{a}_i, \bar{u}_i) & a_i = \bar{a}_i,\ u_i(a) < \bar{u}_i & (3) \\ (c, \bar{a}_i, \bar{u}_i) & a_i = \bar{a}_i,\ u_i(a) = \bar{u}_i & (4) \\ (h, \bar{a}_i, \bar{u}_i) & a_i = \bar{a}_i,\ u_i(a) > \bar{u}_i & (5) \end{array}$$

The first case says that if i experiments and his payoff *does not increase*, then i keeps the previous benchmarks and remains content. The second case says that if i experiments and his payoff *does increase*, he adjusts his benchmark payoff to the new higher level, takes the new strategy as his benchmark strategy, and remains content.

The next three cases deal with the situation in which does not experiment. He becomes watchful, content, or hopeful depending on whether the realized payoff was lower, the same, or higher than his benchmark, where any change in i's payoff must be triggered by a change of strategy by someone else.

Watchful: $z_i = (w, \bar{a}_i, \bar{u}_i)$. Agent i plays his benchmark strategy next period ($a_i = \bar{a}_i$). If the realized payoff $u_i(a)$ is less than his payoff benchmark $\bar{u}_i$ he becomes discontent; if it equals $\bar{u}_i$ he becomes content with the old benchmarks; if it is greater than $\bar{u}_i$ he becomes hopeful with the old benchmarks. These possibilities are shown below:

$$z_i = (w, \bar{a}_i, \bar{u}_i) \longrightarrow \begin{array}{lll} (d, \bar{a}_i, \bar{u}_i) & a_i = \bar{a}_i,\ u_i(a) < \bar{u}_i & (6) \\ (c, \bar{a}_i, \bar{u}_i) & a_i = \bar{a}_i,\ u_i(a) = \bar{u}_i & (7) \\ (h, \bar{a}_i, \bar{u}_i) & a_i = \bar{a}_i,\ u_i(a) > \bar{u}_i & (8) \end{array}$$

Hopeful: $z_i = (h, \bar{a}_i, \bar{u}_i)$. Agent i plays his benchmark strategy ($a_i = \bar{a}_i$): if the realized payoff is lower than $\bar{u}_i$ he becomes watchful with the old benchmarks; if the realized payoff equals $\bar{u}_i$ he becomes content with the old benchmarks; if the realized payoff is greater than $\bar{u}_i$, he becomes content with the realized payoff as the new benchmark.

$$z_i = (h, \bar{a}_i, \bar{u}_i) \longrightarrow \begin{array}{lll} (w, \bar{a}_i, \bar{u}_i) & a_i = \bar{a}_i,\ u_i(a) < \bar{u}_i & (9) \\ (c, \bar{a}_i, \bar{u}_i) & a_i = \bar{a}_i,\ u_i(a) = \bar{u}_i & (10) \\ (c, \bar{a}_i, u_i(a)) & a_i = \bar{a}_i,\ u_i(a) > \bar{u}_i & (11) \end{array}$$

Discontent: $z_i = (d, \bar{a}_i, \bar{u}_i)$. In this case the agent's benchmark strategy and benchmark payoff do not matter: he plays a strategy a_i drawn uniformly at random from A_i.[5] Spontaneously he becomes content with probability $\phi(u_i(a), \bar{u}_i)$, where the *response function* ϕ is bounded away from 0 and 1,

[5] The assumption of a uniform random draw is unimportant. It suffices that every action is chosen with a probability that is bounded away from zero over all possible states of the process.

that is, $\theta \le \phi(u_i, \bar{u}_i) \le 1 - \theta$ for some $\theta > 0$.[6] When agent i becomes content, his current strategy a_i and payoff level $u_i(a)$ serve as his new benchmarks; otherwise he continues to be discontent with the old benchmarks.

$$z_i = (d, \bar{a}_i, \bar{u}_i) \nearrow (c, a_i, u_i(a)) \quad \text{with prob } \phi(u_i(a), \bar{u}_i) \qquad (12)$$
$$z_i = (d, \bar{a}_i, \bar{u}_i) \searrow (d, \bar{a}_i, \bar{u}_i) \quad \text{with prob } 1 - \phi(u_i(a), \bar{u}_i) \qquad (13)$$

The precise form of the response function ϕ is not important for our results, though from a behavioral standpoint it is natural to assume that it is *monotone increasing* in the realized payoff u_i and *monotone decreasing* in the benchmark $\bar{u}_i$: higher values of the former and lower values of the latter mean that the agent is more likely to become content again. Note, however, that there is no *guarantee* that the agent will become content no matter how high u_i is relative to $\bar{u}_i$; in particular he may remain discontent even if his previous benchmark is realized, and may become content even when it is not. Moods are not *determined* by the absolute level of one's payoffs, but moods can change when payoffs change.[7]

To state our main result we shall need two further definitions.

Definition A game G is *interdependent* if any proper subset S of players can influence the payoff of at least one player not in S by some (joint) choice of actions. More precisely, G is *interdependent* if, for every proper subset S and every action-tuple a,

$$\exists i \notin S,\ \exists a'_S \neq a_S \text{ such that } u_i\left(\bar{a}'_S, a_{-S}\right) \neq u_i\left(a_S, a_{-S}\right). \qquad (14)$$

For a randomly generated game G on a finite strategy space A, interdependence holds *generically*, because it holds if there are no payoff ties. Notice, however, that interdependence is a considerably weaker condition: there can be many payoff ties so long as there is enough variation in payoffs that each subgroup can affect the payoff of *someone* not in the group by an appropriate choice of strategies.

Definition Consider a stochastic process $\{X_t\}$ and suppose that each realization of X_t either does or does not have some property P. Given any realization of the process, let p_t be the proportion of times that property

[6] The response functions can differ among agents without changing the results; purely for notational convenience we shall assume that the same ϕ applies to everyone.

[7] One is reminded of the rabbi who instructed the unhappy peasant to put a goat in his house: later he was delighted when the rabbi said he could take it out again.

P holds in the first t periods. *Property P holds at least r of the time* if and only if $\liminf_t p_t \geq r$ for almost all realizations of the process.

Theorem 1 *Let G be an n-person game on a finite joint action space A such that G is interdependent and has at least one pure Nash equilibrium. Given* $\varepsilon' > 0$*, if the players use ITE learning with response function* ϕ *and sufficiently small experimentation probability* $0 < \varepsilon \leq \varepsilon'$*, then a pure Nash equilibrium is played at least* $1 - \varepsilon'$ *of the time.*

The assumption of interdependence is not needed when there are only two players, as we shall show later on in Theorem 2, but when there are more than two players some form of non-genericity is needed, as we shall show by example in Section 5.

4 Proof of Theorem 1: Preliminaries

Before formally proving Theorem 1 let us briefly outline the argument. We begin by observing that states in which someone is not content are inherently unstable: any given player will leave a discontent, hopeful, or watchful state and enter a content state with a probability that is high relative to the experimentation probability ε. Next suppose that the process is in an all-content state but that the benchmark actions do not constitute a Nash equilibrium. Then it takes *only one* person to experiment with the "right" action and the experiment will succeed (yield a higher payoff). Hence the process transits to a state having different benchmarks with probability on the order of ε. If, however, the process is in an all-content state in which the benchmark actions *do* constitute a Nash equilibrium, then it takes at least two experiments (together or in close succession) to cause the benchmarks to change. In other words, the process either transits temporarily to a state with the same benchmarks and quickly reverts to the equilibrium state, or it transits to a state with new benchmarks, where the latter case has probability on the order of ε^2 or less. It follows that, when ε is very small, the process stays in the equilibrium states much longer than in the disequilibrium states. The key point to establish is that the process *enters* an equilibrium state with reasonably high probability starting from an arbitrary initial state. This requires a detailed argument and is the place where the interdependence property is used.

The proof uses the theory of perturbed Markov chains as developed in Chapter 2, which builds on work of Freidlin and Wentzell (1984), Foster and

Young (1990), and Kandori et al. (1993). Suppose that all players in the game G use ITE learning with experimentation probability ε and a given response function ϕ (which will be fixed throughout).[8] Let the probability transition matrix of this process be denoted by P^ε, where for every pair of states $z, z' \in Z$, $P^\varepsilon_{zz'}$ is the probability of transiting in one period from z to z'. We assert that if $P^\varepsilon_{zz'} > 0$, then $P^\varepsilon_{zz'}$ is of order ε^k for some non-negative integer k. To see why, suppose that z is the current state with benchmark strategies $\bar{a}$, and suppose that the vector a is realized next period, resulting in the state z'. If $a \neq \bar{a}$, some subset of $k \geq 1$ content players experimented. The probability of this event is $c\varepsilon^k (1 - \varepsilon)^{n-k}$ where c depends on z' but not on ε. (The other $n - k$ players were either not content in z, or were content and did not experiment.) If $a = \bar{a}$, then no one experimented but someone's mood may have changed; the probability of this event is $c(1 - \varepsilon)^n$ where again c depends on z' but not on ε. Hence in all cases $P^\varepsilon_{zz'}$ is of order ε^k for some integer $k \geq 0$. (In general we shall say that $P^\varepsilon_{zz'}$ *is of order* ε^k, written $P^\varepsilon_{zz'} \approx \varepsilon^k$, if $0 < \lim_{\varepsilon \to 0} P^\varepsilon_{zz'}/\varepsilon^k < \infty$.)

Definition If the transition $z \to z'$ occurs with positive probability ($P^\varepsilon_{zz'} > 0$), the *resistance* of the transition, written $r(z \to z')$, is the unique integer $k \geq 0$ such that $P^\varepsilon_{zz'} \approx \varepsilon^k$.

Let $Z_1, Z_2, \ldots, Z_h$ be the distinct recurrence classes of the Markov chain P^ε. Starting from any initial state, the probability is one that the process eventually enters one of these classes and stays there forever. To characterize the long-run behavior of P^ε, it therefore suffices to examine its long-run behavior when restricted to each of the classes Z_j. Let $P^\varepsilon{}_j$ denote the process restricted to the recurrence class Z_j. This process is irreducible, and the resistances of its transitions are defined just as for P^ε. Hence the restricted process is a *regular perturbed Markov chain*, and we can study its asymptotic behavior for small ε using the theory of large deviations (see Chapter 2 for details).

Given a state $z \in Z_j$, a *tree rooted at z*, or *z-tree*, is a set of $|Z_j| - 1$ directed edges that span the vertex set Z_j, such that from every $z' \in Z_j - \{z\}$ there is a *unique directed path* from z' to z. Denote such a tree by $\mathcal{T}_z$. The *resistance* of $\mathcal{T}_z$ is defined to be the sum of the resistances of its edges:

$$r(\mathcal{T}_z) = \sum_{(z,z') \in \mathcal{T}} r(z \to z'). \tag{15}$$

[8] Players can have different experimentation probabilities provided they go to zero at the same rate. We could assume, for example, that each player i has an experimentation probability $\lambda_i\, \varepsilon > 0$, where the parameter ε is varied while the λ_i are held fixed. This complicates the notation unnecessarily, so in the proofs we shall assume a common rate ε.

The *stochastic potential* of z is defined to be

$$\rho(z) = \min\{r(\mathcal{T}_z) : \mathcal{T}_z \text{ is a tree rooted at } z\}. \tag{16}$$

Let Z_j^- be the subset of all states $z \in Z_j$ that minimize $\rho(z)$. Theorem 4 in Chapter 2 implies the following result.

For each recurrence class Z_j and every $\varepsilon > 0$, let μ_j^ε be the unique stationary distribution of the process P_j^ε. Then for every $z \in Z_j$, $\lim_{\varepsilon \to 0} \mu_j^\varepsilon(z) = \bar{\mu}_j(z)$ exists and the support of $\bar{\mu}_j$ is contained in Z_j^-. (17)

The states z such that $\bar{\mu}_j(z) > 0$ are *stochastically stable*, that is, they have non-vanishing probability when the parameter ε becomes arbitrarily small (Foster and Young 1990; see Chapter 2).

5 Proof of Theorem 1

The proof of Theorem 1 amounts to showing that: (i) every recurrence class Z_j contains at least one all-content state in which the action benchmarks constitute a pure Nash equilibrium of G; (ii) the stochastically stable states are all of this form.

Let Z^o be the subset of states $z = (m, \bar{a}, \bar{u})$ such that $\bar{u}_i = u_i(\bar{a})$ for all agents i. In other words, Z^o is the subset of states such that the agents' benchmark payoffs and benchmark actions are *aligned*. Let $C^o \subset Z^o$ be the subset of such states in which all agents are content. Let E^o be the subset of C^o in which the benchmark actions $\bar{a}$ form a pure Nash equilibrium of G. The first step in the proof (Claim 1 below) will be to show that the only candidates for stochastic stability are states in which everyone is content and benchmarks are aligned (states in C^o). The remainder of the proof will establish that, in fact, the only candidates for stochastic stability are states in E^o.

Definition A *path* in Z is a sequence of transitions $z^1 \to z^2 \to \cdots \to z^m$ such that all states are distinct.

Claim 1 *For every $z \notin C^o$ there exists a zero-resistance path of length at most three from z to some state in C^o.*

Proof If state $z = (m, \bar{a}, \bar{u}) \notin C^o$, then someone is not content and/or someone's benchmark payoff is not aligned with the benchmark actions, that is,

$\bar{u}_i \neq u_i(\bar{a})$ for some player i. I claim that the benchmark action-tuple $\bar{a}$ is played next period with probability $\approx \varepsilon^0$, that is, with a probability that is bounded away from zero for all small ε. Consider the cases: (i) if in state z agent i is content, he plays $\bar{a}_i$ next period with probability $1 - \varepsilon$; (ii) if agent i is hopeful, he plays $\bar{a}_i$ again for sure and waits to see the payoff; (iii) if agent i is watchful he plays $\bar{a}_i$ again for sure and waits to see the payoff; (iv) if agent i is discontent, he plays $\bar{a}_i$ with probability $1/|A_i|$. Therefore $\bar{a}$ is played with probability $\approx \varepsilon^0$.

Notice that, if $\bar{a}$ is played, each discontent agent i *spontaneously becomes content* (and his benchmarks are $\bar{a}_i$, $u_i(\bar{a})$) with probability θ. Assume that this occurs for all discontent agents, and denote the resulting state by z'. Notice that if some player i was hopeful or watchful in z and becomes content in z', then i's new payoff benchmark is $u_i(\bar{a})$. We have therefore shown that, with probability $\approx \varepsilon^0$, $z \to z'$ where z' has action benchmark vector $\bar{a}$, and every content agent has a payoff benchmark that is aligned with $\bar{a}$.

We shall now show that in two more plays of $\bar{a}$, the process reaches a state in C^o. Let us observe first that the transition $z \to z'$ may have caused some players to *become* hopeful, watchful, or discontent, so we cannot assert that $z' \in C^o$. In the *next period*, however, the probability is $\approx \varepsilon^0$ that $\bar{a}$ will again be played and the previously discontent players (if any) will all become content with benchmarks $\bar{a}_i$, $u_i(\bar{a})$. Call this state z''. Since $\bar{a}$ was played twice in succession on the path $z \to z' \to z''$, every hopeful player in z' has become content in z'', every content player in z' is still content, and by construction all the discontent players have become content. Furthermore all of the content players in z'' have benchmarks $\bar{a}_i$, $u_i(\bar{a})$. There remains the possibility that someone who was watchful in z' has just become discontent in z''. However, *in one more transition*, $\bar{a}$ will be played again and *everyone* will become content with the benchmarks $\bar{a}_i$, $u_i(\bar{a})$, all with probability $\approx \varepsilon^0$. We have therefore shown that it takes at most three transitions, each having zero resistance, to go from any state not in C^o to some state in C^o, which establishes Claim 1.

Claim 2 *If $e = (m, \bar{a}, \bar{u}) \in E^o$ and z has action benchmarks that differ from $\bar{a}$, then every path from e to z has resistance at least two.*

Proof Consider any path $e \to z^1 \to z^2 \to \ldots \to z^m = z$. By definition of E^o, everyone in e is content, their actions constitute a pure equilibrium $\bar{a}$, and their benchmark payoffs are aligned with their actions. Hence $r(e \to z^1) \geq 1$, because at least one agent must experiment for the process to exit from e. If

$r(e \to z^1) \geq 2$ we are done. Suppose therefore that $r(e \to z^1) = 1$, that is, the transition involves an experiment by *exactly one agent* (say i). Since $\bar{a}$ is an equilibrium, i's experiment does not lead to a payoff improvement for i. Hence in state z^1 the benchmark actions are still $\bar{a}$, and the benchmark payoffs are still $\bar{u}$. (Note, however, that in z^1 some agents may have become hopeful or watchful, though none is yet discontent.)

Suppose that, in the transition $z^1 \to z^2$, none of the content agents experiments. Then $\bar{a}$ is played, so in z^2 all the hopeful and watchful agents (if any) have *reverted* to a contented mood with benchmarks $\bar{a}, \bar{u}$. But this is the original state e, which contradicts the assumption that a path consists of *distinct* states. We conclude that at least one agent does experiment in the transition $z^1 \to z^2$, which implies that $r(z^1 \to z^2) \geq 1$. Hence the total resistance along the path is at least two, as claimed.

Definition A transition from state z to another state is *easy* if it has the lowest resistance among all transitions out of z. A sequence of transitions $z^1 \to z^2 \to \cdots \to z^m$ is an *easy path* from z^1 to z^m if all states are distinct and all transitions are easy.

Claim 3 *For every state not in E^o, there exists an easy path to some state in E^o.*

Proof Suppose that $z \notin E^o$. If also $z \notin C^o$, then by Claim 1 there exists a zero-resistance path to some state $z^1 \in C^o$, which is obviously an easy path. If $z^1 \in E^o$ we are done. Otherwise it suffices to show that there exists an easy path from z^1 to some state in E^o. The intuitive idea of the proof is as follows. One by one the players experiment and find actions that yield payoff improvements and correspondingly higher aspirations (benchmark payoffs). This process continues until a state in E^o is reached, or some player's aspirations are disappointed. This triggers a sequence in which one player becomes discontent and his flailing around causes all the others to become discontent (this is the step where the interdependence property is invoked). Then with positive probability they simultaneously jump to a Nash equilibrium and become content again. It can be shown that all of these transitions occur with least resistance, that is, they generate easy paths to E^o.

We now give the argument in detail. Let $(\bar{a}, \bar{u})$ be the benchmarks in state z^1, which are aligned in the sense that $\bar{u}_i = u_i(\bar{a})$ for all i, because $z^1 \in C^o$. Since $z^1 \in C^o - E^o$, there is an agent i and an action $a_i \neq \bar{a}_i$ such that $u_i(a_i, \bar{a}_{-i}) > u_i(\bar{a}_i, \bar{a}_{-i}) = \bar{u}_i$. The probability that $(a_i, \bar{a}_{-i})$ is realized next period is $(1 - \varepsilon)^{n-1}\ \varepsilon/(|A_i|-1)$, which occurs when i experiments and chooses a_i, while the others do not experiment. This results in a state z^2 where i is content, i's new action benchmark is a_i, i's new payoff benchmark is $u_i(a_i, \bar{a}_{-i})$, and the others' benchmarks are as before (though their moods may have changed). Note that i's payoff benchmark has *strictly increased*, while the others' payoff benchmarks have stayed the same. Note also that $(1 - \varepsilon)^{n-1}\ \varepsilon/(|A_i|-1) \approx \varepsilon$, so $r(z^1 \to z^2) = 1$. Since all other transitions out of z^1 have resistance at least 1, $z^1 \to z^2$ is an easy path. As we have just seen, it is a *monotone increasing path* (with respect to the payoff benchmarks) in the sense that no one's payoff benchmark decreased and someone's strictly increased.

If $z^2 \in E^o$ we are done. Otherwise there are three possibilities to consider: (i) everyone in z^2 is content; (ii) some are hopeful and no one is watchful; (iii) someone is watchful. (No one can be discontent at this stage, because $z^1 \in C^0$ and it takes at least two periods of disappointing payoffs to become discontent.)

In the first case everyone is content, so evidently i's change of action did not change anyone else's payoff. Hence $z^2 \in C^o$ and we can simply repeat the earlier argument to extend the path by one more transition, $z^2 \to z^3$, having resistance 1. As before, this is an easy and monotone increasing continuation of the path. In the second case there is a zero-resistance (hence easy) transition to a state $z^3 \in C^o$ in which everyone is content, the benchmark payoffs for everyone are at least as high as they were in state z^2, and they are *strictly higher* for those who were hopeful (this happens when everyone in state z^2 plays his action benchmark, an event that has probability $\approx \varepsilon^0$). So again there is an easy and monotone increasing continuation of the path.

We shall consider the third case in a moment. Notice, however, that if the continuation of the path always involves cases (i) and (ii), then it will always be monotone increasing. Since the state space is finite, it must come to an end, which can only happen when it reaches some state in E^o.

We now consider the other case, namely, the path reaches a first transition where some agent becomes *watchful*, but no one is yet discontent. Suppose this happens in the transition $z^k \to z^{k+1}$. Up to this point, transitions have either: (i) involved a single content agent making an experiment that led to a better payoff for himself; or (ii) involved one or more hopeful agents playing their benchmark actions and becoming content with new higher benchmark payoffs (but not both (i) and (ii)). It follows that there are no hopeful agents in state z^k because hopeful agents do not try new actions, so they cannot cause *someone else* to become watchful (which is what happened for the first time in the transition $z^k \rightarrow z^{k+1}$). Thus all agents in z^k are content, $z^k \in C^o$, and in the transition $z^k \rightarrow z^{k+1}$ there is exactly one agent, say i, who experimented and caused the payoff of some other agent, say j, to go down.

Let $\bar{a}^k, \bar{u}^k$ be the benchmark actions and payoffs in state z^k; these are aligned because $z^k \in C^o$ by construction. Let $\bar{a}^{k+1}, \bar{u}^{k+1}$ be the benchmarks in state z^{k+1}. Note that only i's benchmark action and payoff changed between the two states (due to i's successful experiment); agents who became watchful or hopeful in z^{k+1} have not changed their benchmarks yet (they will wait one more period). In the next period the probability is at least $(1-\varepsilon)^{n-1}$ that the current action benchmarks $\bar{a}^{k+1}$ are played again. In this case all the watchful agents experience another disappointing payoff and become discontent, while all the other agents become (or stay) content. Thus the process transits with zero resistance to a state z^{k+2} in which there is at least one discontent agent and there are no hopeful or watchful agents. In state z^{k+2} the benchmarks are partially aligned in the sense that $u_j\left(\bar{a}^{k+1}\right) = \bar{u}_j^{k+1}$ for all agents j who are not discontent.

Let D be the subset of discontent agents in z^{k+2}. To avoid notational clutter let us drop the superscripts on the current benchmarks and denote them by $(\bar{a}, \bar{u})$. By assumption G is interdependent, hence there exists an agent $j \notin D$ and an action-tuple a'_D such that $u_j\left(a'_D, \bar{a}_{N-D}\right) \neq u_j\left(\bar{a}_D, \bar{a}_{N-D}\right) = \bar{u}_j$. We claim that there is a sequence of four (or fewer) easy transitions that make all the agents in $D \cup \{j\}$ discontent.

Case 1 $$u_j\left(a'_D, \bar{a}_{N-D}\right) > u_j\left(\bar{a}_D, \bar{a}_{N-D}\right).$$

Consider the following sequence: in the first and second period the players in D play a'_D and in the third and fourth periods they revert to $\bar{a}_D$, *all the while remaining discontent*. (In each of these periods the players not in D keep playing $\bar{a}_{N-D}$.) This initially raises j's expectations, which are later quashed (the "goat effect" in reverse). The sequence of transitions and play realizations looks like this:

actions	$(a'_D, \bar{a}_{N-D})$		$(a'_D, \bar{a}_{N-D})$		$(\bar{a}_D, \bar{a}_{N-D})$		$(\bar{a}_D, \bar{a}_{N-D})$		
states	z^{k+2}	$\rightarrow$	z^{k+3}	$\rightarrow$	z^{k+4}	$\rightarrow$	z^{k+5}	$\rightarrow$	z^{k+6}
payoffs			$u_j \uparrow$		$\bar{u}_j \uparrow$		$u_j \downarrow$		
moods			j hopeful		j content		j watchful		j discontent

I claim that each of these transitions has zero resistance, so this is an easy path. Indeed, in each transition the players in D play their required actions *and stay discontent*, which has probability at least $(\theta/m)^{|D|}$, where $m = \max_i|A_i|$. Meanwhile, each of the players $i \notin D$ continues playing his benchmark $\bar{a}_i$, which has probability $1 - \varepsilon$ if content and probability 1 if watchful or hopeful. These probabilities are bounded away from zero when ε is small, hence all the transitions have zero resistance. Thus by state z^{k+6}, and possibly earlier, the set of discontent agents has expanded from D to $D \cup \{j\}$ or more.

Case 2

$$u_j\left(a'_D, \bar{a}_{N-D}\right) < u_j\left(\bar{a}_D, \bar{a}_{N-D}\right).$$

In this case it suffices that everyone in D play a'_{D} and stay discontent, while the others play $\bar{a}_{N-D}$. This makes player j discontent in two steps.

Proceeding in this way, we conclude that there is an easy path from z^{k+2} to a state z^d in which *all* agents are discontent. Given any $e \in E^o$, the probability is at least $(\theta/m)^n$ that $z^d \to e$ in one period; indeed this happens if all n agents choose their part of the equilibrium specified by e and spontaneously become content.

We have therefore shown that, from any initial state $z \notin E^o$, there exists an easy path to some state in E^o. This establishes Claim 3.

Recall that, for any state z, $\rho(z)$ is defined to be the *resistance of a least-resistant tree rooted at z*. To establish Theorem 1, it therefore suffices to show the following (see the discussion at the end of Section 3).

Claim 4 *$\forall z \notin E, \exists e \in E^o$ such that $\rho(e) < \rho(z)$.*

Proof Let z be in the recurrence class Z_j, and let $\mathcal{T}_z$ be a least-resistant tree that spans Z_j and is rooted at z. Suppose that $z \notin E$. By Claim 3 there exists an easy path from z to some state $e \in E^o \subset E$. Denote this path by $z \to z^1 \to \cdots \to z^k = e$, and let $\mathcal{P}$ be the set of its k directed edges. We shall construct a new tree that is rooted at e and has *lower resistance* than does $\mathcal{T}_z$.

In $\mathcal{T}_z$, each state $z' \neq z$ has a unique *successor state* $s(z')$; in other words, $z' \to s(z')$ is the unique edge exiting from z'. Adjoin the path $\mathcal{P}$ to the tree $\mathcal{T}_z$; this creates some states with two exiting edges—one from $\mathcal{P}$ and one from $\mathcal{T}_z$. For each such state (*except e*), remove the exiting edge that comes from $\mathcal{T}_z$. The resulting set of edges $\mathcal{S}$ has one more edge than does $\mathcal{T}_z$; in fact, every state (including e) now has exactly one exiting edge, so it is not a tree.

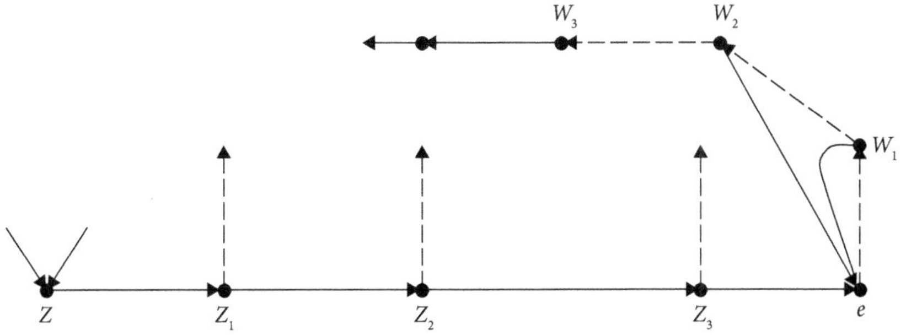

Figure 7.1 Construction of a tree rooted at e from a tree rooted at z by adding edges (solid) and subtracting edges (dashed).

Let us now compare the total resistance, $r(\mathcal{S})$, summed over all the edges in $\mathcal{S}$, with the total resistance, $r(\mathcal{T}_z)$, summed over all the edges in $\mathcal{T}_z$. Since $\mathcal{P}$ is an easy path, each of its transitions $z^j \to z^{j+1}$ has *least resistance* among all transitions out of the state z^j. Hence each edge from $\mathcal{P}$ that replaced an edge from $\mathcal{T}_z$ led to a decrease (or at least no increase) in the resistance, that is

$$r(z^j \to z^{j+1}) \leq r(z^j \to s(z^j)) \quad \text{for } 1 \leq j < k. \tag{18}$$

Furthermore, the "additional" edge $z \to z^1$ has resistance at most 1, since $\mathcal{P}$ is an easy path. It follows that

$$r(\mathcal{S}) \leq r(\mathcal{T}_z) + 1. \tag{19}$$

Next let $e \to w^1 \to w^2 \to \dots \to w^j$ be the unique path in $\mathcal{T}_z$ (and $\mathcal{S}$) leading from e toward z, where w^j is the *first* state on the path such that e and w^j do *not* have the same benchmarks. (There is such a state because e corresponds to an equilibrium and z does not.) From Claim 2 we know that

$$r(e \to w^1) + r(w^1 \to w^2) + \dots + r(w^{j-1} \to w^j) \geq 2. \tag{20}$$

Remove each of these j edges from $\mathcal{S}$,, and adjoin the $j-1$ edges

$$w^1 \to e, w^2 \to e, \dots, w^{j-1} \to e. \tag{21}$$

The result of all of these edge-exchanges is now a tree $\mathcal{T}_z$ that is rooted at e. (See Figure 7.1 for an example.) By construction, each of the states $w^1, w^2, \dots, w^{j-1}$ has the same benchmarks as does e; they differ from e only in that some agents may not be content. Hence

$$r\left(w^1 \to e\right) = r\left(w^2 \to e\right) = \cdots = r\left(w^{j-1} \to e\right) = 0. \qquad (22)$$

Combining (19)–(22) it follows that $r(\mathcal{T}_e) < r(\mathcal{T}_z)$ and hence that $\rho(e) < \rho(z)$. This completes the proof of Claim 4 and thereby the proof of Theorem 1.

6 Non-Generic Payoffs

The interdependence assumption is easy to state, but it is somewhat stronger than necessary. Consider, for example, an n-person game in which the players can be divided into disjoint groups such that the actions of any one group do not affect the payoffs of those outside the group, but the game is interdependent *within* each of these groups. (In effect the game decomposes into two disjoint interdependent games.) If the overall game has a pure equilibrium then so does each of the subgames, and interactive learning will discover it even though the game is not interdependent as a whole.

I shall not attempt to formulate the most general condition under which ITE learning discovers a pure Nash equilibrium; however, *some form of genericity* is needed when there are three or more players (though not when there are two players, as we shall see in Theorem 2 below). Consider the three-person game in Figure 7.2, where each player has two actions. There is a unique pure equilibrium in the lower northeast corner, and a best response cycle on the top square. Note that player 3's payoffs remain unchanged no matter what the other players do.

Suppose that the process starts in a state where player 3 is content. Since her payoffs are constant, no amount of experimenting will produce better results, and nothing the other players do will trigger a change in her mood. *Hence, once player 3 begins in a content state, she remains content and never changes her benchmark action.* If she starts by playing the action corresponding to the top square, no combination of actions by the other two players constitutes a Nash equilibrium, so they keep moving around in a best-response cycle. It follows that there are initial states from which ITE learning never leads to a pure Nash equilibrium even though there is one. (By contrast, if the process begins in a state where player 3 chooses the action corresponding to the lower square, the pure equilibrium will eventually be played with probability one.)

Similar examples can be constructed when there are more than three players, but this is not the case when there are only two players.

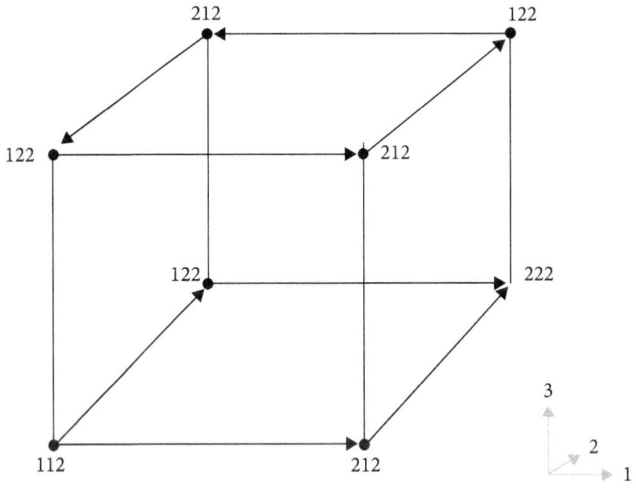

Figure 7.2 A three-person game with non-generic payoffs in which ITE learning does not necessarily lead to Nash equilibrium play. Each triple represents the payoffs to players one, two, and three respectively. Arrows indicate best-response transitions.

Theorem 2 *Let G be a two-person game on a finite joint action space A that has at least one pure Nash equilibrium. Given a small* $\varepsilon' > 0$, *if the players use ITE learning with response function* ϕ *and sufficiently small experimentation probability* $0 < \varepsilon \leq \varepsilon'$, *then a pure Nash equilibrium is played at least* $1 - \varepsilon'$ *of the time.*

Proof Consider a two-person game on a finite joint action space $A = A_1 \times A_2$, where the game possesses at least one pure Nash equilibrium. A *best response path* is a sequence of action-tuples $a^1 \to a^2 \to \cdots \to a^m$ such that the action-tuples are all *distinct*, and for each transition $a^k \to a^{k+1}$ there exists a unique player i such that $a_{-i}^{k+1} = a_{-i}^k$, $a_i^{k+1} \neq a_i^k$, and a_i^{k+1} is a *strict best response* by i to $a^k{}_{-i}$. The sequence is a *best response cycle* if all the states are distinct except that $a^1 = a^m$.

The only part of the proof of Theorem 1 that relied on the interdependence assumption was the proof of Claim 3. We shall show that this claim holds for two players without invoking interdependence, from which the theorem follows. The key idea is that if the players do not continue up a payoff-monotone path, then they fall into a cycle that causes both to become discontent. (When there are more than two players interdependence assures the existence of such a cycle.)

As before, let E^o denote the set of states such that everyone is content, the action benchmarks form a pure Nash equilibrium, and the payoff benchmarks are aligned with the action benchmarks. (For other definitions and notation the reader is referred to the proof of Theorem 1.)

Claim *For every state not in E^o, there exists an easy path to some state in E^o.*

Proof Suppose that $z \notin E^o$. If also $z \notin C^o$, then by Claim 1 (in the proof of Theorem 1) there exists a zero-resistance (hence easy) path to some state $z^1 \in C^o$. If $z^1 \in E^o$ we are done. Otherwise it suffices to show that there exists an easy path from $z^1 \in C^o$ to some state in E^o. Let $(\bar{a}^1, \bar{u}^1)$ be the benchmarks in state z^1, which are aligned by definition of C^o. We now distinguish two cases.

Case 1 *There exists a best response path from $\bar{a}^1$ to some pure Nash equilibrium, say $\bar{a}^1 = a^1 \to a^2 \to \cdots \to a^m$.*

Given such a path, let z^k be the *state* that has action benchmarks a^k, payoff benchmarks $u_i(a^k)$, and both players are content. (Note that the action and payoff benchmarks are aligned.) We shall construct an easy path to $z^m \in E^o$ that mimics the best response path as follows. In state z^1, suppose the relevant player experiments and chooses a best reply to the opponent's current action, which the opponent continues to play next period. In other words, the action-tuple a^2 is played next period with probability $\approx \varepsilon$. With probability $\approx \varepsilon^0$ the action-tuple a^2 is played in each of the next two periods and both players become content. Call this state z^2. In this manner we construct an easy path to the target state $z^m \in E^o$ that mimics the given best response path with the help of some intermediate transitions that have zero resistance.

Case 2 *There exists no best response path from $\bar{a}^1$ to a pure Nash equilibrium.*

Given that there is no best response path from $\bar{a}^1$ to a pure Nash equilibrium, there must exist a best response path from $\bar{a}^1$ that leads to a best response cycle. Denote such a cycle by $b^0 \to b^1 \to \cdots \to b^{m-1} \to b^0 \ldots$ and let the path to it be $\bar{a}^1 = a^1 \to a^2 \to \ldots \to a^j = b^0$. As in case 1 we can construct an easy path that mimics the best response path up to b^0; we need to show that it can be extended as an easy path to a Nash equilibrium.

Along the b-cycle the two players alternate in choosing best responses, say player 1 chooses a strict best response going from b^0 to b^1, player 2 from

b^1 to b^2, and so forth, all indexes being modulo m. Since these are strict best responses and the process cycles, each player's payoff must at some stage *decrease*. Proceeding from b^0, let $b^k \to b^{k+1}$ be the *first* transition in the cycle such that some player's payoff *strictly decreases*, say player 2's. Since this is a best response cycle, player 1's payoff must *strictly increase* in the transition $b^k \to b^{k+1}$. Moreover, in the *preceding* transition, $b^{k-1} \to b^k$, player 2's payoff must *strictly increase* because the players alternate in making best responses. We therefore know that

$$u_1\left(b^{k+1}\right) > u_1\left(b^k\right), u_2\left(b^{k+1}\right) < u_2\left(b^k\right), \text{and } u_2\left(b^k\right) > u_2\left(b^{k-1}\right) \qquad (23)$$

We now consider two possibilities.

Case 2a $u_1\ (b^k) < u_1\ (b^{k-1})$.

By assumption $b^k \to b^{k+1}$ was the first transition (starting from b^0) in which any decrease occurred, and by assumption it occurred for player 2. Hence $k =$ 0 and the hypothesis of case 2a is that $u_1\ (b^0) < u_1\ (b^{m-1})$.

As in case 1 we can construct an easy path (in the full state space) that mimics the transitions along the path $\bar{a}^1 = a^1 \to a^2 \to \ldots \to a^j = b^0$. and then mimics the cycle from b^0 on. Consider the situation when this path *first returns* to b^0, that is, the players play b^0 again after having gone around the cycle once. By construction, the players were all content in the previous state and their benchmarks were aligned. In the transition to b^0, player 1's payoff decreases so he becomes watchful, while player 2's payoff increases so she remains content.

In the next period, the probability is $\approx \varepsilon^0$ that: player 1 plays his current action benchmark b_1^0 again and *becomes discontent,* while player 2 plays action b_2^0 again and remains content. In the next period after that, the probability is $\approx \varepsilon^0$ that player 1 chooses b_1^1 and *remains discontent,* while player 2 does not experiment, chooses $b_2^0 = b_2^1$ again, and remains content. (By assumption, player 1 changed action in the transition $b^0 \to b^1$, hence player 2 *did not* change action, that is, $b_2^1 = b_2^0$.) By (22), player 2's payoff decreases in this transition ($b^0 \to b^1$), so she is now watchful. In the period after that, with probability $\approx \varepsilon^0$ they play b^1 again, player 1 *remains discontent*, and player 2 *becomes discontent.* At this juncture *both* players are discontent. Hence in one more period they will jump to a pure Nash equilibrium and spontaneously become content (with aligned benchmarks), all with probability $\approx \varepsilon^0$. Thus in Case 2a we have constructed an easy path to a state in E^o, that is, to an all-content, aligned Nash equilibrium state.

Case 2b
$$u_1\left(b^k\right) \geq u_1\left(b^{k-1}\right).$$

In this case let us first construct an easy path (in the full state space) that mimics the transitions along the path $\bar{a}^1 = a^1 \rightarrow a^2 \rightarrow \cdots \rightarrow a^j = b^0$, and then mimics the cycle up to the point where b^{k+1} is first played. (Recall that this is the first transition on the cycle where someone's payoff decreases.) At this point player 2 becomes watchful while player 1 remains content. In the next period the probability is $\approx \varepsilon^0$ that b^{k+1} will be played again and that player 2 becomes discontent while player 1 remains content. In the next period after that, the probability is $\approx \varepsilon^0$ that player 2 plays $b_2{}^{k-1}$ and remains discontent, while player 1 plays $b_2{}^{k+1}$. Denote the resulting pair of actions by $\tilde{b} = \left(b_1^{k+1}, b_2^{k-1}\right)$. Again we may distinguish two cases.

Case 2b′
$$u_1\left(b^k\right) \geq u_1\left(b^{k-1}\right) \text{ and } u_1(\tilde{b}) < u_1\left(b^{k+1}\right).$$

In this case player 1 has become watchful in the transition to $\tilde{b}$ while player 2 is still discontent. Hence in one more period the probability is $\approx \varepsilon^0$ that $\tilde{b}$ will be played again and that both players will be discontent. As we have already shown, this leads in one more easy step to an all-content Nash equilibrium, and we are done. It therefore only remains to consider the following.

Case 2b″
$$u_1\left(b^k\right) \geq u_1\left(b^{k-1}\right) \text{ and } u_1(\tilde{b}) \geq u_1\left(b^{k+1}\right).$$

We claim that this case cannot occur. Recall that the players alternate in making best replies around the cycle. Since player 2 best responded in going from b^{k-1} to b^k, player 1 best responded in the previous move. It follows that $b_1{}^{k-1}$ is 1's best response to b_2^{k-1}, from which we deduce that $u_1\left(b^{k-1}\right) \geq u_1(\tilde{b})$ Putting this together with the Case 2b″ assumption we obtain

$$u_1\left(b^k\right) \geq u_1\left(b^{k-1}\right) \geq u_1(\tilde{b}) \geq u_1\left(b^{k+1}\right), \tag{24}$$

which implies that $u_1(b^k) \geq u_1\left(b^{k+1}\right)$, contrary to (22). This concludes the proof of Theorem 2.

7 Extensions

Interactive trial and error learning can be generalized in several ways. One is to assume that players react only to "sizable" changes in payoffs. Given a real number $\tau > 0$, define *ITE learning with payoff tolerance* τ to be the same as before except that: (i) a player becomes *hopeful* only if the gain in payoff relative to the previous benchmark is strictly greater than τ; (ii) a player becomes *watchful* only if the loss in payoff relative to the previous benchmark is strictly greater than τ.

Say that a game is τ*-interdependent* if any proper subset S of players can—by an appropriate choice of joint actions—change the payoff of some player not in S by more than τ. An argument very similar to that of Theorem 1 shows the following: *if a game has a* τ*-equilibrium and is* τ*-interdependent, ITE learning with tolerance* τ *and experimentation rate* ε *leads to* τ*-equilibrium play in at least* $1 - \varepsilon$ *of all time periods provided that* ε *is sufficiently small.*

Extensions of the approach to learning mixed equilibria are not quite as straightforward. The obvious modification to make in this case is to assume that each player computes the *average payoff over a large sample of plays* before changing mood or strategy. If the players are using mixed strategies, however, there is always a risk—due to sample outcome variability—that the realized average payoffs will differ substantially from their expected values, and hence that one or more players changes mood and strategy due to "measurement error" rather than fundamentals. Thus one needs to assume that players only react to *sizable changes in payoff* and that the *sample size is sufficiently large* that sizable changes (due to sample variability) occur with very low probability. Moreover, for our method of proof to work, one would need to know that the game is τ-interdependent for a suitable value of τ, but this does not necessarily hold for the mixed strategy version of the game when the underlying game is τ-interdependent. (Consider for example a 2×2 game in which every two payoffs differ by more than τ. Each player may nevertheless have a mixed strategy that equalizes his own payoffs for all strategies of the opponent, in which case the mixed-strategy version is certainly not τ -interdependent.) Thus, while it may be possible to extend the approach to handle mixed equilibria, the result would be more complex and perhaps not as intuitively appealing as the version described here.

One of the issues that we have not dealt with is how long it takes (in expectation) for the learning process to reach an equilibrium from an arbitrary initial state. The proof of Theorems 1 and 2, which relies on the theory of large deviations in Markov chains, is not very informative on this point. One can compute a rough upper bound on the waiting time by observing that from any state there exists a sequence of transitions, each having probability at most ε, such that the sequence either ends at an equilibrium, or in an all-discontent state from which the process jumps to an equilibrium with probability at most ε^n, n being the number of players. To estimate the expected waiting time more precisely requires knowing how long these sequences are, which depends on the payoff structure of the game. This poses an interesting open problem that we shall not pursue here.

To sum up, interactive trial and error learning is a simple and intuitive heuristic for learning pure equilibria that does not rely on statistical estimation (like regret testing) and does not require observability of the opponents' actions (like the procedure of Hart and Mas-Colell). Even simpler procedures—such as the one described in Chapter 9—work for weakly acyclic games, although these have a fairly special structure. We conclude that there exist simple methods for learning equilibrium even when players know nothing about the structure of the game, who the other players are, or what strategies they are pursuing.

Acknowledgments

I am indebted to Jason Marden, Thomas Norman, Tim Salmon, Christopher Wallace, the referees, and the Associate Editor for helpful comments and suggestions.

References

Babichenko, Y. (2007). Uncoupled automata and pure Nash equilibria. Discussion paper #459. Center for the Study of Rationality, The Hebrew University.

Capra, C. M. (2004), "Mood-driven behavior in strategic interactions," *American Economic Review* (Pap. Proc.) 94: 367–372.

Compte, O. and Postelwaite, A. (2007), Repeated Games and Limited Information Processing. Mimeo, University of Pennsylvania.

Foster, D. P. and Vohra, R. (1999), "Regret in the on-line decision problem," *Games and Economic Behavior* 29: 7–35.

Foster, D. P. and Young, H. P. (1990), "Stochastic evolutionary game dynamics," *Theoretical Population Biology* 38: 219–232.

Foster, D. P. and Young, H. P. (2003), "Learning, hypothesis testing, and Nash equilibrium," *Games and Economic Behavior* 45: 73–96.

Foster, D. P. and Young, H. P. (2006), "Regret testing: Learning to play Nash equilibrium without knowing you have an opponent," *Theoretical Economics* 1: 341–367.

Freidlin, M. and Wentzell, A. (1984), *Random Perturbations of Dynamical Systems* (Berlin: Springer-Verlag).

Germano, F. and Lugosi, G. (2007), "Global convergence of Foster and Young's regret testing," *Games and Economic Behavior* 60: 135–154.

Hart, S. and Mas-Colell, A. (2000), "A simple adaptive procedure leading to correlated equilibrium," *Econometrica* 68: 1127–1150.

Hart, S. and Mas-Colell, A. (2001), "A general class of adaptive strategies," *Journal of Economic Theory* 98: 26–54.

Hart, S. and Mas-Colell, A. (2003), "Uncoupled dynamics do not lead to Nash equilibrium," *American Economic Review* 93: 1830–1836.

Hart, S. and Mas-Colell, A. (2006), "Stochastic uncoupled dynamics and Nash equilibrium," *Games and Economic Behavior* 57: 286–303.

Jehiel, P. (1998), "Learning to play limited forecast equilibria," *Games and Economic Behavior* 22: 274–298.

Kandori, M., Mailath, G., and Rob, R. (1993), "Learning, mutation, and long-run equilibrium in games," *Econometrica* 61: 29–56.

Karandikar, R., Mookherjee, D., Ray, D., and Vega-Redondo, F. (1998), "Evolving aspirations and cooperation," *Journal of Economic Theory* 80: 292–331.

Kirchsteiger, G., Rigotti, L., and Rustichini, A. (2006), "Your morals are your moods," *Journal of Economic Behavior and Organization* 59: 155–172.

Marden, J., Young, H. P., Arslan, G., and Shamma, J. S. (2009), "Payoff-based dynamics for multiplayer weakly acyclic games," *SIAM Journal on Control and Optimization* 48: 373–396.

Smith, K. and Dickhaut, J. (2005), "Economics and emotion: institutions matter," *Games and Economic Behavior* 52: 316–335.

8
Learning in a Black Box

With Heinrich H. Nax, Maxwell N. Burton-Chellew, and Stuart A. West

1 Introduction

Mainstream research in experimental game theory has focused primarily on situations where the structure of the game is known and the actions of other players are observable. In practice, however, many strategic environments are so large and complex that individuals are mostly in the dark about others' actions and how they affect their own payoffs. How do players learn in such situations?

To address this question, we conducted a series of experiments in which subjects played a voluntary contribution game, but they had no information about the nature of the game, and they could not observe the actions or payoffs of the other players.[1] In effect, subjects were inside a "black box" and the only data available to them were the payoffs they received from taking different actions. We tested the key features of learning models that have recently been proposed for such settings, and compared subjects' behaviors in the black box environment with other treatments where players were given increasing amounts of information about the nature of the game and about the actions and/or payoffs of the other players. Although this information permitted more complex adaptive responses, we found that the key features of black-box learning models are still present even when subjects have full information.

There is a growing body of theoretical work that shows how equilibrium (or near-equilibrium) play can evolve from simple, adaptive rules

Originally published Heinrich H. Nax, Maxwell N. Burton-Chellew, Stuart A. West, and H. Peyton Young (2016), "Learning in a Black Box," *Journal of Economic Behavior and Organization* 127: 1–16 Reproduced with permission.

[1] There is a related literature on low-information gambling tasks (Ben Zion et al. 2010; Nevo and Erev 2012; Erev and Haruvy 2013). Prior low- information experiments in games provide subjects with *some* information allowing communication (Colman et al. 2010), or regarding the structure of the game (Rapoport et al. 2002; Weber 2003) or other players' actions (Friedman et al. 2015). http://dx.doi.org/10.1016/j.jebo.2016.04.0060167-2681/© 2016 The Authors. Published by Elsevier B.V. This is an open access article under the CC BY-NC-ND license (http://creativecommons.org/licenses/by-nc-nd/4.0/).

that require no information about what other players are doing or ex ante knowledge of the game. Such rules are said to be *payoff-based* or *completely uncoupled*, because they depend solely on the pattern of an individual's received payoffs, and not on the behaviors or payoffs of anyone else.[2] Classical reinforcement models form one class of examples,[3] but in recent years many other payoff-based rules have been identified that differ from classical reinforcement in certain key respects. These rules go under a variety of names including "win-stay-lose-shift" (Robbins 1952; Nowak et al. 1993), "near-far search" (Motro and Shmida 1995; Thuijsman et al. 1995), "exploration-exploitation" (Eiben and Schippers 1998; Bowling and Veloso 2002), "trial-and-error" (Chapter 7, this volume), and "probe-and-adjust" (Skyrms 2010). Although these rules are broadly similar to reinforcement learning models, they differ by positing important *asymmetries* in the way that subjects respond to different patterns of experienced payoffs. These rules are also quite different from belief based learning models, which typically assume that subjects know the structure of the game and can observe other players' actions.

An essential feature of these rules is that the pattern of received payoffs triggers different modes of search: "experimentation" (present when payoffs are steady or increasing), "exploration" (triggered when payoffs decline) and "inertia" (staying with one's previous choice).[4] In this chapter, we formalize and test five specific behaviors that have been suggested in the literature on payoff-based learning:

(i) *Asymmetric inertia.* Inertia is greater when payoffs are steady or increasing than when they are decreasing.[5]

(ii) *Asymmetric volatility.* Adjustments have higher variance under exploration (when payoffs decline) than under experimentation (when payoffs are steady or increasing).

(iii) *Asymmetric breadth.* Adjustments are broader under exploration than under experimentation.[6]

[2] For details about *uncoupled* and *completely uncoupled* learning rules respectively, seeHart and Mas-Colell (2003, 2006), Foster and Young (2006), and Chapters 7, 9, and 10 in this volume. a

[3] Classical reinforcement models include Thorndike (1898), Bush and Mosteller (1953), Suppes and Atkinson (1959), Herrnstein (1970), Harley (1981), Cross (1983), Roth and Erev (1995), Erev and Rapoport (1998), Nevo and Erev (2012), and Erev and Haruvy (2013).

[4] These asymmetries are crucial for their convergence properties. See chapter 7 (Young 2009), chapter 10 (Marden et al. 2014), as well as Foster and Young, 2006; Germano and Lugosi, 2007; Pradelski and Young, 2012; Huttegger and Skyrms, 2012; Skyrms, 2012; Huttegger et al., 2014).

[5] A separate question is whether inertia is affected by the *size* as well as the direction of recent payoff changes ("surprise triggers change") as proposed by Nevo and Erev (2012). We discuss this issue in more detail in Section 3.

[6] Note that adjustments can be broad (i.e. far from one's current strategy) even though the variance of these adjustments is small; thus (ii) and (iii) are distinct.

(iv) *Reversion.* Under both experimentation and exploration, reversion to one's previous strategy is more likely when payoffs decrease than when they are steady or increasing.

(v) *Directional bias.* When the strategy space can be linearly ordered, the direction of change tends to be reinforced: a direction that leads to higher payoffs tends to be followed by a further change in the same direction next period; a direction that leads to lower payoffs tends to be followed by a change of direction next period.

Various forms of directional bias have been discussed in the experimental literature; see among others Selten and Stoecker (1986), Selten (1998), Bayer et al. (2013), Harstad and Selten (2013), Burton-Chellew et al. (2015), and Nax and Perc (2015). The other four features have been proposed in the following models, some of which are theoretical and some empirical.

1 *Near-far search* (Motro and Shmida, 1995; Thuijsman et al., 1995). This is an empirical model of bee foraging behavior. Bees search for nectar close to their current location (within a given patch of flowers) as long as their payoffs are steady or increasing. Once a drop in the nectar stream occurs they fly far away to a new randomly chosen patch. Thus negative payoffs trigger wide-area search, whereas steady or positive payoffs lead to local search (*asymmetric breadth*, *asymmetric volatility*).

2 *Win-stay-lose-shift* (Robbins, 1952; Nowak et al., 1993). Subjects keep their current strategy if it has steady or increasing payoffs; otherwise they choose a new strategy at random (*asymmetric inertia*).

3 *Probe-and-adjust* (Skyrms, 2010; Huttegger and Skyrms, 2012; Huttegger et al., 2014). Subjects occasionally experiment with different strategies, which they retain if the resulting payoff is higher; otherwise they revert to their previous strategy (*reversion*).

4 *Trial-and-error* (Chapter 7 this volume). Subjects retain their current strategy with high probability if payoffs are steady or increasing. Occasionally they may try a different strategy which is retained if the resulting payoff is higher; otherwise they revert to the previous strategy. When subjects experience a decrease in payoff from their current strategy, they choose a new one at random (*asymmetric inertia*, *reversion*).

We shall show that all five features mentioned earlier are confirmed at high levels of statistical significance in the black box treatment. Moreover, even in the treatments where more information is available, none of the hypotheses can be rejected. We do not claim that these are the only relevant features of payoff-based learning, nor do we attempt to fit a parametric model to the data, as is common in much of the experimental games literature. Rather, our aim is to identify a set of qualitative search behaviors that can be examined in many other settings.

The rest of this chapter contains details of the experimental set-up, the description of the learning model, and the summary of methods and results. An appendix containing experimental instructions, regression tables and figures follows the main text.

2 The Experimental Set-Up

2.1 Data

A total of 236 subjects, in 16 separate sessions involving 12 or 16 subjects making 80 decisions each, participated in our experiments, yielding a total of 18,880 observations. In this section, we shall describe the structure of each repeated game experiment, and the different information treatments. The data underlying this study were first reported in Burton-Chellew and West (2013), which showed that fundamentally different information settings may lead to very similar patterns of play at the aggregate level. The purpose of the present chapter is to identify specific adjustment patterns at the individual level that drive the macro-dynamics.

2.2 Voluntary contribution game

The voluntary contribution game is the standard game in experimental economics to study the provision of public goods (Isaac et al. 1985; Isaac and Walker 1988; Ledyard 1995; Chaudhuri 2011). It proceeds as follows. Each player i in population $N = \{1, \ldots, n\}$ makes a nonnegative, real-numbered contribution, c_i, from a finite budget $B > 0$ (here $B = 40$). The vector of contributions is denoted by $\mathbf{c} = \{c_1, c_2, \ldots, c_n\}$. Given a *rate of return*, $e \geq 1$, the public good is provided in the amount $R(\mathbf{c}) = e \sum_{i \in N} c_i$ and split equally amongst the players (if $e < 1$, $R(\mathbf{c})$ would be a "public bad"). Write $\boldsymbol{\phi}$ for the payoff vector $\{\phi_1, \ldots, \phi_n\}$. Given others' contributions $\mathbf{c}_{-i}$, player i's contribution c_i results in the payoff

$$\phi_i = \frac{e}{n} \sum_{i \in N} c_i + (B - c_i).$$

2.3 Nash equilibria

If the rate of return is *low* ($e < n$), an individual contribution of zero ("free-riding") is the strictly dominant strategy for all players. Similarly, if the rate of return is *high* ($e > n$), an individual contribution of B ("fully contributing") is the strictly dominant strategy for all players. The respective Nash equilibria result in either nonprovision or full provision of the public good.

2.4 Repeated game

In each experimental session, the same population S (with $|S| = 12$ or 16) plays four "phases" where each phase is a twenty times repeated voluntary contribution game. In each period t within a given phase, players in S are randomly matched (as in Andreoni 1988) in groups of four to play the voluntary contribution game. At the start of each period, each subject is given a new budget $B = 40$, of which he can invest any amount; however, subjects cannot invest money carried over from previous periods. Other than that, subjects are free to invest any number of coins between 0 and 40 each periods.

The rate of return is either *low* ($e = 1.6$) or *high* ($e = 6.4$) throughout a given phase. Write $N^t{}_4$ for any of the four-player groups matched at time t. Note that, due to random group rematching, for every player i, the relevant group $N^t{}_4$ (i.e., such that $i \in N^t{}_4$ in period t) typically is a different one in each period t. Given others' contributions $\mathbf{c}^t{}_{-i}$ for all t, each i receives a total of $\phi_i = \sum_{t=1,\ldots,20} \phi_i^t$. The number ϕ_i represents a monetary reward that is paid after the game.

2.5 Information treatments

Each experimental session (involving four phases) is divided into two "stages": phases one and two of the session constitute stage one, phases three and four constitute stage two. Each phase is a twenty-period voluntary contribution game with either the low ($e = 1.6$) or the high ($e = 6.4$) rate of return, and in each stage both rates of return are played. The two stages differ with respect to the information revealed in that treatment.

Before the experiment begins, subjects are told that two separate experiments will be conducted, each stage consisting of two games. At no point before, during, or in between the two separate stages of the experiment are players allowed to communicate. Depending on which treatment is played, the following information is revealed at the start of each stage.[7]

All treatments. During each phase, the same game is repeated for 20 periods. Each player receives 40 coins each period, of which he can invest any amount. The amount not invested goes straight into his private account each period. After investments are made, each player earns a nonnegative return from his investment each period which, at the end of the experiment, he receives together with his uninvested money according to a known exchange rate into real money.

Black box. Subjects have no information about the structure of the game or about other players' actions and payoffs. Subjects play two voluntary contribution games (one with the high and one with the low rate of return). As play proceeds, subjects only know their own contributions and realized payoffs.

Standard. The rules of the game are revealed, including production of the public good, high or low rate of return, and how groups form each period. As the game proceeds, players receive a summary of the relevant contributions in their group at the end of each period.

Enhanced. In addition to the information available in the standard treatment, the payoffs of the other group members are explicitly calculated, and players receive a summary of their own and other players' payoffs in their group at the end of each period.

In each experimental session, every player plays two black box games (one with the high and one with the low rate of return), and either two standard or two enhanced information games (again one high and one low). Either the black box treatment occurs first, or a non-black box treatment occurs. Of the total of 18,880 observations, 9440 are black box, 4640 are standard, and 4800 are enhanced. The order, of high and low rates of return and of treatments, is different in each session. Sessions lasted between fifty and sixty minutes, and subjects earned between £6.20 and £15.50 (mean earnings were £12.40).

2.6 Black box details

Black box is our main treatment, and our main analysis is based on sessions when black box is played first. Recall that we require the recruiting system (ORSEE) to select only those who had not previously participated in public

[7] See Appendix B for instructions. Further details about the experimental instructions such as screenshots can be found in Burton-Chellew and West (2013).

goods experiments. Thus, when black box is played first, subjects are not likely to have prior knowledge of the structure of the game.

We shall reserve the term "black box" for the case when the "no information" treatment is played first. Sessions when it is played after the standard or enhanced treatments will be called "grey box."

In the "black box" treatment, subjects were told that the black box would convert their input (in coins) to an output (in coins) via a mathematical function with a random component. The instructions (see Appendix B) are silent on the question of whether this function might depend on the actions of others. Subjects were not told that they would be engaged in a game, but the instructions do not rule out this possibility either. The asymmetric search behaviors that we are testing are a component of learning rules that have been proposed for highly uncertain environments of this type.

In the "grey box" treatment, subjects are told that a new and separate experiment will be conducted and that all information except for their own payoffs will be withheld. Since play in these sessions was preceded by two voluntary contribution games where subjects received full information about the structure of the game, they might (or might not) think that the grey box treatment has a similar structure. However, they will still be unable to infer others' contributions and the underlying rates of return from the information they receive. We make this distinction between black box and grey box because our payoff-based learning model applies most evidently in the (pure) black box environment where more sophisticated learning models cannot apply due to the complete absence of information about the structure and about others' actions. Nevertheless, we shall find that many of the same learning behaviors are present in both treatments.

3 The Learning Model

In this section, we present and test our behavioral hypotheses based on the analysis of the pure black box data, that is, those sessions when black box was played before the standard or enhanced treatments. There were eight such sessions and 4960 observations.

3.1 Five behavioral hypotheses

Formally, we say that there is "inertia" in period t, that is, between periods t and $t-1$, if $c_i^t = c_i^{t-1}$. Regarding non-inertial adjustments in period t, we distinguish between "experimentation," present after previous payoff patterns were steady or increasing $(\phi_i^{t-1} \geq \phi_i^{t-2})$, and "exploration," which is triggered by payoff decreases $(\phi_i^{t-1} < \phi_i^{t-2})$.[8]

Consider the following example to illustrate our five learning hypotheses. Suppose that over three successive periods, $t-2$, $t-1$ and t, an agent contributes c_i^{t-2}, $c_i^{t-1} = 10$, and $c_i^t = x$ with realized payoffs $\phi_i^{t-2}, \phi_i^{t-1}$ and ϕ_i^t respectively.

(i) *Asymmetric inertia.* $x = 10$ is less likely if $\phi_i^{t-1} < \phi_i^{t-2}$ than if $\phi_i^{t-1} \geq \phi_i^{t-2}$.
(ii) *Asymmetric volatility.* The variance of x is higher if $\phi_i^{t-1} < \phi_i^{t-2}$ than if $\phi_i^{t-1} \geq \phi_i^{t-2}$.
(iii) *Asymmetric breadth.* The expected value of $|x-10|$ is higher if $\phi_i^{t-1} < \phi_i^{t-2}$ than if $\phi_i^{t-1} \geq \phi_i^{t-2}$.
(iv) *Reversion.* Suppose that $x \neq 10$ and that $c_i^{t+1} = y$ is chosen in the next period $(t+1)$. In period $t+1$, $y = 10$ is more likely if $\phi_i^t < \phi_i^{t-1}$ than if $\phi_i^t \geq \phi_i^{t-1}$.
(v) *Directional bias.* Suppose that $x \neq 10$ and that $c_i^{t+1} = y$ is chosen in period $t+1$. If $x > 10$, then, in expectation, $y - x$ is larger when $\phi_i^t \geq \phi_i^{t-1}$ than when $\phi_i^t < \phi_i^{t-1}$. Similarly, if $x < 10$, then, in expectation, $y - x$ is smaller when $\phi_i^t \geq \phi_i^{t-1}$ than when $\phi_i^t < \phi_i^{t-1}$. In other words, the direction of successful adjustments is reinforced.

We find that all five features are confirmed at a statistically significant level (see Tables A1 and A2). The tests of significance were conducted as follows. To test the hypothesis of asymmetric inertia, we used an ordered probit regression of the inertia rate controlling for session, phase, group, period, and individual fixed effects with individual-level clustering. Payoff decreases turn out to have a significant effect on the inertia rate: on average it reduces inertia (i.e., increases search probability) by 12 percent (± 2%): see Table A3.[9] There is significantly less inertia in the first phase than in the second phase, and period fixed effects are negative until period six, suggesting that inertia tends to increase over time. Importantly, asymmetric inertia is already present in the "early" periods (before period 10) of the experiment, as well as in the "later" periods (after period 10).

We also examine the situation in which a subject keeps the same action for two periods in a row and in the second period experiences a payoff shock

[8] The inclusion of stable payoffs in "experimentation" (i.e. not in "exploration") is a key characteristic of trial-and-error learning (see Chapter 7 (Young 2009)).
[9] The range corresponds to the 95% confidence interval.

due to a change of strategy by someone else (whom he knows nothing about due to the black box environment). Again, we find an asymmetric effect on search behavior in the next (third) period: on average a negative payoff shock increases the subsequent probability of search by 13 percent (± 4%) relative to a positive payoff shock. The baseline inertia rate in this setting however is roughly 50 percent higher (± 3%) than when a subject did not keep the same action for the two prior periods.

These findings are broadly consistent with Nevo and Erev's (2012) "surprise triggers change" hypothesis when payoff shocks are negative, but they are inconclusive when payoff shocks are positive. In the Nevo and Erev (2012) experiment, subjects either get very low payoffs (0, or −1) or very high payoffs (10 or −10), whereas in our setting, subjects experience a whole range of payoffs and there is no clear demarcation between "ordinary" and "surprise" payoffs. It is conceivable that if a subject happens to experience a series of small payoffs and then suddenly gets a very large increase, inertia might go down but our dataset does not contain the necessary number of cases to be able to test this possibility.

For asymmetric volatility, we apply Levene's test, a nonparametric test for the equality of variances in different samples. The null hypothesis of equal variances following steady/increasing payoffs versus decreasing payoffs, which would imply no asymmetric volatility, is rejected with 99 percent confidence. The same is true for early and late periods of the game. This holds for both high and low rates of return and for different orders of the games (see Table A4).

We test asymmetric breadth by regressing the absolute values of the noninertial adjustments conditional on steady/increasing payoffs versus payoff declines, controlling for session, phase, group, period, and individual fixed effects with individual-level clustering. The average adjustment following steady/increasing payoffs is smaller than after payoff declines (99 percent confidence; see Table A5). The same is true for early and late periods. Moreover, breadth is significantly larger in the first phase than in the second. Other fixed effects are not significant.

We test for reversion much as we did for asymmetric inertia. Payoff decreases turn out to have significantly positive effects on the reversion rate (see Table A6). Again, the same is true for early and late periods, but the effect is smaller in the first phase than in the second phase, i.e. reversion increases over time. Other fixed effects are not significant.

Finally, we analyze the directional patterns of adjustments. Our hypothesis of directional bias states that (i) if an increase leads to steady/increasing payoffs, the player's next-period contribution will tend to be higher than if it leads to payoff declines; (ii) if a decrease leads to steady/increasing payoffs, the player's contribution next period will tend to be lower than if it leads to payoff declines. To test this hypothesis, we regress the difference between adjustments after steady/increasing payoffs and adjustments after payoff declines, controlling for the direction of the previous adjustment as well as for session, phase, group, period, and individual fixed effects with individual-level clustering. The directional bias is confirmed, for both prior contribution increases and prior contributions decreases, at high levels of significance ($p < 0.01$). The details are given in Table A7.

3.2 Non-black box learning

In this section, we assess our black box findings in light of the data from grey box and from the other two treatments, standard and enhanced. In particular, we investigate whether our learning model describes only black box behavior or whether its components also persist in the other settings, that is, where players gain experience and/or have explicit information about the structure of the game and others' actions (and payoffs).

Figure 8.2 illustrates play in the different treatments, averaged over individuals and sessions. Different information clearly makes a difference in the level of contributions and/or convergence rates under the high rate of return (see also Burton-Chellew and West 2013). However, the basic features of our learning model persist in all sessions and treatments, as is summarized in Figure 8.1.

A noteworthy feature is that convergence to equilibrium is both more pronounced and more similar across treatments when the rate of return is *low* ($e = 1.6$) than when it is *high* ($e = 6.4$). Furthermore, contributions in the low-rate-of-return case are higher in the standard treatment than in the enhanced treatment. Under the high rate of return, contributions tend to polarize over time: some subjects appear to learn the dominant strategy (full contribution) whereas others do not. This contrasts with the learning dynamics under the low rate of return, where contributions gradually shifted from near-random contributions in the early period to low-or-zero contributions in the later period. (See Appendix A, Figure 8.3 for initial versus final contributions in the different treatments.) Hence, providing more information can actually lead to decreasing play of the dominant strategy, which supports the hypothesis that our five behavioral features still have bite with more information. In other

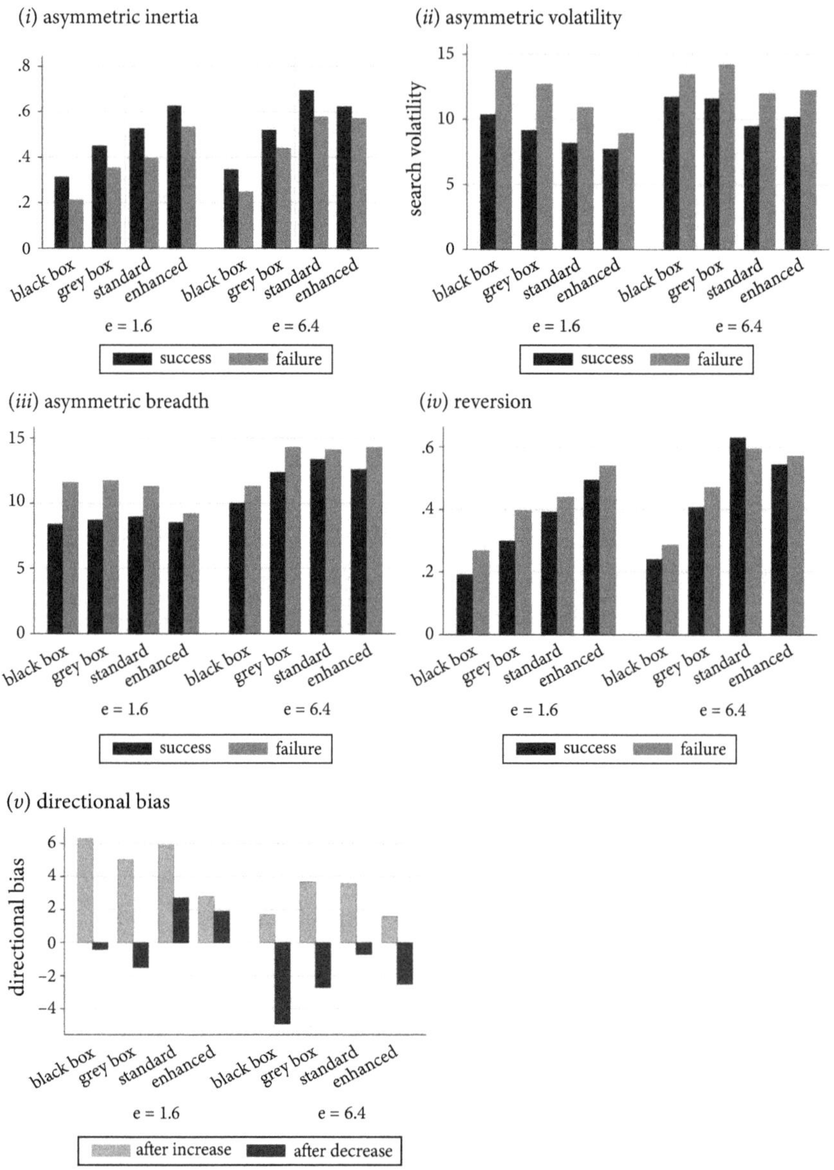

Figure 8.1 All treatments, both rates of return. The bar charts summarize the five learning model components in all treatments for both rates of return separately. For (ii), (iv), and (v), the y-axes are units of contributions; for (i) and (iii), the y-axes are probabilities. We denote the cases of steady or increasing payoffs by "success," and decreasing payoffs by "failure."

Figure 8.2 Mean play all treatments (black box, grey box, standard, enhanced).

words, some subjects continue to act as if they were inside a black box even though they are not. The slow-down of convergence in the high-rate-of-return case, especially under black box, could be the result of the "payoff variability effect" (Myers and Sadler 1960; Busemeyer and Townsend 1993; Erev and Barron 2005; Erev and Haruvy 2013), according to which increased payoff variability may inhibit learning (see Appendix A, Figure 8.4). In the standard and enhanced conditions, it could moreover be the case that subjects did not understand (fully) that they had a dominant strategy, or that relative payoff concerns interfered with equilibrium play. We leave further investigations of this feature to future research.

3.2.1 Black box versus grey box

Next, we shall investigate "grey box" play. Recall that grey box means that black box is played after either enhanced or standard. Even though subjects are explicitly told that a separate experiment is started after the first stage of the experiment, players may or may not make inferences about the game structure. We shall investigate the consequences of this effect in comparison with (pure) black box behavior. Our analysis reveals that, although there are differences in average levels of contribution in the two cases (Figure 8.2), all of the features of our learning model are robust for both rates of return (see also Figure 8.1). In particular, the differences in inertia rates conditional on steady-or-increasing payoff versus decreasing payoff are similarly robust, however, the level of inertia in absolute terms is higher in grey box than in black box (Test 1, Table A8). The levels of asymmetric volatility and asymmetric breadth are not significantly different in the two cases (Test 1, Tables A9 and A10). The differences in reversion are robust, but higher in grey box than in black box (Test 1, Table A11). Directional bias is unchanged in terms of size and significance (Test 1, Table A12). We conclude that grey box differs from black box with regard to the absolute levels of inertia and adjustment, but not qualitatively with respect to any of the five learning model components.

3.2.2 Black box versus standard and enhanced

Finally, we consider whether the standard and enhanced treatments lead to different conclusions. The situation is summarized in Figure 8.1. Panel (i) in Figure 8.1 illustrates that inertia rates decrease after payoff decreases in all treatments. This steady-or-increasing payoffs versus decreasing payoffs differential in inertia rates is statistically significant in all four treatments, and there is higher absolute inertia in non-black box treatment data (see Test 2, Table A8). Asymmetric volatility, though smaller in absolute size in the standard treatment and even smaller in the enhanced treatment, is higher after payoff decreases than after steady or increasing payoffs in all treatments (see panel (ii) in Figure 8.1). Moreover, this difference is significant at the 1 percent level (see Test 2, Table A9). Asymmetric breadth, too, exhibits an analogous differential as in black box (see panel (iii) in Figure 8.1). It continues to be statistically significant at the 1 percent level (see Test 2, Table A10). Panel (iv) in Figure 8.1 illustrates that reversion rates increase after payoff decreases in all treatments except for under the standard treatment when $e = 6.4$, but significant overall (see Test 2, Table A11). The difference is statistically significant in all four treatments, and there are higher absolute reversion rates in non-black box treatment data (see Test 2, Table A11). Finally, panel (v) in Figure 8.1 illustrates the presence of directional biases across all four treatments. We qualitatively confirm the directional bias from the black box data when analyzing the non-black box data at levels that are statistically significant (see Test 2, Table A12).

4 Conclusion

Much of the prior empirical work on learning in games has focused on situations where players have a substantial amount of information about the structure of the game and they can observe the behavior of others as the game proceeds. In this chapter, by contrast, we have examined situations in which players have *no* information about the strategic environment. This takes us to the other end of the information spectrum. Players in such environments must feel their way toward equilibrium based solely on the pattern of their own realized payoffs.

To highlight the potential differences between adaptive learning and best-reply dynamics, we implemented a black box environment in a voluntary contribution game and compared the resulting behavior to play under intermediate and rich information treatments. We identified five key features of the learning dynamics: asymmetric inertia, asymmetric volatility, asymmetric breadth, reversion, and directional bias. Although these components have precursors in both psychology and biology, they have not been given the precise formulation that we propose here. It turns out all five features are validated at high levels of statistical significance in the black box treatment. Moreover, they are present even when players gain more experience and/or have more information about the game. Whether this remains true for other classes of games is an open question for future research.

Appendix A Regression outputs

In the following, a single star indicates $p < 0.05$, double-stars indicate $p < 0.01$.

Table A1 Summary statistics for asymmetric inertia, asymmetric volatility, asymmetric breadth, and reversion (black box).

(a) Learning component	(b)	(c) Payoff decreases	(d) Steady/increasing payoffs	(e) Difference (c) – (d)
(i) *Asymmetric inertia* Probability of zero adjustments	Relative frequency # observations	0.23 520	0.33 722	0.10**
(ii) *Asymmetric breadth* Absolute value of non-zero adjustments	Mean # observations	11.4 1746	9.2 1476	2.2**
(iii) *Asymmetric volatility* Standard deviation of adjustment	Standard deviation # observations	15.6 1746	13.6 1476	2.0**
(iv) *Reversion* Probability of return after search	Relative frequency # observations	0.21 160	0.08 57	0.13**

Table A2 Summary statistics for (v) directional bias (black box).

(a) Mean non-zero adjustment	(b) Steady/increasing payoffs	(c) Payoff decreases	(d) Difference (b) – (c)
After *increase*	–1.6	–6.5	4.9**
After *decrease*	2.5	5.5	–3.0**

Table A3 Asymmetric inertia (black box). We use black box data from phases 1 and 2. We perform an ordered probit regression of absolute adjustments controlling for session, phase, group, period, and individual fixed effects with individual-level clustering. The pseudo-R^2 indicates the level of improvement over the intercept model (not interpretable as percentage of variance explained).

	Coefficient (test statistic)
1 if "payoff decreases"	−0.52 (−10.39)**
1 if "period" < 10	0.22 (1.37)
1 if "phase" = 1	−1.21 (−50.38)**
Periods	Not listed
Individual fixed effects	Not listed
Group fixed effects	Not listed
Session fixed effects	Not listed
Cut	−0.05 (s.e. 0.13)
Observations	4464
Adjusted pseudo-R^2	0.40

Table A4 Asymmetric volatility (black box). We use black box data from phases 1 and 2. We perform Levene's robust variance test, W being the Levene's test statistic.

(a) **Test 1: Asymmetric volatility Null rejected ($W = 39.9, p < 0.01$)**	(b) **Steady/increasing payoffs**	(c) **Payoff decreases**	(d) **Total**
Standard deviation	13.6	15.6	14.8
Mean	+1.2	−1.5	−0.3
Frequency	1476	1746	3222

Table A5 Asymmetric breadth (black box). We use black box data from phases 1 and 2. We perform an OLS regression of absolute adjustments controlling for session, phase, group, period, and individual fixed effects with individual-level clustering.

	Coefficient (test statistic)
1 if "payoff decreases"	2.28 (6.89)**
1 if "period" < 10	1.28 (1.25)

Continued

Table A5 *Continued*

	Coefficient (test statistic)
1 if "phase" = 1	2.75 (2.40)*
Periods	Not significant
Individual fixed effects	Not listed
Group fixed effects	Not listed
Session fixed effects	Not listed
Constant	4.49 (3.07)**
Observations	3222
Adjusted R^2	0.30

Table A6 Reversion (black box). We use black box data from phases 1 and 2. We perform an ordered probit regression of reversion rates controlling for session, phase, group, period, and individual fixed effects with individual-level clustering. The pseudo-R^2 indicates the level of improvement over the intercept model (not interpretable as percentage of variance explained).

	Coefficient (test statistic)
1 if "payoff decreases"	0.72 (8.39)**
1 if "period" < 10	−0.42 (−1.73)*
1 if "phase" = 1	0.08 (1.37)
Periods	Not significant
Individual fixed effects	Not listed
Group fixed effects	Not listed
Session fixed effects	Not listed
Cut	1.29 (s.e. 0.19)
Observations	3252
Adjusted pseudo-R^2	0.26

Table A7 Directional bias (black box). We use black box data from phases 1 and 2. We perform OLS regressions (without constant) to test for the directional bias of adjustments for each directional impulse controlling for phase, group, period, and individual fixed effects with individual-level clustering.

	Coefficient (test statistic)
1 if "up"	−3.48 (1.14)
1 if "up" and "steady or increasing payoff"	4.53 (5.35)**
1 if "down"	9.20 (2.83)**
1 if "down" and "steady or increasing payoff"	−3.07 (3.70)**

1 if "period"<10	−0.25 (−0.19)
Periods	Not significant
Individual fixed effects	Not listed
Group fixed effects	Not listed
Phase fixed effects	Not listed
Session fixed effects	Not listed
Observations	3264
Adjusted R^2	0.11

Table A8 Asymmetric inertia (non-black box). We use non-black box data: phases 3 and 4 for grey box (Test 1), and phases 1–4 for standard and enhanced (Test 2). We perform an ordered probit regression of absolute adjustments controlling for treatment effects as well as session, phase, group, period, and individual fixed effects with individual-level clustering. The pseudo-R^2 indicates the level of improvement over the intercept model (not interpretable as percentage of variance explained).

	Test 1: Grey box coefficient (test statistic)	**Test 2: Standard & enhanced coefficient (test statistic)**
1 if "payoff decreases"	−0.45 (−9.14)**	−0.34 (9.98)**
1 if "period"<10	−0.49 (−3.17)**	−0.35 (3.15)**
1 if "enhanced"	n/a	0.96 (44.85)**
Periods	Not significant	Not significant
Other fixed effects	Not listed	Not listed
Cut	5.05 (s.e. 0.23)	0.52 (s.e. 0.09)
Observations	4032	8464
Adjusted pseudo-R^2	0.39	0.44

Table A9 Asymmetric volatility (non-black box). We use non-black box data: phases 3 and 4 for grey box, phases 1–4 for standard and enhanced. We perform Levene's robust variance tests for asymmetric volatility in grey box (Test 1), and standard and enhanced (Test 2). Recall W is the Levene's test statistic.

	Steady/increasing payoffs	**Decreasing payoffs**	**Total**
Test 1: *Grey box*			
Null rejected ($W = 23.3$, $p < 0.01$)			
Standard deviation	14.6	17.3	16.2
Mean	+2.0	−1.9	-0.1
Frequency	1053	1206	2259

Continued

Table A9 *Continued*

	Steady/increasing payoffs	Decreasing payoffs	Total
Test 2: *Standard and enhanced*			
Null rejected ($W = 9.3, p < 0.01$)			
Standard deviation	14.4	15.8	15.5
Mean	+2.8	–3.6	–0.6
Frequency	1707	1931	3638

Table A10 Asymmetric breadth (non-black box). We use non-black box data: phases 3 and 4 for grey box (Test 1), phases1–4 for standard and enhanced (Test 2). We perform OLS regressions of absolute adjustments controlling for session, phase, group, period, and individual fixed effects with individual-level clustering.

	Test 1: Grey box coefficient (test statistic)	Test 2: Standard and enhanced coefficient (test statistic)
1 if "payoff decreases"	2.09 (5.10)**	1.36 (4.93)**
1 if "period"<10	Not significant	Not significant
Periods	Not significant	Not significant
Other fixed effects	Not listed	Not listed
Constant	19.91 (16.71)**	18.85 (19.68)**
Observations	2259	3638
Adjusted R^2	0.42	0.47

Table A11 Reversion (non-black box). We use non-black box data: phases 3 and 4 for grey box (Test 1), and phases 1–4 for standard and enhanced (Test 2). We perform an ordered probit regression of absolute adjustments controlling for treatment effects as well as session, phase, group, period, and individual fixed effects with individual-level clustering. The pseudo-R^2 indicates the level of improvement over the intercept model (not interpretable as percentage of variance explained).

	Test 1: Grey box coefficient (test statistic)	Test 2: Standard & enhanced coefficient (test statistic)
1 if "payoff decreases"	0.82 (8.17)**	0.60 (8.70)**
1 if "enhanced"	n/a	–0.58 (–11.08)**
1 if "period"<10	–0.04 (0.15)	0.00 (–0.01)
Periods	Not significant	Not significant
Other fixed effects	Not listed	not listed

Cut	2.19 (s.e. 0.24)	1.01 (s.e. 0.17)
Observations	2304	3725
Adjusted pseudo-R^2	0.26	0.22

Table A12 Directional bias (non-black box). We use all the data. We perform OLS regressions (without constant) to test for the directional bias of adjustments for each directional impulse in grey box (Test 1) and standard and enhanced (Test 2) controlling for session, phase, group, period, and individual fixed effects with individual-level clustering.

Test 1: Grey box	**Adjustment: coefficient (test statistic)**
1 if "up"	−3.86 (−1.88)*
1 if "up" and "steady or increasing payoffs"	5.24 (5.58)**
1 if "down"	8.91 (4.50)**
1 if "down" and "steady or increasing payoffs"	−2.41 (2.98)**
1 if "period" < 10	0.48 (0.31)
Period fixed effects	Not significant
Group dummies	Not significant
Other fixed effects	Not listed
Observations	2381
Adjusted R^2	0.18

Test 2: Standard and enhanced	**Adjustment: coefficient (test statistic)**
1 if "up"	−3.04 (−1.84)*
1 if "up" and "steady or increasing payoffs"	3.98 (5.53)**
1 if "down"	8.16 (4.76)**
1 if "down" and "steady or increasing payoffs"	−0.50 (−0.68)
1 if "enhanced"	−1.30 (−0.88)
1 if "period"<10	0.32 (0.27)
Period fixed effects	Not significant
Group dummies	Not significant
Other fixed effects	Not listed
Observations	3938
Adjusted R^2	0.22

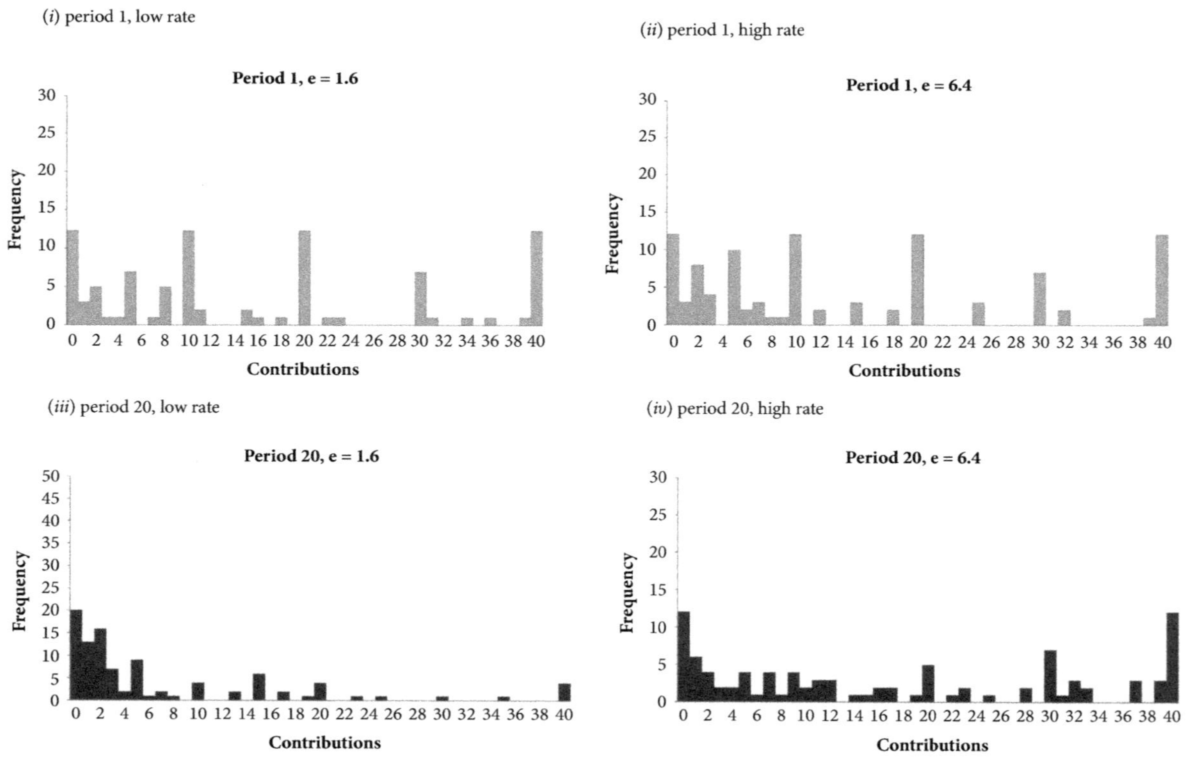

Figure 8.3 Initial and final contributions for both rates of return under black box.

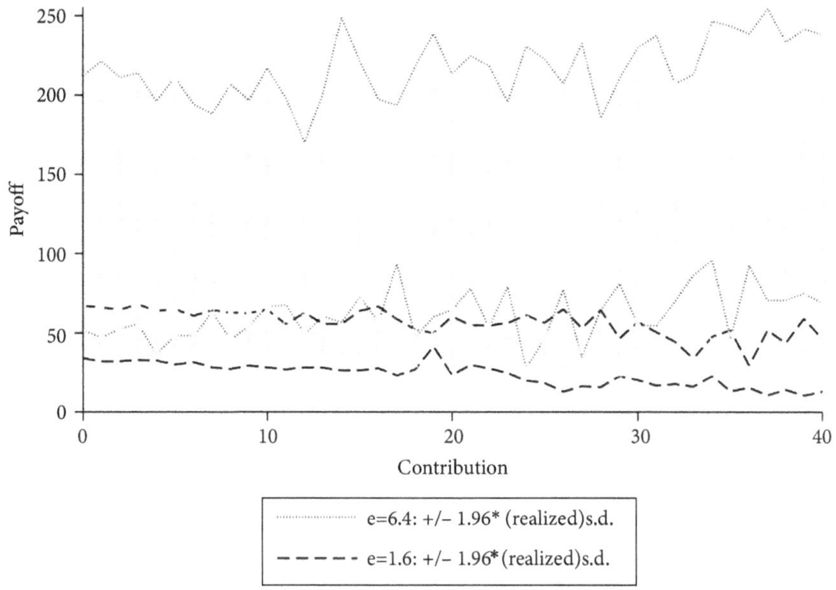

Figure 8.4 Payoff variability (black box).

Appendix B Instructions of the lab experiment

Participants received the following on-screen instructions (in z-Tree) at the start of the game. The set of instructions in standard and enhanced were the same, different instructions were given in black box. In black box, participants had to click an on-screen button saying, "I confirm I understand the instructions" before the game would begin. The same black box instructions were used for both rates of return. The standard/enhanced instructions differ with respect to the relevant numbers for the two rates of return, and the example is adequately modified.

B.1 Black box

The following instructions were used in black box.
Beginning of instruction.

Instructions
Welcome to the experiment. You have been given 40 virtual coins. Each "coin" is worth real money. You are going to make a decision regarding the investment of these "coins." This decision may increase or decrease the number of "coins"

you have. The more "coins" you have at the end of the experiment, the more money you will receive at the end.

During the experiment we shall not speak of £ Pounds or Pence but rather of "Coins." During the experiment your entire earnings will be calculated in Coins. At the end of the experiment the total amount of Coins you have earned will be converted to Pence at the following rate: 100 Coins = 15 Pence. In total, each person today will be given 3200 coins (£4.80) with which to make decisions over 2 economic experiments and their final totals, which may go up or down, will depend on these decisions.

The Decision

You can choose to keep your coins (in which case they will be "banked" into your private account, which you will receive at the end of the experiment), or you can choose to put some or all of them into a **"black box."**

This **"black box"** performs a mathematical function that converts the number of coins inputted into a number of coins to be outputted. The function contains a random component, so if two people were to put the same amount of coins into the **"black box,"** they would not necessarily get the same output. The number outputted may be more or less than the number you put in, but it will never be a negative number, so the lowest outcome possible is to get 0 (zero) back. If you chose to input 0 (zero) coins, you may still get some back from the box.

Any coins outputted will also be "banked" and go into your private account. So, your final income will be the initial 40 coins, minus any you put into the **"black box,"** plus all the coins you get back from the **"black box."**

You will play this game 20 times. Each time you will be given a new set of 40 coins to use. Each game is separate but the **"black box"** remains the same. This means you cannot play with money gained from previous turns, and the maximum you can ever put into the **"black box"** will be 40 coins. And you will never run out of money to play with as you get a new set of coins for each go. The mathematical function will not change over time, so it is the same for all 20 turns. However as the function contains a random component, the output is not guaranteed to stay the same if you put the same amount in each time.

After you have finished your 20 turns, you will play one further series of 20 turns but with a new, and potentially different **"black box."** The two boxes may or may not have the same mathematical function as each other, but the functions will always contain a random component, and the functions will always remain the same for the 20 turns. You will be told when the 20 turns are finished and it is time to play with a new black box.

If you are unsure of the rules please hold up your hand and a demonstrator will help you.

I confirm I understand the instructions (click to confirm)

End of instructions.

B.2 Standard and enhanced

Here, we present the instructions for the rate of return $e = 1.6$. The same instructions apply to standard and enhanced treatments. Equivalent instructions apply for the rate of return $e = 6.4$.

Beginning of instructions.

Instructions

Welcome! You are about to participate in an experimental study of human decision making. Thank you for your participation in our study. Please pay careful attention to the instructions on the following screens. If you wish to return to a previous screen, press the left arrow key. You are now taking part in an economic experiment. If you read the following instructions carefully, you can, depending on your decisions, earn a considerable amount of money. It is therefore very important that you read these instructions with care.

Everybody has received the same instructions. It is prohibited to communicate with the other participants during the experiment. Should you have any questions please ask us by raising your hand. If you violate this rule, we shall have to exclude you from the experiment and from all payments.

During the experiment we shall not speak of £ Pounds or Pence but rather of "Credits." During the experiment your entire earnings will be calculated in Credits. At the end of the experiment the total amount of Credits you have earned will be converted to Pence at the following rate: 100 Credits = 15 Pence. In total, each person today will be given 3200 credits (£4.80) with which to make decisions over two economic experiments and their final totals, which may go up or down, will depend on these decisions.

We are researching the decisions people make.

This part of the experiment is divided into separate rounds. In all, this part of the experiment consists of 20 repeated rounds. In each round the participants are assembled into groups of four. You will therefore be in a group with 3 other participants. The composition of the groups will change at random after each round. **In each round your group will therefore probably consist of different participants.**

In each round the experiment consists of two stages. At the first stage everyone has to individually decide how many credits they would like to contribute to a group project. These decisions have consequences for people's earnings. At the second stage you are informed of the contributions of the three other group members to the project and how many credits you have received from the group project. New groups are then randomly formed and the process repeats itself, with everyone again deciding how many credits they would like to contribute. This process will repeat 20 times.

At the beginning of each round, each participant receives 40 credits. In the following we call this your endowment. Your task is to decide how to use your endowment. You have to decide how many of the 40 credits you want to contribute to a group project and how many of them to keep for yourself. The consequences of your decision are explained in detail in the following slides.

Please note:

- The set-up is anonymous. You will not know with whom you are interacting.
- You will interact with a random set of 3 players in each round.
- Your decisions, and your earnings will remain anonymous to other players, even after the session has ended.

We now provide an animated illustration of a hypothetical scenario to demonstrate how the experiment works. In the following demonstration, we will use these green "disks" to represent "credits." 1 disk equals 1 credit.

Example

Remember! Credits = real money, and the more credits a player has at the end of the experiment, the more money that player will receive.

There are 4 players in each group, For the sake of convenience, we refer to these players as Player A, Player B, Player C, and Player D.

These 4 players then each receive an endowment of 40 credits.

Each of the players can then choose to make a contribution to the group project. They can contribute anything from zero to 40 credits. Non contributed credits are kept in the player's private account. They do this at the same time and anonymously.

Each player is then informed of the decisions of all their group members, although no one will know who the players are and they will randomly change in each round.

After all 4 players have made their decision to contribute or not, and by how much, the resulting total of contributed credits is automatically MULTIPLIED.

In your experiment, in every round, and in this example here, the total will be multiplied by 1.6. So for an imaginary total of 10 credits, this would result in 16 credits.

This new total is then always shared out equally between all 4 players. So, after the multiplication occurs, each player receives one quarter of the credits that are in the group project.

In this example, each player chooses to contribute 20 of their 40 credits. This means the group project has 80 credits; 20 + 20 + 20 + 20 = 80 credits. Which when multiplied by 1.6 results in 128 credits; 80 multiplied by 1.6 = 128 credits. These 128 credits are then shared out equally, giving 32 credits back to each player; 128 divided by 4 = 32 credits each. This gives each of the players a new total. In this case, they all have a new total of 52 credits.

They all started with +40; Contributed −20; and all got +32 in return, **giving them 52 in total.**

That is the end of the demonstration.

Remember, this was just one of many possible scenarios. In the rounds you will now play, all players are free to choose how much they wish to contribute to the pot.

End of instructions.

Acknowledgments

We thank Margherita Comola, Johannes Abeler, Guillaume Hollard, Richard Bradley, Fabrice Etilé, Philippe Jehiel, Thierry Verdier, Vince Crawford, Colin Camerer, Muriel Niederle, Edoardo Gallo, Amnon Rapoport, Guillaume Fréchette, Ozan Aksoy, Margaret Meyer, Roberto Serrano, Arno Riedl, Martin Strobel, Martin Cripps, Madis Ollikainen, Stefano Duca, Aidas Masil-iunas, anonymous referees, a helpful editor, and participants of the Economics and Psychology seminar at MSE Paris, the LSE Choice Group, the 24th Game Theory Workshop at Stony Brook, the CESS seminar at Nuffield College, the 9th Tinbergen Institute Conference and the Economics seminar at the University of Maastricht for helpful discussions, suggestions, and comments.

Nax's research was supported by the Office of Naval Research Grant N00014-09-1-0751 and by the European Commission through the European Research Council Advanced Investigator Grant "Momentum" 324247.

Burton-Chellew's research was supported by the European Research Council. West's research was supported by the European Research Council. Young's research was supported by the Office of Naval Research Grant N00014-09-1-0751 and the Air Force Office of Scientific Research Grant FA9550-09-1-053.

References

Andreoni, J. (1988), "Why free ride? Strategies and learning in public goods experiments," *Journal of Public Economics* 37: 291–304.

Bayer, R.-C., Renner, E., and Sausgruber, R. (2013), "Confusion and learning in the voluntary contributions game," *Experimental Economics* 16, 478–496.

Ben Zion, U., Erev, I., Haruvy, E., and Shavit, T. (2010), "Adaptive behavior leads to under-diversification," *Journal of Economic Psychology* 31: 985–995.

Bowling, M., Veloso, M. (2002), "Multiagent learning using a variable learning rate," *Artificial Intelligence* 136 (2): 215–250.

Burton-Chellew, M. N. and West, S. A., (2013), "Pro-social preferences do not explain human cooperation in public-goods games," *PNAS (Proceedings of the National Academy of Sciences)* 110 (31): 216–221.

Burton-Chellew, M. N., Nax, H. H., and West, S. A. (2015), "Payoff-based learning explains the decline in cooperation in public goods games," *Proceedings of the Royal Society B: Biological Sciences* 1801 (282).

Busemeyer, J. R. and Townsend, J. T. (1993), "Decision field theory: a dynamic-cognitive approach to decision making in an uncertain environment," *Psychological Review* 100: 432–459.

Bush, R. R. and Mosteller, F. (1953), "A stochastic model with applications to learning," *Annals of Mathematical Statistics* 24 (4): 559–585.

Chaudhuri, A. (2011), "Sustaining cooperation in laboratory public goods experiments: a selective survey of the literature," *Experimental Economics* 14: 47–83.

Colman, A. M., Pulford, B. D., Omtzigt, D., and Al Nowaihi, A. (2010), "Learning to cooperate without awareness in multiplayer minimal social situations," *Cognitive Psychology* 61 (3): 201–227.

Cross, J. G. (1983), *A Theory of Adaptive Economic Behavior* (Cambridge: Cambridge University Press).

Eiben, A. E. and Schippers, C. A. (1998), "On evolutionary exploration and exploitation," *Fundamenta Informaticae* 35 (1): 35–50.

Erev, I. and Barron, G. (2005), "On adaptation, maximization, and reinforcement learning among cognitive strategies," *Psychological Review* 112: 912–931.

Erev, I. and Haruvy, E. (2013), "Learning and the economics of small decisions," Handbook of *Experimental Economics* 47: 501–512.

Erev, I. and Rapoport, A. (1998), "Coordination, 'magic,' and reinforcement learning in a market entry game," *Games and Economic Behavior* 23 (2): 146–175.

Foster, D. P. and Young, H. P. (2006,) "Regret testing: learning to play Nash equilibrium without knowing you have an opponent," *Theoretical Economics* 1: 341–367.

Friedman, D., Huck, S., Oprea, R., and Weidenholzer, S. (2015,) "From imitation to collusion: long-run learning in a low-information environment," *Journal of Economic Theory* 155 (C): 185–205.

Germano, F. and Lugosi, G. (2007), "Global Nash convergence of Foster and Young's regret testing," *Games and Economic Behavior* 60: 135–154.
Harley, C. B. (1981), "Learning the evolutionarily stable strategy," *Journal of Theoretical Biology* 89 (4): 611–633.
Harstad, R. M. and Selten, R. (2013), "Bounded-rationality models: tasks to become intellectually competitive," *Journal of Economic Literature* 51 (2): 496–511.
Hart, S. and Mas-Colell, A. (2003), "Uncoupled dynamics do not lead to Nash equilibrium," *American Economic Review* 93: 1830–1836.
Hart, S. and Mas-Colell, A. (2006), "Stochastic uncoupled dynamics and Nash equilibrium," *Games and Economic Behavior* 57: 286–303.
Herrnstein, R. J. (1970), "On the law of effect," *Journal of the Experimental Analysis of Behavior* 13: 243–266.
Huttegger, S. M. and Skyrms, B. (2013), "Emergence of a signaling network with 'probe and adjust,'" In Sterelny, K., Joyce, R., Calcott, B., and Fraser, B. *Cooperation and Its Evolution* (Cambridge, MA: MIT Press).
Huttegger, S. M., Skyrms, B., and Zollman, K. J. S. (2014), "Probe and adjust in information transfer games," *Erkenntnis* 79 (4): 835–853.
Isaac, M. R., McCue, K. F., and Plott, C. (1985), "Public goods provision in an experimental environment," *Journal of Public Economics* 26: 51–74.
Isaac, R. M. and Walker, J. M. (1988), "Group size effects in public goods provision: the voluntary contributions mechanism," *Quarterly Journal of Economics* 103 (1): 179–199.
Ledyard, J. O. (1995), "Public goods: a survey of experimental research," In Kagel, J. H. and Roth, A. E (Eds.), *Handbook of Experimental Economics*, vol. 37, pp. 111–194.
Marden, J.R., Young, H.P., Arslan, G., Shamma, J.S., 2009. Payoff-based dynamics for multiplayer weakly acyclic games. SIAM J. Control Optim. 48 (1), 373–396.
Marden, J.R., Young, H. P., and Pao, L. Y.2014. Achieving Pareto optimality through distributed learning. SIAM J. Control Optim. 52 (5), 2753–2770.
Motro, U. and Shmida, A. (1995), "Near-far search: an evolutionary stable foraging strategy," *Journal of Theoretical Biology* 173: 15–22.
Myers, J. L. and Sadler, E. (1960), "Effects of range of payoffs as a variable in risk taking," *Journal of Experimental Psychology* 60: 306–309.
Nax, H. H. and Perc, M. (2015), "Directional learning and the provisioning of public goods," *Scientific Reports* 5: 8010.
Nevo, I. and Erev, I. (2012), "On surprise, change, and the effect of recent outcomes," *Frontiers of Cognitive Science* 47: 501–512.
Nowak, M., Sigmund, K., et al. (1993), "A strategy of win-stay, lose-shift that outperforms tit-for-tat in the prisoner's dilemma game," *Nature* 364 (6432): 56–58.
Pradelski, B. S. R. and Young, H. P. (2012), "Learning efficient Nash equilibria in distributed systems," *Games and Economic Behavior* 75: 882–897.
Rapoport, A., Seale, D. A., and Parco, J. E. (2002), "Coordination in the aggregate without common knowledge or outcome information," In *Experimental Business Research* (Springer, USA, pp. 69–99).
Robbins, H. (1952), "Some aspects of the sequential design of experiments," *Bulletin of the American Mathematical Society* 58 (5): 527–535.
Roth, A. E. and Erev, I. (1995), "Learning in extensive-form games—experimental data and simple dynamic models in the intermediate term," *Games and Economic Behavior* 8: 164–212.
Selten, R. (1998), "Aspiration adaptation theory," *Journal of Mathematica Psychology* 42: 191–214.

Selten, R. and Stoecker, R. (1986), "End behavior in sequences of finite prisoner's dilemma supergames: a learning theory approach," *Journal of Economic Behavior and Organization* 7: 47–70.

Skyrms, B. (2010), *Signals: Evolution, Learning, and Information* (Oxford: Oxford University Press).

Skyrms, B. (2012), "Learning to signal with probe and adjust," *Episteme* 9 (02): 139–150.

Suppes, P. and Atkinson, A. R. (1959), *Markov Learning Models for Multiperson Situations* (Stanford: Stanford University Press).

Thorndike, E.L. (1898), *Animal Intelligence: An Experimental Study of the Associative Processes in Animals* (New York: Macmillan).

Thuijsman, F., Peleg, B., Amitai, M., and Shmida, A. (1995), "Automata, matching and foraging behavior of bees," *Journal of Theoretical Biology* 175 (3): 305–316.

Weber, R.A. (2003), "'Learning' with no feedback in a competitive guessing game," *Games and Economic Behavior* 44: 134–144.

Young, H. P., 2009. Learning by trial and error. *Games and Economic Behavior* 65, 626–643

Preface to Chapters 9–10
Spontaneous Order by Design

Chapter 7 described a framework in which a large population of dispersed agents with limited information can reach equilibrium without knowing anything about the larger game in which they are embedded. Adaptive behaviors with this property employ some form of experimentation and random search, but they are relatively simple and do not presume any advanced reasoning (or even knowledge) about the strategies and preferences of the other players. They are completely decentralized. Laboratory experiments show that various features of these search rules are present in subjects' actual behavior, as we saw in Chapter 8.

Decentralized search algorithms also have important applications to problems of distributed control in engineering and computer science. The object is to coordinate the behavior of many interacting components without the intervention of a central controlling agent. Unlike applications in the social sciences, the issue here is not whether the components (agents) actually have behaviors that lead to equilibrium; the question is whether the system can be *designed* in such a way that their decentralized, uncoordinated actions lead to a desirable system-wide outcome.

One example of such a system is the routing of information packets in data networks. The objective is to minimize the total time it takes for packets to reach their destinations, but this is difficult to achieve from a centralized perspective because the functional dependence of delay on routing strategies may not be known. Hence in practice the routing problem is often solved by localized strategies that may or may not implement a global optimum.

Another example is the operation of distributed systems of power generation, such as a wind farm. In this case the overall objective is to program the wind turbines so as to maximize total power production, but this is not necessarily achieved by programming each turbine to maximize output in isolation. The reason is that the turbines interact aerodynamically in complex ways, and the behavior of one turbine affects the efficiency of other downwind turbines. In other words, the turbines constitute an interacting system that can be construed as a "game," where the payoffs are defined by the amount of power produced by each turbine.

Why not simply determine the configuration of turbines that maximizes total output given their aerodynamic interactions? The difficulty is that these interactions are often so complex that *even the game designer does not know what the game is.* The challenge, therefore, is to design the behavior of the system components so that they self-organize into a desirable configuration without central direction. This situation bears many similarities to the decentralized learning environments treated in previous chapters, but with the key difference that here the behavioral incentives of the "agents" can be tailored to the application in question.

Chapter 9 treats this design problem in the case where the underlying game is weakly acyclic in better replies. This condition turns out to hold in many applications, such as routing problems where the objective is to minimize some measure of congestion. In weakly acyclic games there exist extremely simple decentralized algorithms that equilibrate the system. Consider the following procedure: at random times each system component (agent) experiments with a new strategy with small probability, which is retained provided it yields a strict payoff improvement. In a weakly acyclic game with a finite number of strategies and players, this completely decentralized procedure leads to a Nash equilibrium with probability one, as Chapter 9 shows.

This result does not imply, however, that the procedure necessarily leads to an efficient (welfare maximizing) equilibrium. Chapter 10 considers variants of the learning procedure that yield efficient outcomes irrespective of whether the outcome is a Nash equilibrium or whether the game is weakly acyclic.[1] The basic idea is similar to trial-and-error learning described in Chapter 7, but is even simpler. At random times, each system component enters one of two behavioral states (content or discontent). Discontent agents are much more likely to experiment than content agents. The transition probability into a content or discontent state does not depend on whether the current realized payoff is higher or lower than the previous payoff (as in trial-and-error learning); it depends only on the *level* of the realized payoff. (Specifically the log probability of becoming content increases linearly in the realized payoff.) Chapter 10 demonstrates that this type of adjustment process maximizes total

[1] A related search procedure that selects efficient Nash equilibria is described in Bary S. R. Pradelski and H. P. Young (2012), "Learning efficient Nash equilibria in distributed systems," *Games and Economic Behavior* 75: 882–897.

system performance with high probability, whether or not the game is weakly acyclic. The process is fully decentralized and does not require any communication between the system's diverse components, nor does it pre-suppose that the system designer knows how these components interact. The approach has many potential applications to computer science and distributed control, including the deployment of mobile sensors, distributed routing of information packets in data networks, distributed power generation, and distributed manufacturing systems.

9

Payoff-Based Dynamics for Multi-Player Weakly Acyclic Games

With Jason R. Marden, Gürdal Arslan, and Jeff S. Shamma

1 Introduction

The objective in distributed cooperative control for multi-agent systems is to enable a collection of "self interested" agents to achieve a desirable "collective" objective. There are two overriding challenges to achieving this objective. The first is complexity. Finding an optimal solution by a centralized algorithm may be prohibitively difficult when there are large numbers of interacting agents. This motivates the use of adaptive methods that enable agents to "self organize" into suitable, if not optimal, collective solutions.

The second challenge is limited information. Agents may have limited knowledge about the status of other agents, except perhaps for a small subset of "neighboring" agents. An example is collective motion control for mobile sensor platforms (e.g., Ganguli, Susca, Martinez, Bullo, and Cortes 2005). In these problems, mobile sensors seek to position themselves to achieve various collective objectives such as rendezvous or area coverage. Sensors can communicate with neighboring sensors, but otherwise do not have global knowledge of the domain of operation or the status and locations of non-neighboring sensors.

A typical assumption is that agents are endowed with a reward or utility function that depends on their own strategies and the strategies of other agents. In motion coordination problems, for example, an agent's utility function typically depends on its position relative to other agents or environmental targets, and knowledge of this function guides local motion adjustments.

In other situations, agents may know nothing about the structure of their utility functions, and how their own utility depends on the actions of other

J. R.Marden, H. P. Young, G. Arslan, and J.S. Shamma (2009). "Payoff-based dynamics for multiplayer weakly acyclic games," SIAM Journal on Control and Optimization 48: 379–396. Reproduced with permission.

Spontaneous Order. H. Peyton Young with Jason R. Marden, Gürdal Arslan, and Jeff S. Shamma, Oxford University Press.
© H. Peyton Young (2024). DOI: 10.1093/oso/9780198892908.003.0009

agents (whether local or far away). In this case, the only course of action is to observe rewards based on experience and "optimize" on a trial and error basis. The situation is further complicated because all agents are trying simultaneously to optimize their own strategies. Therefore, even in the absence of noise, an agent trying the same strategy twice may see different results because of the non-stationary nature of the strategies of other agents.

There are several examples of multi-agent systems that illustrate this situation. In distributed routing for ad hoc data networks (e.g., Borkar and Kumar 2005), routing nodes seek to route packets to neighboring nodes based on packet destinations without knowledge of the overall network structure. The objective is to minimize the delay of packets to their destinations. This delay must be realized through trial and error, since the functional dependence of delay on routing strategies is not known. A similar problem is automotive traffic routing, in which drivers seek to minimize the congestion experienced to get to a desired destination. Drivers can experience the congestion on selected routes as a function of the routes selected by other drivers, but drivers do not know the structure of the congestion function. Finally, in a multi-agent approach to designing manufacturing systems (e.g., Gershwin 1994), it may not be known in advance how performance measures (such as throughput) depend on manufacturing policy. Rather performance can only be measured once a policy is implemented.

Our interest in this chapter is to develop algorithms that enable coordination in multi-agent systems for precisely this "payoff based" scenario, in which agents only have access to (possibly noisy) measurements of the rewards received through repeated interactions with other agents. We adopt the framework of "learning in games." (See Fudenberg and Levine 1998; Young 1998, 2005; Hart 2005 for an extensive overview. See also the recent special issue containing Shoham, Powers, and Grenager (2003), or the survey article by Panait and Luke (2005) for perspectives from machine learning.) Unlike most of the learning rules in this literature, which assume that agents adjust their behavior based on the observed behavior of other agents, we shall assume that agents know only their own past actions and the payoffs that resulted. It is far from obvious that Nash equilibrium can be achieved under such a restriction, but in fact it has recently been shown that such "payoff based" learning rules can be constructed that work in any game (Foster and Young 2006; Germano and Lugosi 2007).

In this chapter we show that there are simpler and more intuitive adjustment rules that achieve this objective for a large class of multi-player games known as "weakly acyclic" games. This class captures many problems of interest in cooperative control (Marden, Arslan, and Shamma 2007a, 2007b). It includes the very special case of "identical interest" games, where each agent receives the same reward. However, weakly acyclic games (and the related concept of potential games) capture other scenarios such as congestion games (Rosenthal 1973) and similar problems such as distributed routing in networks, weapon target assignment, consensus, and area coverage. See Marden, Arslan, and Shamma (2007a), Arslan, Marden, and Shamma (2007) and references therein for a discussion of a learning in games approach to cooperative control problems, but under less stringent assumptions on informational constraints considered in this chapter.

For many multi-agent problems, operation at a pure Nash equilibrium may reflect optimization of a collective objective.[1] We will derive payoff-based dynamics that guarantee asymptotically that agent strategies will constitute a pure Nash equilibrium with arbitrarily high probability. It need not always be the case that at least one Nash equilibrium optimizes a collective objective. Motivated by this consideration, we also discuss the introduction of incentives or tolls in a player's payoff function to assure that there is at least one Nash equilibrium that optimizes a collective objective. Even in this case, however, there may still be suboptimal Nash equilibria.

The remainder of this chapter is organized as follows. Section 2 provides background on finite strategic-form games and repeated games. This is followed by three types of payoff-based dynamics in Section 3 for increasingly general problems. Section 3.1 presents "Safe Experimentation Dynamics" which is restricted to identical interest games. Section 3.2 presents "Simple Experimentation Dynamics" for the more general class of weakly acyclic games but with noise free payoff measurements. Section 3.3 presents "Sample Experimentation Dynamics" for weakly acyclic games with noisy payoff measurements. Section 4 discusses how to introduce tolls and incentives in payoffs so that a Nash equilibrium optimizes a collective objective. Section 5 presents an illustrative example of a traffic congestion game. Finally, Section 6 contains some concluding remarks. An important analytical tool throughout is the method of resistance trees for perturbed Markov chains (see Chapter 2).

[1] Nonetheless, there are varied viewpoints on the role of Nash equilibrium as a solution concept for multi-agent systems. See Shoham, Powers, and Grenager (2003) and Mannor and Shamma (2007).

2 Background

In this section, we will present a brief background of the game theoretic concepts used in the chapter. We refer the readers to Fudenberg and Tirole (1991) and Young (1998, 2005) for a more comprehensive review.

2.1 Finite Strategic-Form Games

Consider a finite strategic-form game with n-player set $\mathcal{P} := \{\mathcal{P}_1, \ldots, \mathcal{P}_n\}$ where each player $\mathcal{P}_i \in \mathcal{P}$ has a finite action set $\mathcal{A}_i$ and a utility function $U_i : \mathcal{A} \to \mathbb{R}$ where $\mathcal{A} = \mathcal{A}_1 \times \ldots \times \mathcal{A}_n$. We will sometimes use a single symbol, e.g., G, to represent the entire game, i.e., the player set, $\mathcal{P}$, action sets, $\mathcal{A}_i$, and utility functions U_i.

For an action profile $a = (a_1, a_2, \ldots, a_n) \in \mathcal{A}$, let a_{-i} denote the profile of player actions *other than* player $\mathcal{P}_i$, i.e.,

$$a_{-i} = \{a_1, \ldots, a_{i-1}, a_{i+1}, \ldots, a_n\}.$$

With this notation, we will sometimes write a profile a of actions as (a_i, a_{-i}). Similarly, we may write $U_i(a)$ as $U_i(a_i, a_{-i})$.

An action profile $a^* \in \mathcal{A}$ is called a *pure Nash equilibrium* if for all players $\mathcal{P}_i \in \mathcal{P}$,

$$U_i\left(a_i^*, a_{-i}^*\right) = \max_{a_i \in \mathcal{A}_i} U_i\left(a_i, a_{-i}^*\right) \tag{1}$$

Furthermore, if the above condition is satisfied with a unique maximizer for every player $\mathcal{P}_i \in \mathcal{P}$, then a^* is called a *strict (Nash) equilibrium.*

In this chapter we will consider three classes of games: identical interest games, potential games, and weakly acyclic games. Each class of games has a connection to general cooperative control problems and multi-agent systems for which there is some global utility or potential function $\phi : \mathcal{A} \to \mathbb{R}$ that a global planner seeks to maximize (Marden, Arslan, and Shamma 2007a).

2.1.1 Identical interest games

The most restrictive class of games that we will review in this chapter is identical interest games. In such a game, the players' utility functions $\{U_i\}_{i=1}^n$ are chosen to be the same. That is, for some function $\phi : \mathcal{A} \to \mathbb{R}$,

$$U_i(a) = \phi(a),$$

for every $\mathcal{P}_i \in \mathcal{P}$ and for every $a \in \mathcal{A}$. It is easy to verify that all identical interest games have at least one pure Nash equilibrium, namely any action profile a that maximizes $\phi(a)$.

2.1.2 Potential games

A significant generalization of an identical interest game is a potential game. In a potential game, the change in a player's utility that results from a unilateral change in strategy equals the change in the global utility. Specifically, there is a function $\phi : \mathcal{A} \to \mathbb{R}$ such that for every player $\mathcal{P}_i \in \mathcal{P}$, for every $a_{-i} \in \mathcal{A}_{-i}$, and for every $a_i', a_i'' \in \mathcal{A}_i$,

$$U_i\left(a_i', a_{-i}\right) - U_i(a_i'', a_{-i}) = \phi\left(a_i', a_{-i}\right) - \phi\left(a_i'', a_{-i}\right).$$

When this condition is satisfied, the game is called an exact potential game with the potential function ϕ.[2] It is easy to see that, in potential games, any action profile maximizing the potential function is a pure Nash equilibrium, hence every potential game possesses at least one such equilibrium. An example of an exact potential game is illustrated in Figure 9.1.

2.1.3 Weakly acyclic games

Consider any finite game G with a set A of action profiles. A *better reply path* is a sequence of action profiles $a^1, a^2, \ldots, a^L$ such that for each successive pair a^j, a^{j+1} there is exactly one player such that $a_i^j \neq a_i^{j+1}$ and for that player $U_i(a^{j+1}) > U_i(a^j)$. In other words, one player moves at a time, and each time a player moves he increases his own utility.

Suppose now that G is a potential game with potential function ϕ. Starting from an arbitrary action profile $a \in \mathcal{A}$, construct a better reply path $a = a^1, a^2, \ldots, a^L$ until it can no longer be extended. Note first that such a path cannot cycle back on itself, because ϕ is strictly increasing along the path. Since $\mathcal{A}$ is finite, the path cannot be extended indefinitely. Hence, the last element in a maximal better reply path from any joint action, a, must be a Nash equilibrium of G.

	L	R
U	0, 0	−1, 1
D	1, −1	0, 0

Payoffs

	L	R
U	0	1
D	1	2

Potential

Figure 9.1 An example of a two player exact potential game.

[2] There are weaker notions of potential games such as ordinal or weighted potential games. Rather than discuss each variation specifically, we will discuss a more general framework, weakly acyclic games, in the ensuing section. Any potential game, whether exact, ordinal, or weighted, is a weakly acyclic game.

	L	C	R
U	0, 0	0.1, 0	1, 1
M	1, 0	0, 1	0, 0
D	0, 1	1, 0	0, 0

Figure 9.2 An example of a two player weakly acyclic game.

This idea may be generalized as follows. The game G is *weakly acyclic* in better replies if for any $a \in \mathcal{A}$, there exists a better reply path starting at a and ending at some pure Nash equilibrium of G.[3] Potential games are special cases of weakly acyclic games. An example of a two player weakly acyclic game is illustrated in Figure 9.2. Notice that the illustrated game is not a potential game.

2.2 Repeated games

In a repeated game, at each time $t \in \{0, 1, 2, \dots\}$, each player $\mathcal{P}_i \in \mathcal{P}$ simultaneously chooses an action $a_i(t) \in \mathcal{A}_i$ and receives the utility $U_i(a(t))$ where $a(t) := (a_1(t), \dots, a_n(t))$. Each player $\mathcal{P}_i \in \mathcal{P}$ chooses his action $a_i(t)$ at time t according to a probability distribution $p_i(t)$, which we will refer to as the *strategy* of player $\mathcal{P}_i$ at time t. A player's strategy at time t can rely only on observations from times $\{0, 1, 2, \dots, t-1\}$. Different learning algorithms are specified by both the assumptions on available information and the mechanism by which the strategies are updated as information is gathered. For example, if a player knows the functional form of his utility function and is capable of observing the actions of all other players at every time step, then the strategy adjustment mechanism of player $\mathcal{P}_i$ can be written in the general form

$$p_i(t) = F_i(a(0), \dots, a(t-1); U_i).$$

An example of a learning algorithm, or strategy adjustment mechanism, of this form is the well-known fictitious play (Monderer and Shapley 1996). For a detailed review of learning in games, we direct the reader to Weibull (1995), Samuelson (1997), Fudenberg and Levine (1998), Hofbauer and Sigmund (1998), and Young (1998, 2005).

In this chapter we deal with the issue of whether players can learn to play a pure Nash equilibrium through repeated interactions under the most restrictive observational conditions: players *only* have access to (i) the action they

[3] This version of weak acyclicity is less demanding than weak acyclicity under best replies, which was the version considered in Chapter 2.

played and (ii) the utility (possibly noisy) they received. In this setting, the strategy adjustment mechanism of player $\mathcal{P}_i$ takes on the form

$$p_i(t) = F_i(\{a_i(0), U_i(a(0)) + v_i(0)\}, \ldots, \{a_i(t-1), U_i(a(t-1)) + v_i(t-1)\}), \tag{2}$$

where the $v_i(t)$ are zero mean, independent and identically distributed (i.i.d.) random variables.

3 Payoff-Based Learning Algorithms

In this section, we will introduce three simple payoff-based learning algorithms. The first, called *Safe Experimentation*, guarantees convergence to a pure optimal Nash equilibrium in any identical interest game. Such an equilibrium is optimal because each player's utility is maximized. The second learning algorithm, called *Simple Experimentation*, guarantees convergence to a pure Nash equilibrium in any weakly acyclic game. The third learning algorithm, called *Sample Experimentation*, guarantees convergence to a pure Nash equilibrium in any weakly acyclic game even when utility measurements are corrupted with noise.

3.1 Safe Experimentation dynamics for identical interest games

3.1.1 Constant exploration rates

Before introducing the learning dynamics, we introduce the following function. Let

$$U_i^{\max}(t) := \max_{0 \le \tau \le t-1} U_i(a(\tau))$$

be the maximum utility that player $\mathcal{P}_i$ has received up to time $t-1$.

We will now introduce the Safe Experimentation dynamics for identical interest games.

1. **Initialization**: At time $t = 0$, each player randomly selects and plays any action, $a_i(0)$. This action will be initially set as the player's *baseline action* at time $t = 1$ and is denoted by $a_i^b(1) = a_i(0)$.
2. **Action Selection**: At each subsequent time step, each player selects his baseline action with probability $(1 - \epsilon)$ or experiments with a new random action with probability ϵ, i.e.:

- $a_i(t) = a^b{}_i(t)$ with probability $(1 - \epsilon)$
- a_i(t) is chosen randomly (uniformly) over $\mathcal{A}_i$ with probability ϵ

The variable ϵ will be referred to as the player's *exploration rate.*

3. **Baseline Strategy Update**: Each player compares the actual utility received, $U_i(a(t))$, with the maximum received utility $U_i^{\max}(t)$ and updates his baseline action as follows:

$$a_i^b(t+1) = \begin{cases} a_i(t), & U_i(a(t)) > U_i^{\max}(t) \\ a_i^b(t), & U_i(a(t)) \leq U_i^{\max}(t) \end{cases}$$

   Each player updates his maximum received utility regardless of whether or not Step 2 involved exploration.

4. Return to step 2 and repeat.

The reason that this learning algorithm is called "Safe" Experimentation is that the utility evaluated at the baseline action, $U(a^b(t))$, is non-decreasing with respect to time.

Theorem 3.1 *Let G be a finite n-player identical interest game in which all players use the Safe Experimentation dynamics. Given any probability p < 1, if the exploration rate $\epsilon > 0$ is sufficiently small, then for all sufficiently large times t, a(t) is an optimal Nash equilibrium of G with at least probability p.*

Proof Since G is an identical interest game, let the utility of each player be expressed as $U : \mathcal{A} \to \mathbb{R}$ and let $\mathcal{A}^*$ be the set of "optimal" Nash equilibrium of G, i.e.,

$$\mathcal{A}^* = \left\{a^* \in \mathcal{A} : U(a^*) = \max_{a \in \mathcal{A}} U(a)\right\}$$

For any joint action, $a(t)$, the ensuing joint action will constitute an optimal Nash equilibrium with at least probability

$$\left(\frac{\epsilon}{|\mathcal{A}_1|}\right)\left(\frac{\epsilon}{|\mathcal{A}_2|}\right)\cdots\left(\frac{\epsilon}{|\mathcal{A}_n|}\right)$$

where $|\mathcal{A}_i|$ denotes the cardinality of the action set of player $\mathcal{P}_i$. Therefore, an optimal Nash equilibrium will eventually be played with probability 1 for any $\epsilon > 0$.

Suppose an optimal Nash equilibrium is first played at time t^*, i.e., $a(t^*) \in \mathcal{A}^*$ and $a(t^* - 1) \notin \mathcal{A}^*$. Then the baseline joint action must remain constant from that time onward, i.e., $a^b(t) = a(t^*)$ for all $t > t^*$. An optimal Nash equilibrium will then be played at any time $t > t^*$ with at least probability $(1 - \epsilon)^n$. Since $\epsilon > 0$ can be chosen arbitrarily small, and in particular such that $(1 - \epsilon)^n > p$ this completes the proof.

3.1.2 Diminishing exploration rates

In the Safe Experimentation dynamics, the exploration rate ϵ was defined as a constant. Alternatively, one could let the exploration rate vary to induce desirable behavior. One example would be to let the exploration rate decay, such as $\epsilon_t = (1/t)^{1/n}$. This would induce exploration at early stages and reduce exploration at later stages of the game. The theorem and proof hold under the following conditions for the exploration rate:

$$\lim_{t\to\infty} \epsilon_t = 0$$

$$\lim_{t\to\infty} \prod_{\tau=1}^{t} \left[1 - \left(\frac{\epsilon_\tau}{|\mathcal{A}_1|}\right)\left(\frac{\epsilon_\tau}{|\mathcal{A}_2|}\right)\cdots\left(\frac{\epsilon_\tau}{|\mathcal{A}_n|}\right)\right] = 0$$

3.2 Simple Experimentation dynamics for weakly acyclic games

We will now introduce the Simple Experimentation dynamics for weakly acyclic games. These dynamics will allow us to relax the assumption of identical interest games.

1. **Initialization**: At time $t = 0$, each player randomly selects and plays any action, $a_i(0)$. This action will be initially set as the player's *baseline action* at time 1, i.e., $a_i^b(1) = a_i(0)$. Likewise, the player's *baseline utility* at time 1 is initialized as $u_i^b(1) = U_i(a(0))$.
2. **Action Selection**: At each subsequent time step, each player selects his baseline action with probability $(1 - \epsilon)$ or experiments with a new random action with probability ϵ; i.e.,
   - $a_i(t) = a_i^b(t)$ with probability $(1 - \epsilon)$
   - $a_i(t)$ is chosen randomly (uniformly) over $\mathcal{A}_i$ with probability ϵ

   The variable ϵ will be referred to as the player's *exploration rate*. Whenever $a_i(t) \neq a_i^b(t)$, we will say that player $\mathcal{P}_i$ *experimented*.
3. **Baseline Action and Baseline Utility Update**: Each player compares the utility received, $U_i(a(t))$, with his baseline utility, $u_i^b(t)$, and updates his baseline action and utility as follows:

- If player $\mathcal{P}_i$ *experimented* (i.e., $a_i(t) \neq a_i^b(t)$) and if $U_i(a(t)) > u_i^b(t)$ then

$$\begin{aligned} a_i^b(t+1) &= a_i(t), \\ u_i^b(t+1) &= U_i(a(t)). \end{aligned}$$

- If player $\mathcal{P}_i$ *experimented* and if $U_i(a(t)) \leq u_i^b(t)$ then

$$\begin{aligned} a_i^b(t+1) &= a_i^b(t), \\ u_i^b(t+1) &= u_i^b(t). \end{aligned}$$

- If player $\mathcal{P}_i$ *did not experiment* (i.e., $a_i(t) = a_i^b(t)$) then

$$\begin{aligned} a_i^b(t+1) &= a_i^b(t), \\ u_i^b(t+1) &= U_i(a(t)). \end{aligned}$$

4. Return to Step 2 and repeat.

As before, these dynamics require only utility measurements, and hence almost no information regarding the structure of the game.

Theorem 3.2 *Let G be a finite n-player weakly acyclic game in which all players use the Simple Experimentation dynamics. Given any probability $p < 1$, if the exploration rate $\epsilon > 0$ is sufficiently small, then for all sufficiently large times t, $a(t)$ is a Nash equilibrium of G with at least probability p.*

The remainder of this subsection is devoted to the proof of Theorem 3.2. The proof relies on the theory of resistance trees for perturbed Markov chains developed in the Appendix to Chapter 2 (Young 1993).

Define the *state* of the dynamics to be the pair $[a, u]$, where a is the baseline joint action and u is the baseline utility vector. We will omit the superscript b to avoid cumbersome notation.

Partition the state space into the following three sets. First, let X be the set of states $[a, u]$ such that $u_i \neq U_i(a)$ for at least one player $\mathcal{P}_i$. Let E be the set of states $[a, u]$ such that $u_i = U_i(a)$ for all players $\mathcal{P}_i$ and a is a Nash equilibrium. Let D be the set of states $[a, u]$ such that $u_i = U_i(a)$ for all players $\mathcal{P}_i$ and a is a disequilibrium (not a Nash equilibrium). These are all the states.

Claim 3.1

a. *Any state $[a, u] \in X$ transitions to a state in $E \cup D$ in one period with probability $O(1)$.*

b. *Any state $[a, u] \in E \cup D$ transitions to a different state $[a', u']$ with probability at most $O(\varepsilon)$.*

Proof For any $[a, u'] \in X$, there exists at least one player $\mathcal{P}_i$ such that $u'_i \neq U_i(a)$. If all players repeat their part of the joint action profile a, which occurs with probability $(1 - \epsilon)^n$, then $[a, u']$ transitions to $[a, u]$, where $u_i = U_i(a)$ for all players $\mathcal{P}_i$. Thus the process moves to $[a, u] \in E \cup D$ with prob O(1). This proves statement (a). As for statement (b), any state in $E \cup D$ transitions back to itself whenever no player experiments, which occurs with probability at least O(1).

Claim 3.2 *For any state $[a, u] \in D$, there is a finite sequence of transitions to a state $[a^*, u^*] \in E$, where the transitions have the form:*[5]

$$[a, u] \underset{O(\epsilon)}{\rightarrow} [a^1, u^1] \underset{O(\epsilon)}{\rightarrow} \cdots \underset{O(\epsilon)}{\rightarrow} [a^*, u^*]$$

where $u_i^k = U_i(a^k)$ for all i and for all $k > 0$, and each transition occurs with probability $O(\epsilon)$.

Proof Such a sequence is guaranteed by weak acyclicity. Since a is not an equilibrium, there is a better reply path from a to some equilibrium a^*, say a, $a^1, a^2, \ldots, a^*$.

At $[a, u]$ the appropriate player $\mathcal{P}_i$ experiments with probability ϵ, chooses the appropriate better reply with probability $1/|\mathcal{A}_i|$, and no one else experiments. Thus the process moves to $[a^1, u^1]$ where $u_i^1 = U_i(a^1)$ for all players $\mathcal{P}_i$ with probability $O(\epsilon)$ (more precisely, $O(\epsilon(1 - \epsilon)^{n-1})$). Notice that for the deviator $\mathcal{P}_i$, $U_i(a^1) > U_i(a)$, therefore $u_i^1 = U_i(a^1)$. For the non-deviator, say player $\mathcal{P}_j$, $u_j^1 = U_j(a^1)$ since $a_j^1 = a_j$. Thus $[a^1, u^1] \in D \cup E$. In the next period, the appropriate player deviates and so forth.

Claim 3.3 *For any equilibrium $[a^*, u^*] \in E$, any path from $[a^*, u^*]$ to another state $[a, u] \in E \cup D$, $a \neq a^*$, that does not loop back to $[a^*, u^*]$ must be one of the following two forms:*

[5] We will use the notation $z \rightarrow z'$ to denote the transition from state z to state z'. We use $z \underset{O(\epsilon)}{\rightarrow} z'$ to emphasize that this transition occurs with probability of order ϵ.

1. $[a^*, u^*] \underset{O(\epsilon)}{\rightarrow} [a^*, u'] \underset{O(\epsilon^k)}{\rightarrow} [a', u''] \rightarrow \ldots \rightarrow [a, u]$ *where* $k \geq 1$;

2. $[a^*, u^*] \underset{O(\epsilon^k)}{\rightarrow} [a', u''] \rightarrow \ldots \rightarrow [a, u]$ *where* $k \geq 2$.

Proof The path must begin by either one player experimenting or more than one player experimenting. Case (2) results if more than one player experiments. Case (1) results if exactly one agent, say agent $\mathcal{P}_i$, experiments with an action $a'_i \neq a^*$ and all other players continue to play their part of a^*. This happens with probability $(\epsilon/|\mathcal{A}_i|)(1-\epsilon)^{n-1}$. In this situation, player $\mathcal{P}_i$ cannot be better off, meaning that $U_i(a'_i, a^*_{-i}) \leq U_i(a^*)$, since by assumption a^* is an equilibrium. Hence the baseline action next period remains a^* for all players, though their baseline utilities may change. Denote the next state by $[a^*, u']$. If in the subsequent period all players continue to play their part of the action a^* again, which occurs with probability $(1-\epsilon)^n$, then the state reverts back to $[a^*, u^*]$ and we have a loop. Hence the only way the path can continue without a loop is for one or more players to experiment in the next stage, which has probability $O(\epsilon^k)$, $k \geq 1$. This is exactly what case (1) alleges.

Proof [Theorem 3.2] This is a finite aperiodic Markov process on the state space $\mathcal{A} \times \bar{U}_1 \times \cdots \times \bar{U}_n$ where $\bar{U}_i$ denotes the (finite) range of $U_i(\cdot)$. Furthermore, from every state there exists a positive probability path to a Nash equilibrium. Hence, every recurrence class has at least one Nash equilibrium. We will now show that within any recurrence class, the trees rooted at the Nash equilibria have lowest resistance. It follows that the a priori probability of being in a Nash equilibrium can be made arbitrarily close to 1 (see the Appendix to Chapter 2 (Young 1993).

The proof proceeds by constructing trees whose vertices consist of all possible states $E \cup D \cup X$. We then show that the minimum resistance among all such trees occurs only at trees rooted in E.

Suppose, by way of contradiction, that a minimum resistance tree T is rooted at a vertex v that is not in E. If $v \in X$, it is easy to construct a new tree that has lower resistance. Namely, by Claim 3.1a, there is a 0-resistance one-hop path P from v to some state $[a, u] \in E \cup D$. Add the edge of P to T and subtract the edge in T that exits from the vertex $[a, u]$. This results in an $[a, u]$-tree T'. It has lower resistance than T because the added edge has zero resistance while the subtracted edge has resistance greater than or equal to 1 because of Claim 3.1b. This argument is illustrated in Figure 9.3, where the dashed edge

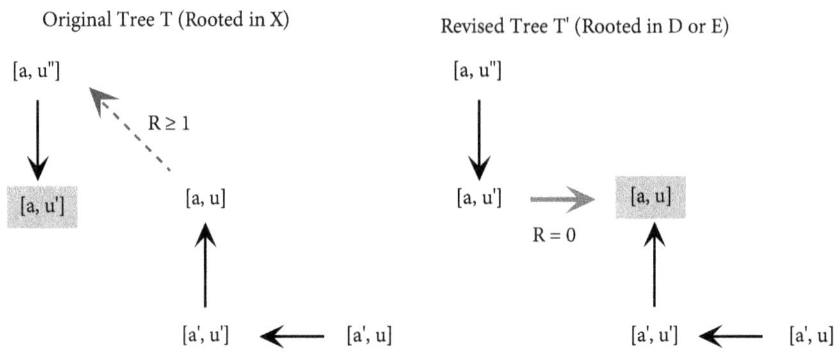

Figure 9.3 Construction of alternative to tree rooted in X.

of strictly positive resistance is removed and replaced with the grey edge of zero resistance.

Suppose next that $v = [a, u] \in D$ but not in E. Construct a path P as in Claim 3.2 from $[a, u]$ to some state $[a^*, u^*] \in E$. As above, construct a new tree T' rooted at $[a^*, u^*]$ by adding the edges of P to T and taking out the redundant edges (the edges in T that exit from the vertices in P). The nature of the path P guarantees that the edges taken out have total resistance at least as high as the resistances of the edges put in. This is because the entire path P lies in $E \cup D$, each transition on the path has resistance 1, and, from Claim 3.2b, the resistance to leave any state in $E \cup D$ is at least 1.

To construct a new tree that has strictly lower resistance, we will inspect the effect of removing the exiting edge from $[a^*, u^*]$ in T. Note that this edge must fit either case (1) or case (2) of Claim 3.3.

In case (2), the resistance of the exiting edge is at least 2, which is larger than any edge in P. Hence the new tree has strictly lower resistance than T, which is a contradiction. This argument is illustrated in Figure 9.4. A new path is created from the original root $[a, u] \in D$ to the equilibrium $[a^*, u^*] \in E$ (grey edges). Redundant (dashed) edges emanating from the new path are removed. In case (2), the redundant edge emanating from $[a^*, u^*]$ has a resistance of at least 2.

In case (1), the exiting edge has the form $[a^*, u^*] \to [a^*, u']$ which has resistance 1 where $u^* \neq u'$. The next edge in T, say $[a^*, u'] \to [a', u'']$, also has at least resistance 1. Remove the edge $[a^*, u'] \to [a', u'']$ from T, and put in the edge $[a^*, u'] \to [a^*, u^*]$. The latter has resistance 0 since $[a^*, u'] \in X$. This results in a tree T'' that is rooted at $[a^*, u^*]$ and has strictly lower resistance than does T, which is a contradiction. This argument is illustrated in Figure 9.5. As in Figure 9.4, a new (grey) path is constructed and redundant (dashed) edges are

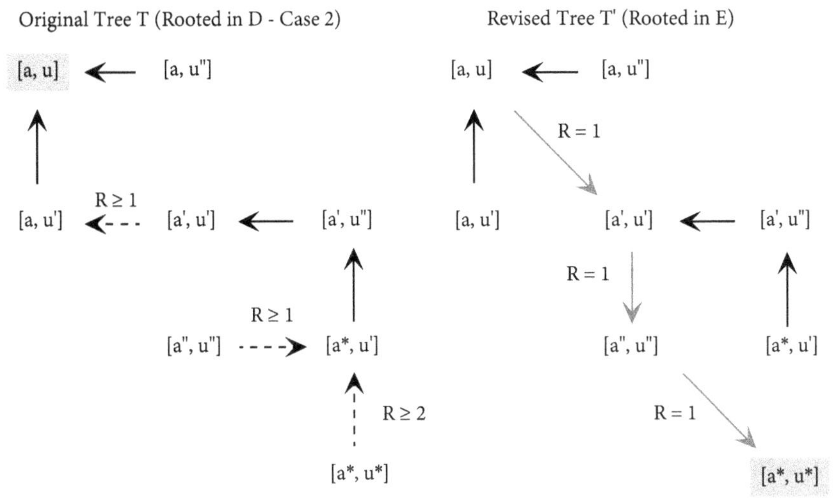

Figure 9.4 Construction of alternative to tree rooted in *D* for Case (2).

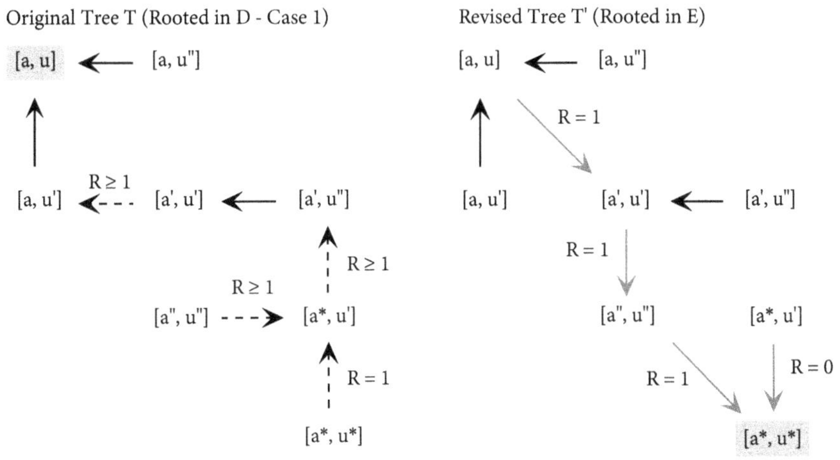

Figure 9.5 Construction of alternative to tree rooted in *D* for Case (1).

removed. The difference is that the edge $[a^*, u'] \to [a', u'']$ is removed and replaced with $[a^*, u'] \to [a^*, u^*]$.

To recap, a minimum resistance tree cannot be rooted at any state in X or D, rather, only at a state in E. Therefore, when ϵ is sufficiently small, the long-run probability on E can be made arbitrarily close to 1, and in particular larger than any specified probability p. This concludes the proof of Theorem 3.2.

3.3 Sample Experimentation dynamics for weakly acyclic games with noisy utility measurements

3.3.1 Noise-free utility measurements

In this section we will focus on developing payoff-based dynamics for which the limiting behavior exhibits that of a pure Nash equilibrium with arbitrarily high probability in any finite weakly acyclic game *even in the presence of utility noise*. We will show that a variant of the so-called Regret Testing algorithm (Foster and Young 2006) accomplishes this objective for weakly acyclic games with noisy utility measurements.

We now introduce Sample Experimentation dynamics.

1. **Initialization**: At time $t = 0$, each player randomly selects and plays any action, $a_i(0) \in \mathcal{A}_i$. This action will be initially set as the player's *baseline action*, $a_i^b(1) = a_i(0)$.
2. **Exploration Phase**: After the baseline action is set, each player engages in an *exploration phase* over the next m periods. The length of the exploration phase need not be the same or synchronized for each player, but we will assume that they are for the proof. For convenience, we will double index the time of the actions played as

$$\check{a}(t_1, t_2) = a(mt_1 + t_2)$$

where t_1 indexes the number of the exploration phase and t_2 indexes the actions played in that exploration phase. We will refer to t_1 as the *exploration phase time* and t_2 as the *exploration action time*. By construction, the exploration phase time and exploration action time satisfy $t_1 \geq 1$ and $\mathrm{m} \geq t_2 \geq 1$. The baseline action will only be updated at the end of the exploration phase and will therefore only be indexed by the exploration phase time.

During the exploration phase, each player selects his baseline action with probability $(1 - \epsilon)$ or experiments with a new random action with probability ϵ. That is, for any exploration phase time $t_1 \geq 1$ and for any exploration action time satisfying $\mathrm{m} \geq t_2 \geq 1$,

- $\check{a}_i(t_1, t_2) = a_i^b(t_1)$ with probability$(1 - \epsilon)$,
- $\check{a}_i(t_1, t_2)$ is chosen randomly (uniformly) over $\left(\mathcal{A}_i \backslash a_i^b(t_1)\right)$ with probability ϵ.

Again, the variable ϵ will be referred to as the player's *exploration rate*.

3. **Action Assessment**: After the exploration phase, each player evaluates the average utility received when playing each of his actions during the exploration phase. Let $n_i^{a_i}(t_1)$ be the number of times that player $\mathcal{P}_i$ played action a_i during the exploration phase at time t_1. The average utility for action a_i during the exploration phase at time t_1 is

$$\hat{V}_i^{a_i}(t_1) = \begin{cases} \frac{1}{n_i^{a_i}(t_1)} \sum_{t_2=1}^{m} I\{a_i = \check{a}_i(t_1, t_2)\} U_i(\check{a}(t_1, t_2)), & n_i^{a_i}(t_1) > 0; \\ U_{\min} & n_i^{a_i}(t_1) = 0, \end{cases}$$

where $I\{\cdot\}$ is the usual indicator function and $U_{\min}$ satisfies

$$U_{\min} < \min_i \min_{a \in \mathcal{A}} U_i(a).$$

In words, $U_{\min}$ is less than the smallest payoff any agent can receive.

4. **Evaluation of Better Response Set**: Each player compares the average utility received when playing his baseline action, $\hat{V}_i^{a_i^b(t)}(t_1)$, with the average utility received for each of his other actions, $\hat{V}_i^{a_i}(t_1)$, and finds all played actions which performed δ better than the baseline action. The term δ will be referred to as the players' *tolerance level*. Define $\mathcal{A}_i^*(t_1)$ to be the set of actions that outperformed the baseline action as follows:

$$\mathcal{A}_i^*(t_1) = \left\{ a_i \in \mathcal{A}_i : \hat{V}_i^{a_i}(t_1) \geq \hat{V}_i^{a_i^b(t_1)}(t_1) + \delta \right\}. \tag{3}$$

5. **Baseline Strategy Update**: Each player updates his baseline action as follows:
   - If $A_i^*(t_1) = \emptyset$, then $a_i^b(t_1 + 1) = a_i^b(t_1)$.
   - If $A_i^*(t_1) \neq \emptyset$, then
   - With probability ω, set $a_i^b(t_1 + 1) = a_i^b(t_1)$. (We will refer to ω as the player's inertia.)
   - With probability $1 - \omega$, randomly select $a_i^b(t_1 + 1) \in \mathcal{A}_i^*(t_1)$ with uniform probability.
6. Return to Step 2 and repeat.

For simplicity, we will first state and prove the desired convergence properties using noiseless utility measurements. The setup for the noisy utility measurements will be stated afterward.

Before stating the following theorem, we define the constant $\alpha > 0$ as follows. If $U_i(a^1) \neq U_i(a^2)$ for any joint actions $a^1, a^2 \in \mathcal{A}$ and any player $\mathcal{P}_i \in \mathcal{P}$, then $|U_i(a^1) - U_i(a^2)| > \alpha$. In words, if any two joint actions result in different utilities at all, then the difference would be greater than α.

Theorem 3.3 *Let G be a finite n-player weakly acyclic game in which all players use the Sample Experimentation dynamics. For any*

- *probability $p < 1$,*
- *tolerance level $\delta \in (0, \alpha)$,*
- *inertia $\omega \in (0, 1)$, and*
- *exploration rate ε satisfying $\min\{(\alpha-\delta)/4, \delta/4, 1-p\} > (1-(1-\epsilon)^n) > 0$,*

if the exploration phase length m is sufficiently large, then for all sufficiently large times $t > 0$, $a(t)$ is a Nash equilibrium of G with at least probability p.

The remainder of this subsection is devoted to the proof of Theorem 3.3.

We will assume for simplicity that utilities are between −1/2 and 1/2, i.e., $|U_i(a)| \leq 1/2$ for any player $\mathcal{P}_i \in \mathcal{P}$ and any joint action $a \in \mathcal{A}$.

We begin with a series of useful claims. The first claim states that for any player $\mathcal{P}_i$ the average utility for an action $a_i \in \mathcal{A}_i$ during the exploration phase can be made arbitrarily close (with high probability) to the actual utility the player would have received provided that all other players never experimented. This can be accomplished if the experimentation rate is sufficiently small and the exploration phase length is sufficiently large.

Claim 3.4 *Let a^b be the joint baseline action at the start of an exploration phase of length m. For*

- *any probability $p < 1$,*
- *any $\delta^* > 0$, and*
- *any exploration rate $\epsilon > 0$ satisfying $\delta^*/2 \geq (1 - (1 - \epsilon)^{n-1}) > 0$,*

if the exploration phase length m is sufficiently large then

$$\Pr\left[\left|\hat{V}_i^{a_i} - U_i\left(a_i, a_{-i}^b\right)\right| > \delta^*\right] < 1 - p$$

Proof Let $n_i(a_i)$ represent the number of times player $\mathcal{P}_i$ played action a_i during the exploration phase. In the following discussion, *all probabilities and expectations are conditioned on $n_i(a_i) > 0$*. We omit making this explicit for the sake of notational simplicity. The event $n_i(a_i) = 0$ has diminishing probability as the exploration phase length m increases, and so this case will not affect the desired conclusions for increasing phase lengths.

For an arbitrary $\delta^* > 0$,

$$\begin{aligned}
&\mathbf{Pr}\left[|\hat{V}_i^{a_i} - U_i(a_i, a_{-i}^b)| > \delta^*\right] \\
&\quad \le \mathbf{Pr}\left[|\hat{V}_i^{a_i} - E\{\hat{V}_i^{a_i}\}| + |E\{\hat{V}_i^{a_i}\} - U_i(a_i, a_{-i}^b)| > \delta^*\right] \\
&\quad \le \underbrace{\mathbf{Pr}\left[|\hat{V}_i^{a_i} - E\{\hat{V}_i^{a_i}\}| > \delta^*/2\right]}_{(*)} + \underbrace{\mathbf{Pr}\left[|E\{\hat{V}_i^{a_i}\} - U_i(a_i, a_{-i}^b)| > \delta^*/2|\right]}_{(**)}.
\end{aligned}$$

First, let us focus on (**). We have

$$E\{\hat{V}_i^{a_i}\} - U_i\left(a_i, a_{-i}^b\right) = \left[1 - (1-\epsilon)^{n-1}\right]\left[E\left\{U_i(a_i, a_{-i}(t)) \mid a_{-i}(t) \neq a_{-i}^b\right\} - U_i\left(a_i, a^b\right)\right]$$

which approaches 0 as $\epsilon \downarrow 0$. Therefore, for any exploration rate ϵ satisfying $\delta^*/2 > (1 - (1-\varepsilon)^{n-1}) > 0$, we know that

$$\Pr\left[|E\{\hat{V}_i^{a_i}\} - U_i(a_i, a_{-i}^b)| > \delta^*/2\right] = 0.$$

Now we will focus on (*). By the weak law of large numbers, (*) approaches 0 as $n_i(a_i) \uparrow \infty$. This implies that for any probability $\bar{p} < 1$ and any exploration rate $\epsilon > 0$, there exists a sample size $n^*(a_i)$ such that if $n_i(a_i) > n_i^*(a_i)$ then

$$\Pr\left[|\hat{V}_i^{a_i} - E\{\hat{V}_i^{a_i}\}| > \rho/2\right] < 1 - \bar{p}.$$

Lastly, for any probability $\bar{p} < 1$ and any fixed exploration rate, there exists a minimum exploration length $\underline{m} > 0$ such that for any exploration length $m > \underline{m}$,

$$\Pr\left[n_i(a_i) \ge n_i^*(a_i)\right] \ge \bar{p}.$$

In summary, for any fixed exploration rate ϵ satisfying $\delta^*/2 \ge (1 - (1-\epsilon)^{n-1}) > 0$, (*) + (**) can be made arbitrarily close to 0, provided that the exploration length m is sufficiently large. This concludes the proof of Claim 3.4.

Claim 3.5 *Let a^b be the joint baseline action at the start of an exploration phase of length m. For any*

- *probability $p < 1$,*
- *tolerance level $\delta \in (0, \alpha)$, and*

- *exploration rate* $\epsilon > 0$ *satisfying* $min\{(\alpha - \delta)/4, \delta/4\} \geq (1 - (1 - \epsilon)^{n-1}) > 0$,
- *if the exploration length m is sufficiently large, then each player's better response set* $\mathcal{A}_i^*$ *will contain* ***only and all*** *actions that are a better response to the joint baseline action, i.e.,*

$$a_i^* \in \mathcal{A}_i^* \Leftrightarrow U_i\left(a_i^*, a_{-i}^b\right) > U_i\left(a^b\right),$$

with at least probability p.

Proof Suppose a^b is not a Nash equilibrium. For some player $\mathcal{P}_i \in \mathcal{P}$, let $a^*{}_i$ be a strict better reply to the baseline joint action, i.e., $U_i\ (a^*{}_i, a^b{}_{-i}) > U_i\ (a^b)$ and let $a_i{}^w$ be a non-better reply to the baseline joint action, i.e., $U_i\ (a_i{}^w, a^b{}_{-i}) \leq U_i\ (a^b)$.

Using Claim 3.4, for any probability $\bar{p} < 1$ and any exploration rate $\epsilon > 0$ satisfying $\min\{(\alpha - \delta)/4, \ \delta/4\} \geq (1 - (1 - \epsilon)^{n-1}) > 0$ there exists a minimum exploration length $\underline{m} > 0$ such that for any exploration length $m > \underline{m}$ the following expressions are true:

$$\Pr\left[\left|\hat{V}_i^{a_i^b} - U_i\left(a_i^b, a_{-i}^b\right)\right| < \delta^*\right] \geq \bar{p} \tag{4}$$

$$\Pr\left[\left|\hat{V}_i^{a_i^*} - U_i\left(a_i^*, a_{-i}^b\right)\right| < \delta^*\right] \geq \bar{p} \tag{5}$$

$$\Pr\left[\left|\hat{V}_i^{a_i^w} - U_i\left(a_i^w, a_{-i}^b\right)\right| < \delta^*\right] \geq \bar{p} \tag{6}$$

where $\delta^* = \min\{(\alpha - \delta)/2, \delta/2\}$. Rewriting equation 4 we obtain

$$\Pr\left[\left|\hat{V}_i^{a_i^b} - U_i\left(a_i^b, a_{-i}^b\right)\right| < \delta^*\right] \leq \Pr\left[\hat{V}_i^{a_i^b} - U_i\left(a_i^b, a_{-i}^b\right) < (\alpha - \delta)/2\right],$$

and rewriting equation 5 we obtain

$$\begin{aligned}\Pr\left[\left|\hat{V}_i^{a_i^*} - U_i\left(a_i^*, a_{-i}^b\right)\right| < \delta^*\right] &\leq \Pr\left[\hat{V}_i^{a_i^*} - U_i\left(a_i^*, a_{-i}^b\right) > -(\alpha - \delta)/2\right] \\ &\leq \Pr\left[\hat{V}_i^{a_i^*} - \left(U_i\left(a_i^b, a_{-i}^b\right) + \alpha\right) > -(\alpha - \delta)/2\right] \\ &= \Pr\left[\hat{V}_i^{a_i^*} - U_i\left(a_i^b, a_{-i}^b\right) > (\alpha + \delta)/2\right],\end{aligned}$$

meaning that

$$\Pr\left[a_i^* \in \mathcal{A}_i^*\right] \geq \bar{p}^2,$$

Similarly, rewriting equation 4 we obtain

$$\Pr\left[\left|\hat{V}_i^{a_i^b} - U_i\left(a_i^b, a_{-i}^b\right)\right| < \delta^*\right] \leq \Pr\left[\hat{V}_i^{a_i^b} - U_i\left(a_i^b, a_{-i}^b\right) > -\delta/2\right],$$

and rewriting equation 6 we obtain

$$\begin{aligned}\Pr\left[\left|\hat{V}_i^{a_i^w} - U_i\left(a_i^w, a_{-i}^b\right)\right| < \delta^*\right] &\leq \Pr\left[\hat{V}_i^{a_i^w} - U_i\left(a_i^w, a_{-i}^b\right) < \delta/2\right],\\ &\leq \Pr\left[\hat{V}_i^{a_i^w} - U_i\left(a_i^b, a_{-i}^b\right) < \delta/2\right],\end{aligned}$$

meaning that

$$\Pr\left[a_i^w \notin \mathcal{A}_i^*\right] \geq \bar{p}^2.$$

Since $\bar{p}$ can be chosen arbitrarily close to 1, the proof of claim 3.5 is complete.

Proof [Theorem 3.3]. The evolution of the baseline actions from phase to phase is a finite aperiodic Markov process on the state space of joint actions, $\mathcal{A}$. Furthermore, since G is weakly acyclic, from every state there exists a better reply path to a Nash equilibrium. Hence, every recurrent class has at least one Nash equilibrium. We will show that these dynamics can be viewed as a perturbation of a certain Markov chain whose recurrent classes are restricted to Nash equilibria. We will then appeal to Theorem 4 in Chapter 2 to obtain the desired result.

We begin by defining an "unperturbed" process on baseline actions. For any $a^b \in \mathcal{A}$, define the *true* better reply set as

$$\bar{\mathcal{A}}_i^*\left(a^b\right) = \left\{a_i : U_i\left(a_i, a_{-i}^b\right) > U_i\left(a^b\right)\right\}$$

Now define the transition process from $a^b(t_1)$ to $a^b(t_1+1)$ as follows:

- If $\bar{\mathcal{A}}_i^*\left(a^b(t_1)\right) = \emptyset$, then $a_i^b(t_1+1) = a_i^b(t_1)$.
- If $\bar{\mathcal{A}}_i^*\left(a^b(t_1)\right) \neq \emptyset$, then

  - With probability ω, set $a_i^b(t_1+1) = a_i^b(t_1)$.
  - With probability $1-\omega$, randomly select $a_i^b(t_1+1) \in \bar{\mathcal{A}}_i^*(t_1)$ with uniform probability.

This is a special case of a so-called "better reply process with finite memory and inertia." The joint actions of this process converge to a Nash equilibrium with probability 1 in any weakly acyclic game [Young 2005, Theorem 6.2]. Therefore, the recurrence classes of this unperturbed process are precisely the set of pure Nash equilibria.

The above unperturbed process closely resembles the Baseline Strategy Update process described in Step 5 of Sample Experimentation dynamics. The difference is that the above process uses the true better reply set, whereas Step 5 uses a better reply set constructed from experimentation over a phase. However, by Claim 3.5, for any probability $\bar{p} < 1$, acceptable tolerance level δ, and acceptable exploration rate ϵ, there exists a minimum exploration phase length $\underline{m}$ such that for any exploration phase length $m > \underline{m}$, each player's better response set will contain only and all actions that are a strict better response with at least probability $\bar{p}$.

With parameters selected according to Claim 3.5, the transitions of the baseline joint actions in Sample Experimentation dynamics follow that of the above unperturbed better reply process with probability $\bar{p}$ arbitrarily close to 1. Since the recurrence classes of the unperturbed process are only Nash equilibria, we conclude that as $\bar{p}$ approaches 1, the probability that the baseline action for sufficiently large t_1 will be a (pure) Nash equilibrium can be made arbitrarily close to 1. By making the exploration probability ϵ sufficiently small, we can also conclude that the joint action during exploration phases, i.e., $a(mt_1 + t_2)$, will also be a Nash equilibrium with probability arbitrarily close to 1.

3.3.2. Noisy utility measurements

Suppose that each player receives a noisy measurement of his true utility, i.e.,

$$\tilde{U}_i\left(a_i, a_{-i}\right) = U_i\left(a_i, a_{-i}\right) + \nu_i,$$

where ν_i is an i.i.d. random variable with zero mean. In the regret testing algorithm with noisy utility measurements, the average utility for action a_i during the exploration phase at time t_1 is now

$$\hat{V}_i^{a_i}(t_1) = \begin{cases} \frac{1}{n_i^{a_i}(t_1)} \sum_{t_2=1}^{m} I\{a_i = \check{a}_i(t_1, t_2)\} \tilde{U}_i(\check{a}(t_1, t_2)), & n_i^{a_i}(t_1) > 0; \\ U_{\min} & n_i^{a_i}(t_1) = 0. \end{cases}$$

A straightforward modification of the proof of Theorem 3.3 leads to the following theorem.

Theorem 3.4 *Let G be a finite n-player weakly acyclic game where players' utilities are corrupted with a zero mean noise process. If all players use the Sample Experimentation dynamics, then for any*

- *probability* $p < 1$,
- *tolerance level* $\delta \in (0, \alpha)$,
- *inertia* $\omega \in (0, 1)$, *and*
- *exploration rate* ε *satisfying* $\min\{(\alpha - \delta)/4, \delta/4, 1 - p\} > (1 - (1 - \epsilon)^n) > 0$,

if the exploration phase length m is sufficiently large, then for all sufficiently large times $t > 0$, $a(t)$ *is a Nash equilibrium of G with at least probability p.*

3.3.3 Comment on length and synchronization of players' exploration phases

In the proof of Theorem 3.3, we assumed that all players' exploration phases were synchronized and of the same length. This assumption was used to ensure that when a player assessed the performance of a particular action, the baseline action of the other players remained constant. Because of the players' inertia this assumption is unnecessary. The general idea is as follows: a player will repeat his baseline action regardless of his better response set with positive probability because of his inertia. Therefore, if all players repeat their baseline action a sufficient number of times, which happens with positive probability, then the joint baseline action would remain constant long enough for any player to evaluate an accurate better response set for that particular joint baseline action.

4 Influencing Nash Equilibria in Resource Allocation Problems

In this section we will derive an approach for influencing the Nash equilibria of a resource allocation problem using the idea of marginal cost pricing. We will illustrate the setup and our approach on a congestion game which is an example of a resource allocation problem.

4.1 Congestion game setup

We consider a transportation network with a finite set R of road segments (or resources) that needs to be shared by a set of selfish drivers labeled as $D := \{d_1, ..., d_n\}$. Each driver has a fixed origin/destination pair connected through multiple routes. The set of all routes available to driver d_i is denoted by $\mathcal{A}_i$. A route $a_i \in \mathcal{A}_i$ consists of multiple road segments, therefore, $a_i \subset R$. Player $\mathcal{P}_i$ taking route a_i incurs a cost c_r for each road segment $r \in a_i$. The utility of driver d_i taking route a_i is defined as the negative of the total cost incurred, i.e.,

$U_i = -\sum_{r \in a_i} c_r$. Of course, the utility of each driver will depend on the routes chosen by other drivers.

If we assume that the cost incurred in a road segment depends *only* on the total number of drivers sharing that road, then drivers are anonymous, and this leads to a *congestion game* (Rosenthal 1973). The utility of driver d_i is now stated more precisely as

$$U_i(a) = -\sum_{r \in a_i} c_r(\sigma_r(a)),$$

where $a := (a_1, ..., a_n)$ is the profile of routes chosen by all drivers and $\sigma_r(a)$ is the total number of drivers using the road segment r.

It is known that a congestion game admits the following potential function,

$$\hat{\phi}(a) = \sum_{r \in R} \sum_{k=1}^{\sigma_r(a)} c_r(k).$$

Unfortunately, this potential function lacks practical significance for measuring the effectiveness of a routing strategy in terms of the overall congestion.

4.2 Congestion game with tolls setup

One approach for equilibrium manipulation is to influence drivers' utilities with tolls (Sandholm 2002). In a congestion game with tolls, a driver's utility takes the form

$$U_i(a) = -\sum_{r \in a_i} c_r(\sigma_r(a)) + t_r(\sigma_r(a)),$$

where $t_r(k)$ is the toll imposed on route r if there are k users.

Suppose that the global planner is interested in minimizing the total congestion experienced by all drivers on the network, which can be evaluated as

$$T_c(a) := \sum_{r \in R} \sigma_r(a) c_r(\sigma_r(a)).$$

It has been shown that there exists a set of tolls such that the potential function associated with the congestion game with tolls is aligned with the total congestion experienced by all drivers on the network (Marden, Arslan, and Shamma 2008, Proposition 4.1).

Proposition 4.1 *Consider a congestion game of any network topology. If the imposed tolls are set as*

$$t_r(k) = (k-1)[c_r(k) - c_r(k-1)], \forall k \geq 1,$$

then the total negative congestion experienced by all drivers, $\phi_c(a) = -T_c(a)$, is a potential function for the congestion game with tolls.

This tolling scheme results in drivers' local utility functions being aligned with the global objective of minimal total congestion.

Now suppose that the global planner is interested in minimizing a *more general* measure,[6]

$$\phi(a) := \sum_{r \in R} f_r(\sigma_r(a))c_r(\sigma_r(a)), \tag{7}$$

where $f_r : \{0,1,2,\ldots\} \to \mathbb{R}$ is any arbitrary function. An example of an objective function that fits within this framework and may be practical for general resource allocation problems is

$$\phi(a) = \sum_{r \in R} c_r(\sigma_r(a)).$$

We will now show that there exists a set of tolls, $t_r(\bullet)$, such that the potential function associated with the congestion game with tolls will be aligned with the global planner's objective function of the form given in equation (7).

Proposition 4.2 *Consider a congestion game of any network topology. If the imposed tolls are set as*

$$t_r(k) - (f_r(k) - 1)c_r(k) - f_r(k-1)c_r(k-1), \quad \forall k \geq 1,$$

then the global planner's objective, $\phi_c(a) = -\phi(a)$, is a potential function for the congestion game with tolls.

Proof Let $a^1 = \{a_i^1, a_{-i}\}$ and $a^2 = \{a_i^2, a_{-i}\}$. We will use the shorthand notation $\sigma_r^{a^1}$ to represent $\sigma_r(a^1)$. The change in utility incurred by driver d_i in changing from route a_i^2 to route a_i^1 is

$$\begin{aligned} U_i(a^1) - U_i(a^2) &= -\sum_{r \in a_i^1} (c_r(\sigma_r^{a^1}) + t_r(\sigma_r^{a^1})) + \sum_{r \in a_i^2} (c_r(\sigma_r^{a^2}) + t_r(\sigma_r^{a^2})), \\ &= -\sum_{r \in a_i^1 \backslash a_i^2} (c_r(\sigma_r^{a^1}) + t_r(\sigma_r^{a^1})) + \sum_{r \in a_i^2 \backslash a_i^1} (c_r(\sigma_r^{a^2}) + t_r(\sigma_r^{a^2})). \end{aligned}$$

[6] In fact, if $c_r(\sigma_r(a)) \neq 0$ for all a, then equation (7) is equivalent to $\sum_{r \in R} f_r(\sigma_r(a))$.

The change in the total negative congestion from the joint action a^2 to a^1 is

$$\phi_c(a^1) - \phi_c(a^2) = - \sum_{r\in(a_i^1 \cup a_i^2)} (f_r(\sigma_r^{a^1})c_r(\sigma_r^{a^1}) - f_r(\sigma_r^{a^2})c_r(\sigma_r^{a^2})).$$

Since

$$\sum_{r\in(a_i^1 \cup a_i^2)} (f_r(\sigma_r^{a^1})c_r(\sigma_r^{a^1}) - f_r(\sigma_r^{a^2})c_r(\sigma_r^{a^2})) = 0$$

the change in the total negative congestion is

$$\phi_c(a^1) - \phi_c(a^2) = \\ - \sum_{r\in a_i^1 \setminus a_i^2} (f_r(\sigma_r^{a^1})c_r(\sigma_r^{a^1}) - f_r(\sigma_r^{a^2})c_r(\sigma_r^{a^2})) - \sum_{r\in a_i^2 \setminus a_i^1} (f_r(\sigma_r^{a^1})c_r(\sigma_r^{a^1}) - f_r(\sigma_r^{a^2})c_r(\sigma_r^{a^2})).$$

Expanding the first term, we obtain

$$\begin{aligned}
&\sum_{r\in a_i^1 \setminus a_i^2} (f_r(\sigma_r^{a^1})c_r(\sigma_r^{a^1}) - f_r(\sigma_r^{a^2})c_r(\sigma_r^{a^2})) \\
&\quad = \sum_{r\in a_a^1 \setminus a_i^2} (f_r(\sigma_r^{a^1})c_r(\sigma_r^{a^1}) - (f_r(\sigma_r^{a^1} - 1))c_r(\sigma_r^{a^1} - 1)), \\
&\quad = \sum_{r\in a_i^1 \setminus a_i^2} (f_r(\sigma_r^{a^1})c_r(\sigma_r^{a^1}) - ((f_r(\sigma_r^{a^1}) - 1)c_r(\sigma_r^{a^1}) - t_r(\sigma_r^{a^1}))) \\
&\quad = \sum_{r\in a_i^1 \setminus a_i^2} (c_r(\sigma_r^{a^1}) + t_r(\sigma_r^{a^1})).
\end{aligned}$$

Therefore,

$$\begin{aligned}
\phi_c(a^1) - \phi_c(a^2) &= - \sum_{r\in a_i^1 \setminus a_i^2} (c_r(\sigma_r^{a^1}) + t_r(\sigma_r^{a^1})) + \sum_{r\in a_i^2 \setminus a_i^1} (c_r(\sigma_r^{a^2}) + t_r(\sigma_r^{a^2})), \\
&= U_i(a^1) - U_i(a^2).
\end{aligned}$$

By implementing the tolling scheme set forth in Proposition 4.2, we guarantee that all action profiles that minimize the global planner's objective are equilibria of the congestion game with tolls.

In the special case that $f_r(\sigma_r(a)) = \sigma_r(a)$, then Proposition 4.2 produces the same tolls as in Proposition 4.1

5 Illustrative Example—Congestion Game

We will consider a discrete representation of the congestion game setup considered in Braess' Paradox (Braess 1968). In our setting, there are 1000 vehicles

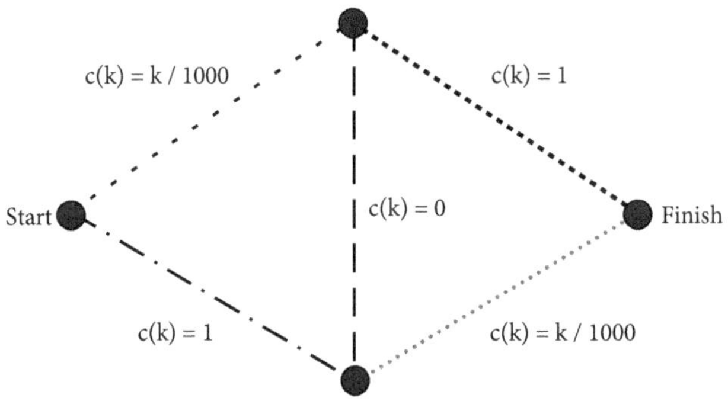

Figure 9.6 Congestion game setup.

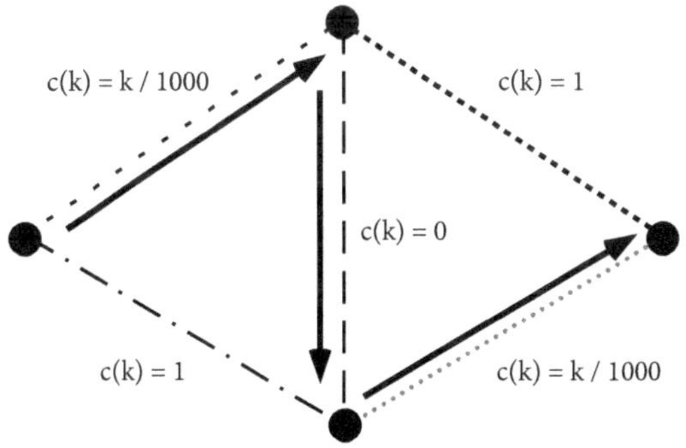

Figure 9.7 Illustration of Nash equilibrium in proposed congestion game.

that need to traverse through the network. The network topology and associated congestion functions are illustrated in Figure 9.6. Each vehicle can select one of the four possible paths to traverse across the network.

The reason for using this setup as an illustration of the learning algorithms and equilibrium manipulation approach developed in this chapter is that the Nash equilibrium of this particular congestion game is easily identifiable. The unique Nash equilibrium is when all vehicles take the route as highlighted in Figure 9.7. At this Nash equilibrium each vehicle has a utility of 2 and the total congestion is 2000.

Since a potential game is weakly acyclic, the payoff-based learning dynamics in this chapter are applicable learning algorithms for this congestion game. In a congestion game, a payoff-based learning algorithm means that drivers have access *only* to the actual congestion experienced. Drivers are unaware of the congestion level on any alternative routes. Figure 9.8 shows the evolution of drivers on routes when using the Simple Experimentation dynamics. This simulation used an experimentation rate of $\epsilon = 0.25\%$. One can observe

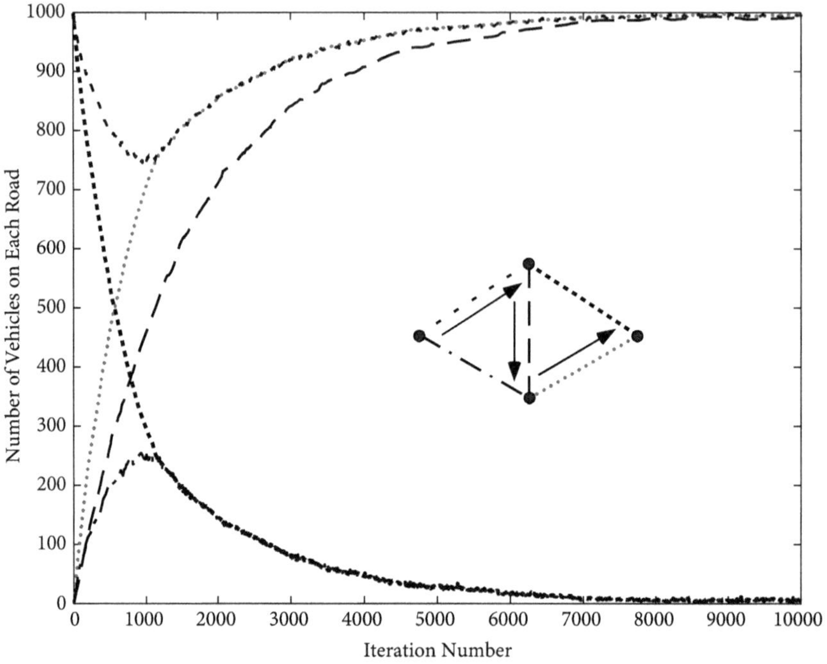

Figure 9.8 Evolution of number of vehicles on each road using Simple Experimentation dynamics: The number of vehicles on the roads highlighted by arrows approach 1,000 while the number of vehicles on all remaining roads approach 0.

that the vehicles' collective behavior does indeed approach that of the Nash equilibrium.

In this congestion game, it is also easy to verify that this vehicle distribution does not minimize the total congestion experience by all drivers over the network. The distribution that minimizes the total congestion over the network is when half the vehicles occupy the top two roads and the other half occupy the bottom two roads. The middle road is irrelevant.

One can employ the tolling scheme developed in the previous section to locally influence vehicle behavior to achieve this objective. In this setting, the new cost functions, i.e., congestion plus tolls, are illustrated in Figure 9.9.

Figure 9.10 shows the evolution of drivers on routes when using the Simple Experimentation dynamics. This simulation used an experimentation rate of $\epsilon = 0.25\%$. When using this tolling scheme, the vehicles' collective behavior approaches the new Nash equilibrium which now minimizes the total congestion experienced on the network. The total congestion experienced on the network is now approximately 1500.

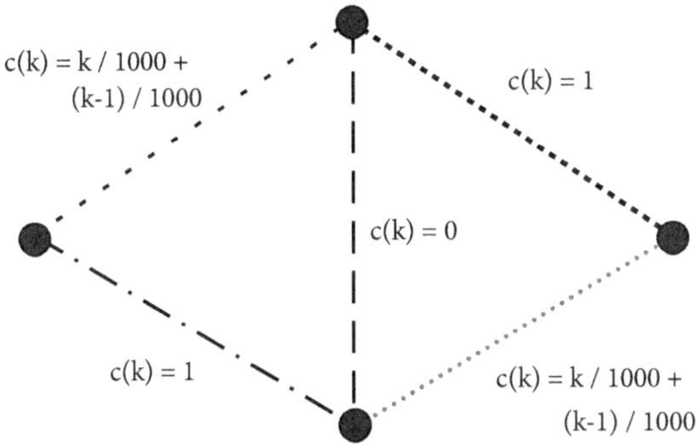

Figure 9.9 Congestion game setup with tolls to minimize total congestion.

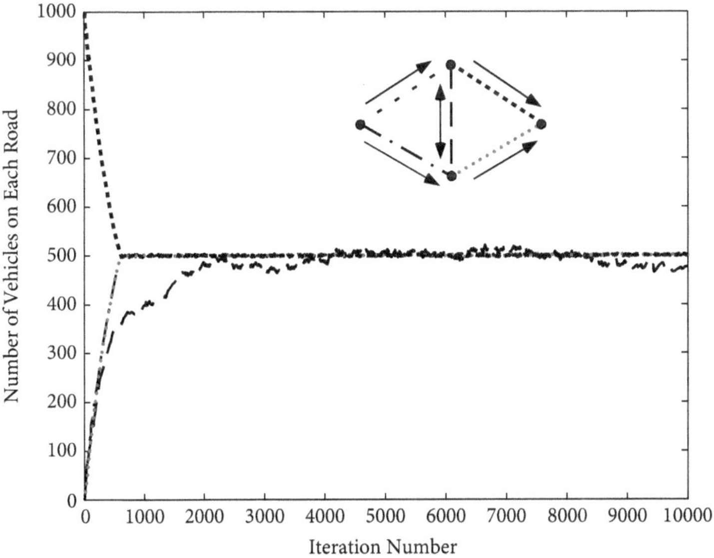

Figure 9.10 Evolution of number of vehicles on each road using Simple Experimentation dynamics with optimal tolls: The number of vehicles on the middle road fluctuates around 500 while the number of vehicles on all remaining roads stabilizes at around 500.

There are other tolling schemes that would have resulted in the desired allocation. One approach is to assign an infinite cost to the middle road, which is equivalent to removing it from the network. Under this scenario, the unique Nash equilibrium is for half the vehicles to occupy the top route and half the bottom, which would minimize the total congestion on the network. Therefore, the existence of this extra road, even though it has zero cost, resulted in the unique Nash equilibrium having a higher total congestion. This is Braess' Paradox (Braess 1968).

The advantage of the tolling scheme set forth in this chapter is that it gives a systematic method for influencing the Nash equilibria of any congestion game. We would like to highlight that this tolling scheme only guarantees that the action profiles that maximize the desired objective function are Nash equilibria of the new congestion game with tolls. However, it does not guarantee the lack of suboptimal Nash equilibria.

In many applications, players may not have access to their true utility, but do have access to a noisy measurement of their utility. For example, in the traffic setting, this noisy measurement could be the result of accidents or weather conditions. We will revisit the original congestion game (without tolls) as illustrated in Figure 9.6. We will now assume that a driver's utility measurement takes on the form

$$\widetilde{U}_i(a) = -\Sigma_{r \in a_i} c_r(\sigma_r(a)) + \nu_i$$

where ν_i is a random variable with zero mean and variance of 0.1. We will assume that the noise is driver specific rather than road specific.

Figure 9.11 shows a comparison of the evolution of drivers on routes when using the Simple and Sample Experimentation dynamics. The Simple Experimentation dynamics simulation used an experimentation rate $\epsilon = 0.25\%$. The Sample Experimentation dynamics simulation used an exploration rate $\epsilon = 0.25\%$, a tolerance level $\delta = 0.002$, an exploration phase length $m = 500{,}000$, and inertia $\omega = 0.85$. As expected, the noisy utility measurements influenced vehicle behavior more in the Simple Experimentation dynamics than the Sample Experimentation dynamics.

6 Concluding Remarks

We have introduced Safe Experimentation dynamics for identical interest games, Simple Experimentation dynamics for weakly acyclic games with noise-free utility measurements, and Sample Experimentation dynamics for weakly acyclic games with noisy utility measurements. For all three settings, we have shown that for sufficiently large times, the joint action taken by players will constitute a Nash equilibrium. Furthermore, we have shown how to guarantee that a collective objective in a congestion game is a (non-unique) Nash equilibrium. An important, but unaddressed topic in this work, is to

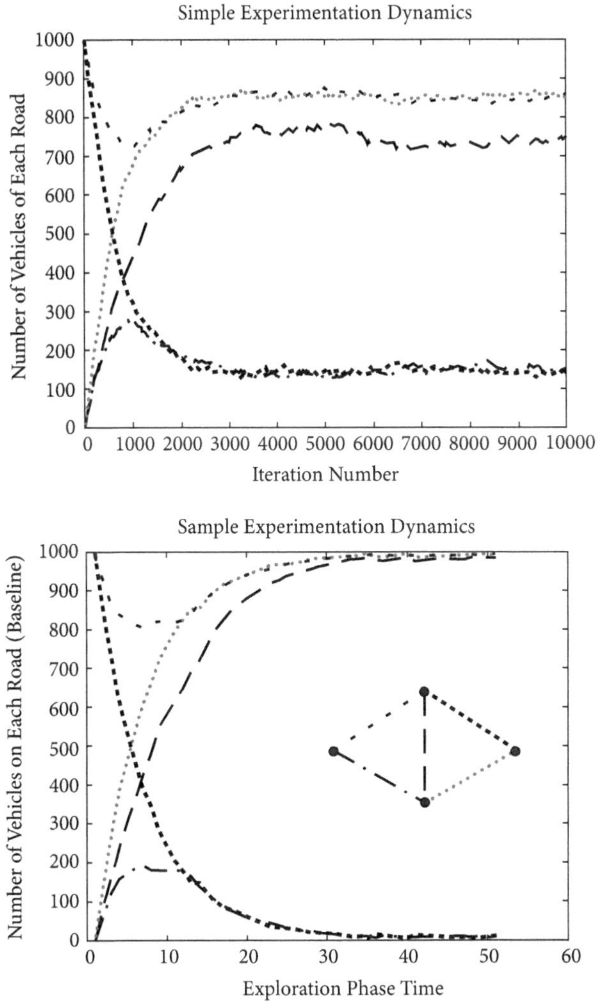

Figure 9.11 Comparison of evolution of number of vehicles on each road using Simple Experimentation dynamics and Sample Experimentation dynamics (baseline) with noisy utility measurements. The number of vehicles on the route (upper left, middle, lower right) dominates the number of vehicles on all remaining roads in both settings.

characterize the resulting convergence rates. It is likely that tools regarding mixing times of Markov chains (Montenegro and Tetali 2006) will be relevant.

Our motivation has been that in many engineered systems, the functional forms of utility functions are not available, and so players must adjust their strategies through an adaptive process using only payoff measurements.

In the dynamic processes defined here, there is no explicit cooperation or communication between players. On the one hand, this lack of explicit coordination offers an element of robustness to a variety of uncertainties in the strategy adjustment processes. Nonetheless, an interesting future direction would be to investigate to what degree explicit coordination through limited communications could be beneficial.

Acknowledgments

Research supported by NSF grant #ECS-0501394, ARO grant #W911NF-04-1-0316, and AFOSR grant #FA9550-05-1-0239.

References

Arslan, G., Marden, J. R., and Shamma, J. S. (2007), "Autonomous vehicle-target assignment: a game theoretical formulation," *ASME Journal of Dynamic Systems, Measurement and Control* 129: 584–596.

Borkar, V. S. and Kumar, P. R. (2003), "Dynamic Cesaro-Wardrop equilibration in networks," *IEEE Transactions on Automatic Control* 48 (3): 382–396.

Braess, D. (1968), "Uber ein paradoxen der verkehrsplanning," *Unternehmensforschung* 12: 258–268.

Foster, D. P. and Young, H. P. (2006), "Regret testing: Learning to play Nash equilibrium without knowing you have an opponent," *Theoretical Economics* 1: 341–367.

Fudenberg, D. and Tirole, J. (1991), *Game Theory* (Cambridge, MA: MIT Press).

Fudenberg, D. and Levine, D. K. (1998), *The Theory of Learning in Games* (Cambridge, MA: MIT Press).

Ganguli, A., Susca, S., Martinez, S., Bullo, F., and Cortes, J. (2005), "On collective motion in sensor networks: sample problems and distributed algorithms," in *Proceedings of the 44th IEEE Conference on Decision and Control*, Seville, Spain, pp. 4239–4244, December.

Germano, F. and Lugosi, G. (2007), "Global Nash convergence of Foster and Young's regret testing," *Games and Economic Behavior* 60: 135–154.

Gershwin, S. B. (1994), *Manufacturing Systems Engineering* (Prentice-Hall).

Hart, S. (2005), "Adaptive heuristics," *Econometrica* 73 (5): 1401–1430.

Hofbauer, J. and Sigmund, K. (1998), *Evolutionary Games and Population Dynamics* (Cambridge: Cambridge University Press).

Mannor, S. and Shamma, J. S. (2007), "Multi-agent learning for engineers," *Artificial Intelligence* 171: 365-377.

Marden, J. R., Arslan, G., and Shamma, J. S. (2007a), "Connections between cooperative control and potential games illustrated on the consensus problem," in Proceedings of the 2007 European Control Conference (ECC '07), July.

Marden, J. R., Arslan, G., and Shamma, J. S. (2007b), "Regret based dynamics: Convergence in weakly acyclic games," in Proceedings of the 2007 International Conference on Autonomous Agents and Multiagent Systems (AAMAS), Honolulu, Hawaii, May.

Marden, J. R., Arslan, G., and Shamma, J. S. (2009), "Joint strategy fictitious play with inertia for potential games," IEEE Transactions on Automatic Control54: 208–220.
Monderer, D. and Shapley, L.S. (1996), "Fictitious play property for games with identical interests," *Journal of Economic Theory* 68: 258–265.
Montenegro, R. and Tetali, P. (2006), *Mathematical Aspects of Mixing Times in Markov Chains* (Boston, MA: Now Publishers Inc).
Panait, L. and Luke, S. (2005), "Cooperative multi-agent learning: The state of the art," *Autonomous Agents and Multi-Agent Systems* 11: 387–434.
Rosenthal, R. W. (1973), "A class of games possessing pure-strategy Nash equilibria," *International Journal of Game Theory* 2: 65–67.
Samuelson, L. (1997), *Evolutionary Games and Equilibrium Selection* (Cambridge, MA: MIT Press).
Sandholm, W. (2002), "Evolutionary implementation and congestion pricing," *Review of Economic Studies* 69 (3): 667–689.
Shoham, Y., Powers, R., and Grenager, T. (2003), "If multi-agent learning is the answer, what is the question?" *Foundations of Multi-Agent Learning* 171: 365–377. special issue ed. R. Vohra and M. Wellman.
Weibull, J.W. (1995), *Evolutionary Game Theory* (Cambridge, MA: MIT Press).
[24] Young, H. P. (1993), "The evolution of conventions,". *Econometrica*, 61 (1): 57–84, January 1993.
Young, H. P. (1998), *Individual Strategy and Social Structure* (Princeton, NJ: Princeton University Press).
Young, H. P. (2005), *Strategic Learning and its Limits* (Oxford: Oxford University Press).

10
Achieving Pareto Optimality through Distributed Learning

With Jason R. Marden and Lucy Y. Pao

1 Introduction

Game theory has important applications to the design and control of multiagent systems. This design choice requires two steps. First, the system designer must model the system components as "agents" embedded in an interactive, game-theoretic environment. This step involves defining a set of choices and a local objective function for each agent. Second, the system designer must specify the agents' behavioral rules, i.e., the way in which they react to local conditions and information. The goal is to complete both steps in such a way that the agents' behavior leads to desirable system-wide behavior even though the agents themselves do not have access to the information needed to determine the state of the system.[1]

The existing literature primarily focuses on distributed learning algorithms that are suitable for implementation in engineering systems. Accordingly, most of the results in distributed learning are concerned with the attainment of (pure) Nash equilibria for particular classes of games that arise in such systems. For example, Stankovic, Johansson, and Stipanovice (2010), Zhu and Martinez (2010a), and Durr, Stankovic, and Johansson (2011) establish learning algorithms that converge to Nash equilibria for games that are relevant to mobile sensor networks. Arslan, Marden, and Shamma (2007), Marden, Arslan, and Shamma (2009a, 2009b), and Marden, Young, Arslan, and Shamma (2009)

Originally published J. R. Marden, H. P. Young, and L. Y. Pao (2014), "Achieving Pareto optimality through distributed learning," *SIAM J. Control and Optimization* 52(5), 2753–2770. Reproduced with permission.

[1] See among others Komali and MacKenzie 2006; Arslan, Marden, and Shamma 2007; Johari 2007; Marden, Arslan, and Shamma 2009a; Chasparis and Shamma 2011; Gopalakrishnan, Marden, and Wierman 2011; Li and Marden 2013, 2014; Marden 2012.

Spontaneous Order. H. Peyton Young with Jason R. Marden and Lucy Y. Pao, Oxford University Press.
 DOI: 10.1093/oso/9780198892908.003.0010

establish learning algorithms that converge to Nash equilibria for potential games (Monderer and Shapley 1996).[2]

There are, however, two limitations to this body of work. First, as highlighted above, most results in this domain focus on convergence to a Nash equilibrium, which may be very inefficient in achieving the system-level objective. Characterizing this inefficiency is a highly active research area in algorithmic game theory (Nisan, Roughgarden, Tardos, and Vazirani 2007). The second limitation of this framework is that it is frequently impossible to represent the interactions of a given system as a potential game, because engineering systems possess inherent constraints on the types of objective functions that can be assigned to the agents. These constraints are a byproduct of the information available to different components of the system. Furthermore, in many complex systems, the relationship between the behavior of the components and overall system performance is not known with any precision.

One example of a system that exhibits these challenges is the control of a wind farm to maximize total power production (Marden, Ruben, and Pao 2013). Controlling an array of turbines in a wind farm is fundamentally more challenging than controlling a single turbine. The reason is the aerodynamic interactions among the turbines, which render many of the single-turbine control algorithms *highly inefficient* for optimizing total power production (Johnson and Thomas 2009). Here, the goal is to establish a *distributed* control algorithm that enables the individual turbines to adjust their behavior based on local conditions, so as to maximize total system performance. One way to handle this large-scale coordination problem is to model the interactions of the turbines in a game theoretic environment. However, the space of admissible utility functions for the individual turbines is limited because of the following informational constraints:

(i) no turbine has access to the actions[3] of other turbines, due to the lack of a suitable communication system;
(ii) no turbine has access to the functional relationship between the total power generated and the action of the other turbines. The reason is that the aerodynamic interactions between the turbines is poorly understood from an engineering standpoint.

These limitations restrict the ability of the designer to represent the interaction framework as a potential game. For example, one of the common design

[2] See also Krstic, Frihauf, Krieger and Basar 2010; Frihauf, Krstic and Basar 2010, 2011; and Chapter 9.
[3] A turbine's action is called an *axial induction factor*. The axial induction factor indicates the amount of power the turbine extracts from the wind.

approaches is to assign each turbine an objective function that measures the turbine's marginal contribution to the power production of the wind farm, that is, the difference between the total power produced when the turbine is active and the total power produced when the turbine is inactive (Wolpert and Tumor 1999; Marden, Arslan, and Shamma 2009a). This assignment ensures that the resulting interaction framework is a potential game and that the action profile which optimizes the potential function also optimizes the total power production of the wind farm. Calculating this marginal contribution may not be possible due to lack of knowledge about the aerodynamic interactions, hence the existing literature does not provide suitable control algorithms for this situation.

The contribution of this chapter is to demonstrate the existence of simple, completely decentralized learning algorithms that lead to efficient system-wide behavior *irrespective* of the game structure. We measure the efficiency of an action profile by the sum of the agents' utility functions. In a wind farm, this sum is precisely equal to the total power generated. Our main result is the development of a simple payoff-based learning algorithm that guarantees convergence to an efficient action profile whenever the underlying game has finite action sets and generic payoffs. This result holds whether or not this efficient action profile is a Nash equilibrium. It therefore differs from the approach of Pradelski and Young (2012), which shows how to achieve constrained efficiency *within* the set of Nash equilibrium outcomes.

In the prior literature, the principal approaches to distributed or networked optimization are subgradient methods,[4] consensus-based methods,[5] and two-step consensus-based methods.[6] A key difference between our proposed algorithm and these earlier methods is that the latter depend heavily on the particular structure of the objective function and also on the form of the underlying network. For example, common requirements are that the objective function is convex and the communication graph is connected. These stand in sharp contrast to our approach, where there is no communication between the agents and the objective function is completely general.

2 Background

Let G be a finite strategic-form game with n agents. The set of agents is denoted by $N := \{1, \ldots, n\}$. Each agent $i \in N$ has a finite action set $\mathcal{A}_i$ and a utility

[4] Tsitsiklis, Bertsekas, and Athans 1986; Solodov 1998; Blatt, Hero, and Gauchman 2008; Nedic,Olshevsky, Ozdaglar, and Tsitsiklis 2009; Zhu and Martinez 2010b; Lobel and Ozdaglar 2011.

[5] Tsitsiklis and Athans 1984; Jadbabaie, Lin, and Morse 2003; Hatano and Mesbahi 2005.

[6] Androulakis and Reklaitis 1999; Boyd, Parikh, Chu, Peleato, and Eckstein 2010.

function $U_i : \mathcal{A} \to \mathbb{R}$, where $\mathcal{A} = \mathcal{A}_1 \times \cdots \times \mathcal{A}_n$ denotes the joint action set. We shall henceforth refer to a finite strategic-form game simply as "a game." Given an action profile $a = (a_1, a_2, ..., a_n) \in \mathcal{A}$, let a_{-i} denote the profile of agent actions *other than* agent i, that is., $a_{-i} = (a_1, ..., a_{i-1}, a_{i+1}, ..., a_n)$. With this notation, we shall sometimes denote a profile a of actions by (a_i, a_{-i}) and $U_i(a)$ by $U_i(a_i, a_{-i})$. We shall also let $\mathcal{A}_{-i} = \prod_{j \neq i} \mathcal{A}_j$ denote the set of possible collective actions of all agents other than agent i. The *welfare* of an action profile $a \in \mathcal{A}$ is defined as

$$W(a) = \sum_{i \in N} U_i(a).$$

An action profile that optimizes the welfare will be denoted by $a^{\text{opt}} \in \arg\max_{a \in \mathcal{A}} W(a)$.

2.1 Repeated games

We shall assume that a given game G is repeated one each period $t \in \{0, 1, 2, ...\}$. In period t, the agents simultaneously choose actions $a(t) = (a_1(t), ..., a_n(t))$ and receive payoffs $U_i(a(t))$. Agent $i \in N$ chooses the action $a_i(t)$ according to a probability distribution $p_i(t) \in \Delta(\mathcal{A}_i)$, which is the simplex of probability distributions over $\mathcal{A}_i$. We shall refer to $p_i(t)$ as the *strategy* of agent i at time t. We adopt the convention that $p_i^{a_i}(t)$ is the probability that agent i selects action a_i at time t according to the strategy $p_i(t)$. An agent's strategy at time t relies only on observations from times $\{0, 1, 2, ..., t-1\}$.

Different learning algorithms are specified by the agents' information and the mechanism by which their strategies are updated as information is gathered. Suppose, for example, that an agent knows his own utility function and is capable of observing the actions of all other agents at every time step but does not know their utility functions. Then the strategy adjustment mechanism of a given agent i can be written in the form

$$p_i(t) = F_i(a(0), ..., a(t-1); U_i). \tag{1}$$

Such an algorithm is said to be *uncoupled* (Hart and Mas-Colell 2003, 2006).

In this chapter we ask whether agents can learn to play the welfare maximizing action profile under even more restrictive observational conditions. In particular, we shall assume that agents *only* have access to: (i) the action they played and (ii) the payoff they received. In this setting, the strategy adjustment mechanism of agent i takes the form

$$p_i(t) = F_i\left(\{a_i(\tau), U_i(a(\tau))\}_{\tau=0,\ldots,t-1}\right). \tag{2}$$

Such a learning rule is said to be *completely uncoupled* or *payoff-based* (Foster and Young 2006). Recent work has shown that for finite games with generic payoffs there exist completely uncoupled learning rules that lead to Pareto optimal Nash equilibria (Pradelski and Young 2012; see also Fudenberg and Maskin 1986; Young 2009; Arieli and Babichenko 2012). Here we exhibit a different class of learning procedures that lead to Pareto optimal outcomes whether or not they are Nash equilibria.[7]

3 A Payoff based Algorithm for Maximizing Welfare

Our proposed algorithm is a variant of the approach described in Chapter 7 and is closely related to the method introduced by Pradelski and Young (2012). The key feature of these learning rules is that each agent possesses an internal state variable (called a 'mood) that affects the agent's behavior, where changes in mood are triggered by particular patterns in the agent's realized payoffs. The key difference between our algorithm and the one in Pradelski and Young (2012) is that the latter guarantees convergence to a Pareto Nash equilibrium, whereas our proposed algorithm converges to a Pareto (efficient) action profile irrespective of whether or not this action profile is a Nash equilibrium. Furthermore, our algorithm uses fewer state variables than the method in Pradelski and Young (2012).

At each point in time an agent's *state* can be represented as a triple $[\bar{a}_i, \bar{u}_i, m_i]$, where

- The *benchmark action* is $\bar{a}_i \in \mathcal{A}_i$.
- The *benchmark payoff* is $\bar{u}_i$, which is in the range of $U_i(\cdot)$.
- The *mood* is m_i, which can take on two values: *content* (C) and *discontent* (D).

The learning algorithm produces a sequence of action profiles $a(1), \ldots, a(t)$, where the behavior of an agent i in each period $t = 1, 2, \ldots$, is conditioned on agent i's underlying benchmark payoff $\bar{u}_i(t)$, benchmark action $\bar{a}_i(t)$, and mood $m_i(t) \in \{C, D\}$.

[7] Such a result might seem reminiscent of the Folk Theorem, which specifies conditions under which an efficient action profile can be implemented as an equilibrium of a repeated game (Fudenberg and Maskin 1986). In the present context, however, we are interested in whether agents can learn to play an efficient action profile without having any information about the game as a whole or what the other agents are doing. Hence they cannot condition their behavior on the observed behavior of others, which is a key requirement of most repeated game equilibria.

We divide the dynamics into the following two parts: the agent dynamics and the state dynamics. Without loss of generality we shall focus on the case where agent utility functions are strictly bounded between 0 and 1, i.e., for any agent $i \in N$ and action profile $a \in \mathcal{A}$ we have $1 > U_i(a) \geq 0$. Consequently, for any action profile $a \in \mathcal{A}$, the welfare function satisfies $n > W(a) \geq 0$.

Agent Dynamics: Fix an experimentation rate $\epsilon > 0$ and constant $c > n$. Let $[\bar{a}_i, \bar{u}_i, m_i]$ be the current state of agent i.

- **Content** ($m_i = C$): In this state, the agent chooses an action a_i according to the following probability distribution

$$p_i^{a_i} = \begin{cases} \dfrac{\epsilon^c}{|\mathcal{A}_i|-1} & \text{for } a_i \neq \bar{a}_i \\ 1-\epsilon^c & \text{for } a_i = \bar{a}_i \end{cases} \tag{3}$$

where $|\mathcal{A}_i|$ represents the cardinality of the set $\mathcal{A}_i$.

- **Discontent** ($m_i = D$): In this state, the agent chooses an action a_i according to the following probability distribution:

$$p_i^{a_i} = \frac{1}{|\mathcal{A}_i|} \text{ for every } a_i \in \mathcal{A}_i \tag{4}$$

Note that the benchmark action and utility play no role in the agent dynamics when the agent is discontent.

State Dynamics: Once the agent selects an action $a_i \in \mathcal{A}_i$ and receives the payoff $u_i = U_i(a_i, a_{-i})$, where a_{-i} is the action selected by all agents other than agent i, the state is updated as follows:

- **Content** ($m_i = C$): If $[a_i, u_i] = [\bar{a}_i, \bar{u}_i]$ the new state is determined by the transition

$$[\bar{a}_i, \bar{u}_i, C] \xrightarrow{[\bar{a}_i, \bar{u}_i]} [\bar{a}_i, \bar{u}_i, C]. \tag{5}$$

If $[a_i, u_i] \neq [\bar{a}_i, \bar{u}_i]$ the new state is determined by the transition

$$[\bar{a}_i, \bar{u}_i, C] \xrightarrow{[a_i, u_i]} \begin{cases} [a_i, u_i, C] & \text{with prob } \epsilon^{1-u_i} \\ [a_i, u_i, D] & \text{with prob } 1-\epsilon^{1-u_i}. \end{cases}$$

- **Discontent** ($m_i = D$): If the selected action and received payoff are $[a_i, u_i]$, the new state is determined by the transition

$$[\bar{a}_i, \bar{u}_i, D] \xrightarrow{[a_i,u_i]} \begin{cases} [a_i, u_i, C] & \text{with prob } \epsilon^{1-u_i} \\ [a_i, u_i, D] & \text{with prob } 1-\epsilon^{1-u_i}. \end{cases}$$

To ensure that the dynamics converge to an efficient action profile, we require the following notion of interdependence in the game structure (see Chapter 7 (Young 2009)).

Definition 1 (Interdependence). *An n-person game G on the finite action space $\mathcal{A}$ is interdependent if, for every $a \in \mathcal{A}$ and every proper subset of agents $J \subset N$, there exists an agent $i \notin J$ and a choice of actions $a'_J \in \prod_{j \in J} \mathcal{A}_j$ such that $U_i\left(a'_J, a_{-J}\right) \neq U_i\left(a_J, a_{-J}\right)$.*

Roughly speaking, the interdependence condition states that it is not possible to divide the agents into two distinct subsets that do not mutually interact with one another.

These dynamics induce a Markov process over the finite state space $Z = \prod_{i \in N}(\mathcal{A}_i \times \mathcal{U}_i \times M)$, where $\mathcal{U}_i$ denotes the finite range of $U_i\,(a)$ over all $a \in \mathcal{A}$ and $M = \{C, D\}$ is the set of moods. We shall denote the transition probability matrix by P^ϵ for each $\epsilon > 0$. Computing the stationary distribution of this process is challenging because of the large number of states and the fact that the underlying process is not reversible. Accordingly, we shall focus on characterizing the *support* of the limiting stationary distribution, whose elements are as the *stochastically stable states* (Young 1993). More precisely, a state $z \in Z$ is stochastically stable if and only if $\lim_{\epsilon \to 0^+} \mu(z, \epsilon) > 0$ where $\mu(z, \epsilon)$ is the stationary distribution of the process P^ϵ for a fixed $\epsilon > 0$.

Theorem 1 Let G be an interdependent n-person game on a finite joint action space $\mathcal{A}$. Under the dynamics defined above, a state $z = [\bar{a}, \bar{u}, m] \in Z$ is stochastically stable if and only if the following conditions are satisfied:

(i) *The action profile $\bar{a}$ optimizes $W(\bar{a}) = \sum_{i \in N} U_i(\bar{a})$.*

(ii) *The benchmark actions and payoffs are aligned, i.e., $\bar{u}_i = U_i\,(\bar{a})$ for all i.*

(iii) *The mood of each agent is content, i.e., $m_i = C$ for all i.*

4 Proof of Theorem 1

The proof relies on the theory of resistance trees for regular perturbed Markov decision processes as described in the Appendix to Chapter 2. For convenience we briefly review the key elements that we shall need to establish Theorem 1. Let P^0 denote the probability transition matrix of a finite state Markov chain on the state space Z. Consider a "perturbed" process P^ϵ on Z where the "size" of the perturbations can be indexed by a scalar $\epsilon > 0$. The process P^ϵ is a *regular perturbed Markov process* if P^ϵ is ergodic for all sufficiently small $\epsilon > 0$ and P^ϵ approaches P^0 at an exponentially smooth rate, that is,

$$\forall z, z' \in Z, \quad \lim_{\epsilon \to 0^+} P^\epsilon_{zz'} = P^0_{zz'},$$

and

$$\forall z, z' \in Z, \quad P^\epsilon_{zz'} > 0 \text{ for some } \epsilon > 0 \Rightarrow 0 < \lim_{\epsilon \to 0^+} \frac{P^\epsilon_{zz'}}{\epsilon^{r(z \to z')}} < \infty,$$

where $r(z \to z')$ is a nonnegative real number called the *resistance* of the transition $z \to z'$. (Note in particular that if $P^0_{zz'} > 0$ then $r(z \to z') = 0$.)

Let the recurrence classes of P^0 be denoted by $E_1, E_2, \ldots, E_M$. For each pair of distinct recurrence classes E_i and E_j, $i \neq j$, an *ij-path* is defined to be a sequence of distinct states $\zeta = (z_1 \to z_2 \to \ldots \to z_m)$ such that $z_1 \in E_i$ and $z_m \in E_j$. The *resistance* of this path is the sum of the resistances of its edges, that is,

$$r(\zeta) = r(z_1 \to z_2) + r(z_2 \to z_3) + \ldots + r(z_{m-1} \to z_m).$$

Let $\rho_{ij} = \min r(\zeta)$ be the least resistance over all *ij*-paths ζ. Note that ρ_{ij} must be positive for all distinct i and j, because there exists no path of zero resistance between distinct recurrence classes.

Now construct a complete directed graph with M vertices, one for each recurrence class. The vertex corresponding to class E_j will be called j. The weight on the directed edge $i \to j$ is ρ_{ij}. A *j-tree* T is a set of $M - 1$ directed edges such that, from every vertex different from j, there is a unique directed path in the tree to j. The resistance of such a tree is the sum of the resistances on the $M - 1$ edges that compose it. The *stochastic potential*, γ_j, of the recurrence class E_j is the minimum resistance over all trees rooted at j. The following result provides a simple criterion for determining the stochastically stable states (see Chapter 2, Theorem 4).

Let P^ϵ be a regular perturbed Markov process, and for each $\epsilon > 0$ let μ^ϵ be the unique stationary distribution of P^ϵ. Then $\lim_{\epsilon \to 0} \mu^\epsilon$ *exists and the limiting distribution μ^0 is a stationary distribution of P^0. The stochastically stable* *states (i.e., the support of μ^0) are precisely those states contained in the recurrence classes with minimum stochastic potential.*[8]

It can be verified that the dynamics introduced above define a regular perturbed Markov process. The proof of Theorem 1 proceeds by a series of lemmas. Let C^0 be the subset of states in which each agent is content and the benchmark action and utility are aligned. That is, if $[\bar{a}, \bar{u}, m] \in C^0$ then $u_i = U_i(\bar{a})$ and $m_i = C$ for each agent $i \in N$. Let D^0 represent the set of states in which everyone is discontent. That is, if $[\bar{a}, \bar{u}, m] \in D^0$ then $\bar{u}_i = U_i(\bar{a})$ and $m_i = D$ for each agent $i \in N$.

The first lemma provides a characterization of the recurrence classes of the unperturbed process P^0.

Lemma 1 *The recurrence classes of the unperturbed process P^0 are D^0 and all singletons $z \in C^0$.*

Proof The set of states D^0 represents a single recurrence class of the unperturbed process since the probability of transitioning between any two states $z_1, z_2 \in D^0$ is $O(1)$ and when $\epsilon = 0$ there is no possibility of exiting from D^0.[9] Any state $[\bar{a}, \bar{u}, C] \in C^0$ is a recurrent class of the unperturbed process, because all agents will continue to play their baseline action at all future times.

We will now show that the states D^0 and all singletons $z \in C^0$ represent the *only* recurrent states. Suppose that a proper subset of agents $S \subset N$ is discontent and the benchmark actions and benchmark utilities of all other agents are a_{-S} and u_{-S} respectively. By interdependence, there exists an agent $j \notin S$ and an action tuple $a'_S \in \prod_{i \in S} \mathcal{A}_i$ such that $u_j \neq U_j(a'_S, a_{-S})$. This situation cannot be a recurrence class of the unperturbed process because the agent set S will eventually play action a'_S with probability 1, thereby causing agent j to become discontent. Agent set S will eventually play action a'_s with probability 1, because each agent $i \in S$ is discontent and hense selects actions uniformly for the action set $\mathcal{A}_i$. Consequently, at each subsequent period, the action a'_s will be played with probability $1 / |\mathcal{A}_s|$. Note that once the agent set S selects action a'_s, the payoff of agent j will be different from agent j's baseline utility, i.e., $\bar{u}_j \neq U_j(a'_S, \bar{a}_{-S})$, thereby causing j to become discontent. This process can be

[8] In section 6, we illustrate how to compute the resistances and stochastic potential of each state in several concrete examples.

[9] The notation $O(1)$ refers to transition probabilities that are bounded away from 0. For the situation highlighted above, the probability of the transition $z_1 \to z_2$ is $1/|\mathcal{A}|$. The notation $O(\epsilon)$ refers to transition probabilities that are on the order of ϵ.

repeated to show that all agents will eventually become discontent with probability $O(1)$; hence any state that consists of a partial collection of discontent agents $S \subset N$ is not a recurrence class of the unperturbed process.

Lastly, consider a state $[a, u, C]$ where all agents are content but there exists at least one agent i whose benchmark action and benchmark utility are not aligned, i.e., $u_i \neq U_i(a)$. For the unperturbed process, at the ensuing time step the action profile a will be played and agent i will become discontent since $u_i \neq U_i(a)$. Since one agent is discontent, all agents will eventually become discontent. This completes the proof of Lemma 1.

We know from Chapter 2, Theorem 4 that the computation of the stochastically stable states can be reduced to an analysis of rooted trees on the vertex set consisting solely of the recurrence classes. We denote the collection of states D^0 by a single variable D to represent this single recurrence class since the exit probabilities are the same for all states in D^0. By Lemma 1, the set of recurrence classes consists of the singleton states in C^0 and also the singleton state D. Accordingly, we represent a state $z \in C^0$ by just $[\bar{a}, \bar{u}]$ and drop the extra notation highlighting that the agents are content. We now reiterate the definition of edge resistance.

Definition 2 (Edge resistance). *For every pair of distinct recurrence classes w and z, let $r(w \to z)$ denote the total resistance of the least-resistance path that starts in w and ends in z. We call $w \to z$ an edge and $r(w \to z)$ the resistance of the edge.*

Let $z = [\bar{a}, \bar{u}]$ and $z' = [\bar{a}', \bar{u}']$ be any two distinct states in C^0. The following observations will be useful.

(i) The resistance of the transition $z \to D$ satisfies

$$r(z \to D) = c.$$

To see this, consider any state $z \in C^0$. In order to trasition out of the state z, at least one agent needs to experiment, which happens with a probability $O(\epsilon^c)$. This experimenting agent will become discontent at the ensuing step with probability $O(1)$. Given this event, Lemma 1 implies that all agents will become discontent with probability $O(1)$. Hence, the resistance of the transition $z \to D$ equals c.

(ii) The resistance of the transition $D \to z$ satisfies

$$r(D \to z) = \sum_{i \in N} (1 - u_i) = n - W(a).$$

According to the state dynamics, transitioning from discontent to content requires that each agent must accept the benchmark payoff $\bar{u}_i$, which has a resistance $(1 - \bar{u}_i)$. Consequently, the resistane associated with this transition is $\sum_{i \in N}(1 - \bar{u}_i) = n - W(\bar{a})$.

(iii) The resistance of the transition $z \to z'$ satisfies

$$c \leq r\left(z \to z'\right) < 2c.$$

This holds because $r(z \to z') \leq r(z \to D) + r(D \to z')$ by the definition of edge resistance. Therefore, each transition of minimum resistance includes at most one agent who experiments.

The following lemma characterizes the stochastic potential of the states in C^0. Before stating this lemma, we define a *path* $\mathcal{P}$ over the states $D \cup C^0$ to be a sequence of edges of the form

$$\mathcal{P} = \left\{z^0 \to z^1 \to \ldots \to z^m\right\},$$

where each z^k for $k \in \{0, 1, \ldots, m\}$ is in $D \cup C^0$. The *resistance* of a path $\mathcal{P}$ is the sum of the resistance of each edge in the path, i.e., $R(\mathcal{P}) = \sum_{k=1}^{m} r\left(z^{k-1} \to z^k\right)$.

Lemma 2 The stochastic potential of any state $z = [\bar{a}, \bar{u}]$ in C^0 is

$$\gamma(z) = c\left(\left|C^0\right| - 1\right) + \sum_{i \in N} (1 - \bar{u}_i). \tag{6}$$

Proof We first prove that (5) is an upper bound for the stochastic potential of z by constructing a tree rooted at z with the prescribed resistance. To that end, consider the tree T with the following properties:

P-1: The edge exiting each state $z' \in C^0 \backslash [z]$ is of the form $z' \to D$. The total resistance associated with these edges is $c\left(\left|C^0\right| - 1\right)$.

P-2: The edge existing the state D is of the form $D \to z$. The resistance associated with this edge is $\sum_{i \in N}(1 - \bar{u}_i)$.

The tree T is rooted at z and has total resistance $c\left(\left|C^0\right| - 1\right) + \sum_{i \in N} (1 - \bar{u}_i)$. It follows that $\gamma(z) = c\left(\left|C^0\right| - 1\right) + \sum_{i \in N}(1 - \bar{u}_i)$, hence (5) holds as an inequality. It remains to be shown that the right-hand side of (5) is also a lower bound for the stochastic potential.

We argue this by contradiction. Suppose there exists a tree T rooted at z with resistance $R(T) < c\left(|C^0| - 1\right) + \sum_{i \in N}(1 - \bar{u}_i)$. Since the tree T is rooted at z we know that there exists a path $\mathcal{P}$ from D to z of the form

$$\mathcal{P} = \left\{D \to z^1 \to z^2 \to \ldots \to z^m \to z\right\},$$

where $z^k \in C^0$ for each $k \in \{1, \ldots, m\}$. We claim that the resistance associated with this path of $m + 1$ transitions satisfies

$$R(\mathcal{P}) \geq mc + \sum_{i \in N} (1 - \bar{u}_i).$$

The term mc comes from applying observation (iii) to the last m transitions on the path $\mathcal{P}$. The term $\sum_{i \in N}(1 - \bar{u}_i)$ comes from the fact that each agent needs to accept $\bar{u}_i$ as the benchmark payoff at some point during the transitions.

Construct a new tree T' still rooted at z by removing the edges in $\mathcal{P}$ and adding the following edges:

- $D \to z$ which has resistance $\sum_{i \in N}(1 - \bar{u}_i)$.
- $z^k \to D$ for each $k \in \{1, \ldots, m\}$ which has total resistance mc.

The new tree T' is still rooted at z and has a total resistance that satisfies $R(T') \leq R(T)$. Note that if the path $\mathcal{P}$ was of the form $D \to z$, then this augmentation does not alter the tree structure.

Now suppose that there exists an edge $z' \to z''$ in the tree T' for some states $z', z'' \in C^0$. By observation (iii) the resistance of this edge satisfies $r(z' \to z'') \geq c$. Construct a new tree T'' by removing the edge $z' \to z''$ and adding the edge $z' \to D$, which has a resistance c. This new tree T'' is rooted at z, and its resistance satisfies

$$\begin{aligned} R(T'') &= R(T') + r(z' \to D) - r(z' \to z'') \\ &\leq R(T') \\ &\leq R(T). \end{aligned}$$

Repeat this process until we have constructed a tree T^* for which no such edges exist. Note that the tree T^* satisfies properties P-1 and P-2 and consequently has a total resistance $R(T^*) = c\left(|C^0| - 1\right) + \sum_{i \in N}(1 - \bar{u}_i)$. Since by construction $R(T^*) \leq R(T)$ we have a contradiction. This completes the proof of Lemma 2.

We will now prove Theorem 1 by analyzing the minimum resistance trees using the above lemmas. We first show that the state D is not stochastically stable. Suppose, by way of contradiction, that there exists a minimum resistance tree T rooted at the state D. Then there exists an edge in the tree T of the form $z \to D$ for some state $z \in C^0$. The resistance of this edge is c and the resistance of the opposing edge $D \to z$ (not in the tree T) is strictly less than n. Note that the only way the resistance of this opposing edge $D \to z = [\bar{a}, \bar{u}]$ could equal n is if $\bar{u}_i = 0$ for all $i \in N$. By interdependence, this is not possible for all $z \in C^0$.

Create a new tree T' rooted at z by removing the edge $z \to D$ from T and adding the edge $D \to z$. Therefore

$$\begin{aligned} R(T') &= R(T) + r(D \to z) - r(z \to D) \\ &\leq R(T) + n - c \\ &< R(T). \end{aligned}$$

Hence T is not a minimum resistance tree. This contradiction shows that the state D is not stochastically stable. It follows that all the stochastically stable states are contained in the set C^0.

From Lemma 2 we know that a state $z = [\bar{a}, \bar{u}]$ in C^0 is stochastically stable if and only if

$$\bar{a} \in \arg\min_{a^* \in \mathcal{A}} \left\{ c\left(|C^0| - 1\right) + \sum_{i \in N} (1 - U_i(a^*)) \right\},$$

equivalently

$$\bar{a} \in \arg\max_{a^* \in \mathcal{A}} \left\{ \sum_{i \in N} U_i(a^*) \right\}.$$

Therefore, a state is stochastically stable if and only if the action profile is efficient. This completes the proof of Theorem 1.

5 Relaxing Interdependence

In this section we focus on whether the interdependence condition in Definition 1 can be relaxed while ensuring that the stochastically stable states remain efficient. Recall that a game is interdependent if it is not possible to

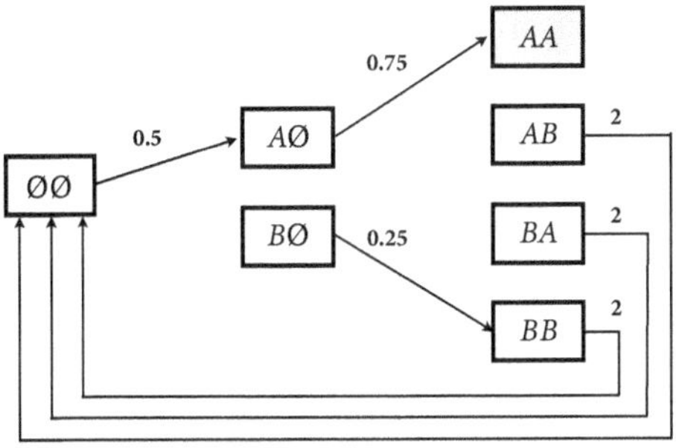

Figure 10.1 Illustration of the minimum resistance tree rooted at the state AA.

partition the agents into two distinct groups S and $N \backslash S$ that do not mutually interact with one another. One way that this condition can fail is that the game can be broken into two completely separate sub-games that can be analyzed independently. In this case our algorithm ensures that in each sub-game the only stochastically stable states are the efficient action profiles. Hence, this remains true in the full game.

In general, however, some version of interdependence is needed. To see why, consider the following two-player game:

	A	*B*
A	1/2, 1/4	1/2, 0
B	1/4, 0	1/4, 3/4

Here, the row agent affects the column agent but the reverse is not true. Consequently, the recurrent states of the unperturbed process are {*AA*, *AB*, *BA*, *BB*, *A*∅, *B*∅, ∅∅} where: *A*∅ is the state where agent 1 is content with action profile *A* and agent 2 is discontent; ∅∅ is the state where both agents are discontent. We claim that the action profile *AA*, which is not efficient, is stochastically stable. This can be deduced from Figure 10.1 (here we choose $c = n = 2$). The illustrated resistance tree has minimum stochastic potential because each edge in the given tree has minimum resistance among the edges exiting from that vertex. Consequently, this inefficient action profile *AA* is stochastically stable.

At first glance, this example merely demonstrates that our proposed algorithm does not guarantee convergence to the efficient action profile for all finite strategic-form games. However, it turns out that this example also establishes that there does not exist a distributed learning algorithm that guarantees convergence to the efficient action profile. The following proposition makes this precise.

Proposition 1 *There exists no uncoupled learning algorithm that leads to an efficient action profile for all finite strategic-form games.*[10]

Proof. We will prove this proposition by contradiction. Suppose that there exists an uncoupled learning algorithm of the form (1) that leads to an efficient action profile for all finite strategic-form games. When considering the two-player game highlighted above, this algorithm will lead behavior to the action profile (BB). However, from player 1's perspective, this game is equivalent to a simple one-player game with payoffs:

A	1/2
B	1/4

This learning algorithm leads to the behavior A in this one-player game. Therefore, we have a contradiction since the same learning algorithm is not able to ensure that behavior leads to A for the one-player setting and (BB) for the two-player setting.

6 Illustrations

In this section, we provide two simulations that illustrate the mechanics of our proposed algorithm. In the first subsection, we apply our results to a prisoner's dilemma game and provide a detailed analysis of the stochastically stable states. In the second subsection, we simulate our algorithm on a three-player game that exhibits many of the same challenges associated with the prisoner's dilemma game.

6.1 Prisoner's dilemma

Consider the following prisoner's dilemma game, highlighted in Figure 10.2, where all players' utilities are scaled between 0 and 1. It is easy to verify that these payoffs satisfy the interdependence condition. Consequently, our

[10] Here, we use the term "lead to" to mean either convergence, almost sure convergence, or convergence in the sense of stochastic stability. The authors would like to acknowledge conversations with Yakov Babichenko which led to this result.

	A	*B*
A	3/4, 3/4	0,4/5
B	4/5, 0	1/3, 1/3

Figure 10.2 A two-player strategic-form game in which both player 1 (row player) and player 2 (column player) choose either A or B. (B, B) is the unique pure Nash equilibrium.

algorithm guarantees that the action profile *AA* is the only stochastically stable state. We will now verify this by computing the resistances for each of the transitions. The recurrence classes of the unperturbed process are (*AA*, *AB*, *BA*, *BB*, $\phi\phi$), where the agents are content for the four listed action profiles and $\phi\phi$ corresponds to the scenario where both agents are discontent. (For notational simplicity, we omit the baseline utilities for each of the four action profiles.)

Consider the transition $AA \to BB$. Its resistance is

$$r(AA \to BB) = c + (1 - 1/3) + (1 - 1/3) = c + 4/3.$$

The term c comes from the fact that we have only one experimenter. The term $2(1 - 1/3)$ results from the fact that both agents 1 and 2 need to accept the new benchmark payoff of 1/3 to make this transition. For the sake of concreteness, let $c = n = 2$ for the remainder of this section. The resistances of all possible transitions are shown in Table 10.1. Each entry in the table represents the resistance going from the row-state to the column-state.

Table 10.1 Evaluation of resistances for prisoner's dilemma game.

	AA	*AB*	*BA*	*BB*	$\phi\phi$
AA	.	2 + (1–4/5) + (1–0) = 16/5	2 + (1–4/5) + (1–0) = 16/5	2 + 2(1–1/3) = 10/3	2
AB	2 + 2(1–3/4) = 5/2	.	2 + (1–4/5) + (1–0) = 16/5	2 + 2(1–1/3) = 10/3	2
BA	2 + 2(1–3/4) = 5/2	2 + (1–4/5) + (1–0) = 16/5	.	2 + 2(1–1/3) = 10/3	2
BB	2 + 2(1–3/4) = 5/ 2	2 + (1–4/5) + (1–0) = 16/5	2 + (1–4/5) + (1–0) = 16/5	.	2
$\phi\phi$	2(1–3/4) = 1/2	(1–4/5) + (1–0) = 6/5	(1–4/5) + (1–0) = 6/5	2(1–1/3) = 4/3	

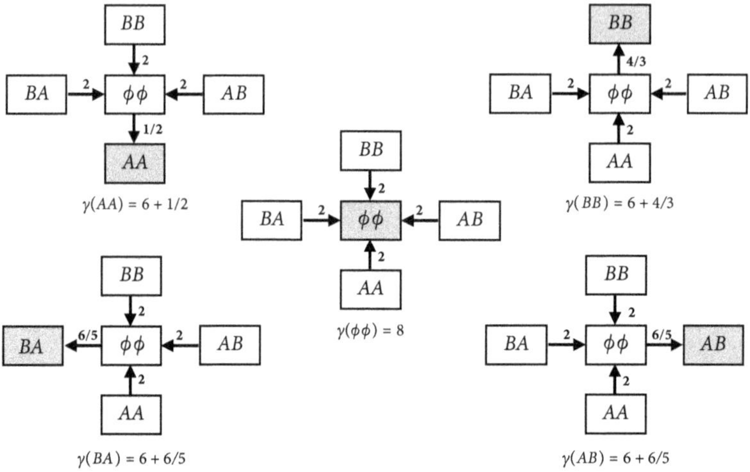

Figure 10.3 Stochastic potential for each state in the prisoner's dilemma game.

The stochastic potential of each of the five states can be evaluated by analyzing the trees rooted at each state. These are shown in Figure 10.3. Note that each of the minimum resistance trees has the very simple structure identified in Lemma 2. It is evident that *AA* has minimum stochastic potential, hence it is the unique stochastically stable state.

Figure 10.4 illustrates a simulation of our learning algorithm on the prisoner's dilemma game defined above. The figure on the left-hand side of Figure 10.4 shows a single run of the algorithm using parameters $\varepsilon = 0.001$, $c = 2$, and an initial state $[\bar{a}_i = B, \bar{u}_i = 1/3, m_i = C]$ for each agent $i \in \{1, 2\}$. Observe that the cumulative empirical frequency distribution eventually puts most of the weight on the efficient joint action *AA*, as predicted by Theorem 1. The table on the right-hand side of Figure 10.4 identifies how the experimentation rate ε affects both the efficiency of the limiting behavior and the underlying convergence rate. The second column shows how the expected welfare varies with ε. The third column shows the average hitting time, i.e., the number of iterations it takes for the system to transition from the initial state highlighted above to the optimal state $[\bar{a}_i = A, \bar{u}_i = 3/4, m_i = C]$ for each $i \in N$. The average hitting time, as with the forthcoming data, was computed from 100,000 runs of the algorithms starting from the specified initial condition. The key point is that nearly optimal behavior is attained in a reasonable length of time when ε is small but not extremely small. For example, if we let $\varepsilon = 0.01$, the emergent system behavior is 97.2 percent of the optimal welfare and the expected hitting time is approximately 6,000 iterations.

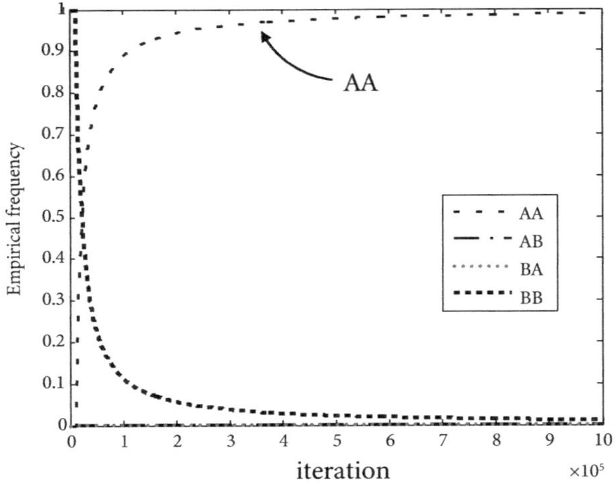

ϵ	Expected Welfare (at stationary distribution)	Average Hitting Time (AA and all agents content)
0.1	0.98 (65.6% of opt)	138
0.05	1.39 (92.9% of opt)	380
0.02	1.44 (95.7% of opt)	1707
0.01	1.46 (97.2% of opt)	5934
0.001	1.49 (99.4% of opt)	512460

Figure 10.4 Simulation results for the two- player prisoner's dilemma game. The figure on the left shows a single run of the algorithm using parameters $\varepsilon = 0.001$, c = 2, and an initial state $[\bar{a}_i = B, \bar{u}_i = 1/3, m_i = C]$ for each agent $i \in N$. The table on the right highlights how the hitting time and the expected welfare associated with the stationary distribution depends on the experimentation rate ε. Observe that nonlimiting values of ϵ achieve desirable behavior with respect to both the efficiency of the limiting behavior and the hitting time to reach a content state at the efficient joint action AA.

For comparative purposes, the following table shows the average hitting time for the other states (*AB*, *BA*, *BB*) when starting from the state *AA*:[11]

Average hitting times starting from state AA content

ε	AA	AB (Content)	BA (Content)	BB (Content)
0.1	X	312	303	246
0.05	X	1509	1514	1277
0.02	X	14921	15046	14076
0.01	X	93913	94413	99152
0.001	X	5.06 E+7	5.04 E+7	8.34 E+7

Note that the hitting times associated with the desirable state *AA* are drastically different than the hitting times associated with the undesirable states (*BA*, *AB*, *BB*). For example, when $\varepsilon = 0.01$, the average hitting time for the state *AA* when starting from the state *BB* is approximately 6,000 as shown in Figure 10.4. Alternatively, the average hitting time for the state *BB* when starting from the state *AA* is approximately 100,000. Accordingly, the rate of transitioning from undesirable states to desirable states is much greater than the rate of transitioning from desirable states to undesirable states.

The following table provides a more direct comparison of the average hitting times associated with the four states (*AA*, *AB*, *BA*, *BB*) by setting the initial state to be discontent:

Average hitting times starting from discontent state

ε	AA (Content)	AB (Content)	BA (Content)	BB (Content)
0.1	78	266	270	206
0.05	158	1320	1337	1065
0.02	426	13826	13815	12989
0.01	911	91082	89584	93326
0.001	1.13 E+4	5.02 E+7	5.03 E+7	8.18 E+7

Observe that the average hitting time for the state *AA* is significantly shorter than the average hitting time for the other states (*BA*, *AB*, *BB*) even for intermediate values of ε.

The next table highlights the stationary distribution associated with the main states of our algorithm (*AA*, *AB*, *BA*, *BB*) and the discontent state as a function of ε. As predicted by Theorem 1, the majority of the mass of the stationary distribution is on the desirable state *AA* as ε goes to 0. Furthermore,

[11] Here, we use the term "lead to" to mean either convergence, almost sure convergence, or convergence in the sense of stochastic stability. The authors would like to acknowledge conversations with Yakov Babichenko which led to this result.

observe that the mass of the stationary distribution on the desirable state *AA* is significantly greater than the mass of the stationary distribution associated with the other states even for intermediate values of ε.

Stationary distribution associated with main states

ε	AA (Content)	AB (Content)	BA (Content)	BB (Content)	Discontent
0.1	0.6216	0.124	0.124	0.0912	0.0391
0.05	0.7406	0.091	0.091	0.061	0.0165
0.02	0.8523	0.0551	0.0551	0.0327	0.0048
0.01	0.9065	0.0361	0.0361	0.0195	0.0008
0.001	0.9812	0.0078	0.0078	0.0031	0.0001

We also point out that the earlier algorithms in this literature (Chapter 7 (Young 2009); Pradelski and Young 2012) produce quite different results as they lead to the inefficient action profile *BB*. The reason for these different asymptotic behaviors stems from the structural differences between the algorithms. In particular, Young (2009) and Pradelski and Young (2012) make use of four mood variables for the agents (content, discontent, hopeful, and watchful) instead of the two mood variables as in our approach. The transitional moods (hopeful and watchful) have the effect of making pure Nash equilibria more attractive than alternative action profiles, including efficient nonequilibrium profiles.

6.2 A three-player version of the prisoner's dilemma game

Consider the three-player version of the prisoner's dilemma game in Figure 10.5 where all players' utilities are scaled between 0 and 1. The efficient action profile is *AAA* and the total system welfare is 2.85. Figure 10.6 illustrates a simulation of our learning algorithm on this three-player prisoner's dilemma

	A	*B*
A	0.95, 0.95, 0.95	0, 0.99, 0
B	0.99, 0, 0	0.1, 0, 0

A

	A	*B*
A	0, 0, 0.99	0, 0, 0.1
B	0, 0.1, 0	0.2, 0.2, 0.2

B

Figure 10.5 A three-player strategic-form game in which player 1 (row player), player 2 (column player), and player 3 {box player, i.e., the player that chooses if we are in the payoff matrix on the left or right} choose either A or B. Observe that (B, B, B) is the unique pure Nash equilibrium.

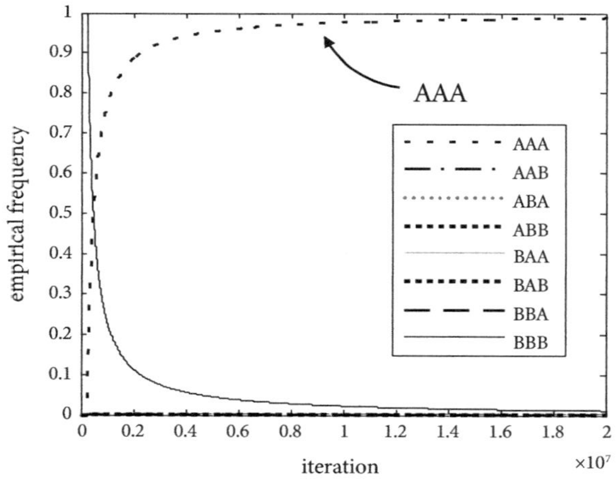

ε	Expected Welfare (at stationary distribution)	Average Hitting Time (AAA and all agents content)
0.2	2.21 (77.6% of opt)	115
0.1	2.69 (94.4% of opt)	513
0.05	2.82 (98.8% of opt)	3070
0.025	2.84 (99.7% of opt)	22168
0.01	2.85 (100% of opt)	334990

Figure 10.6 Simulation results for the three-player prisoner's dilemma game. The figure on the left shows a single run of the algorithm using parameters $\varepsilon = 0.01$, $c = 3$, and an initial state $[\bar{a}_i = B, \varepsilon_i = 0.2, m_i = C]$ for each agent $i \in N$. The table on the right highlights how the hitting time and the expected wel fare associated with the stationary distribution depends on the experimentation rate ε. Observe that nonlimiting values of ε achieve desirable behavior with respect to both the efficiency of the limiting behavior and the hitting time to reach a content state at the efficient joint action (AAA).

game. The figure on the left in Figure 10.6 shows a single run of the algorithm using the parameters $\varepsilon = 0.01$, $c = 3$, and the initial state $[\bar{a}_i = B, \bar{u}_i = 0.2, m_i = \mathrm{C}]$ for each $i \in N$. Observe that the cumulative empirical frequency distribution puts most of the weight on the efficient joint action *AAA*. The table on the right in Figure 10.6 shows how the experimentation rate ε affects both the efficiency of the limiting behavior and the underlying convergence rate. As with the two-player prisoner's dilemma game, the average hitting time was computed from 100,000 runs of the algorithms starting from the specified initial condition. Once again, observe that nonlimiting values of ε achieve nearly optimal behavior within a reasonable length of time. For example, if we let $\epsilon = 0.05$, the emergent system behavior is 98.8 percent of the optimal welfare and the expected hitting time is approximately 3,000 iterations.

7 Conclusion

Most of the distributed learning algorithms in the prior literature focus on reaching a Nash equilibrium for particular game structures, such as potential games. However, there are many engineering applications that cannot be represented by potential games, in which case these algorithms are inadequate. This chapter establishes the existence of simple learning algorithms that lead to efficiency for arbitrary game structures. The methodology raises a number of important theoretical and practical issues. The convergence rate, i.e., the time it takes (in expectation) to reach an efficient outcome, represents one of the most important issues. Clearly, as $\varepsilon \to 0$, the hitting times associated with this learning algorithm could be prohibitively long. However, Figures 10.4 and 10.6 demonstrate that nonlimiting values of ε can lead to nearly optimal system behavior in a reasonable length of time. Characterizing the precise relationship between the experimentation rate ε and the time it takes to achieve a given efficiency bound is a challenging open question. Another key issue is the role of information. In particular, would providing agents with more information improve the transient behavior? If so, what information should be communicated to the agents? We conjecture that the answers to both of these questions depend on the payoff structure of the game under consideration; however, detailed analyses of these issues would take us beyond the scope of this chapter and are left for future investigation.

Acknowledgments

This research was supported by AFOSR grants#FA9550-09-1-0538 and FA9550-12-1-0359, ONR grants #N00014-09-1-0751 and N00014-12-0643, and the Center for Research and Education in Wind The conference version of this work appeared in Proceedings of the 51st IEEE Conference on Decision and Control, 2012.

References

Androulakis, I. and Reklaitis, G. (1999), "Approaches to asynchronous decentralized decision making," *Computational Chemical Engineering* 23: 330–354.

Arieli, I. and Babichenko, Y. (2012), "Average testing and Pareto efficiency," *Journal of Economic Theory* 147: 2376–2398.

Arslan, G., Marden, J. R., and Shamma, J. S. (2007), "Autonomous vehicle-target assignment: a game theoretical formulation," *ASME Journal of Dynamic Systems, Measurement and Control* 129: 584–596.

Blatt, D., Hero, A. O., and Gauchman, H. (2008), "A convergent incremental gradient method with a constant step size," *SIAM Journal of Optimization* 18: 29–51.

Boyd, S., Parikh, N., Chu, E., Peleato, B., and Eckstein, J. (2010), "Distributed optimization and statistical learning via the alternating direction method of multipliers," *Foundations and Trends in Machine Learning* 3: 1–122.

Chasparis, G. and Shamma, J. (2011), "Distributed dynamic reinforcement of efficient outcomes in multiagent coordination and network formation," Discussion Paper, Department of Electrical Engineering, Georgia Tech.

Durr, H. B, Stankovic, M. S., and Johansson, K. H. (2011), "Distributed positioning of autonomous mobile sensors with application to coverage control," *Proceedings of the 2011 American Control Conference.*

Foster, D. and Young, H. (1990), "Stochastic evolutionary games dynamics," Journal of Theoretical Population Biology 38: 219–232.

Foster, D. and Young, H. (2006), "Regret testing: Learning to play Nash equilibrium without knowing you have an opponent," Theoretical Economics 1: 341–367.

Frihauf, P., Krstic, M., and Basar, T. (2010), "Nash equililbrium seeking for games with non-quadratic payoffs," in *Proceedings of the 2010 IEEE Conference on Decision and Control.*

Frihauf, P., Krstic, M., and Basar, T. (2011), "Nash equilibrium seeking with infinitely many players," in *Proceedings of the 2011 IEEE American Control Conference.*

Fudenberg, D. and Maskin, E. (1986), "The folk theorem in repeated games with discounting or with incomplete information," *Econometrica* 54: 533–554.

Gopalakrishnan, R., Marden, J. R., and Wierman, A. (2011), "An architectural view of game theoretic control," *ACM SIGMETRICS Performance Evaluation Review* 38 (3): 31–36.

Hart, S. and Mas-Colell, A. (2003), "Uncoupled dynamics do not lead to Nash equilibrium," *American Economic Review* 93 (5): 1830–1836.

Hart, S. and Mas-Colell, A. (2006), "Stochastic uncoupled dynamics and Nash equilibrium," *Games and Economic Behavior* 57 (2): 286–303.

Hatano, Y. and Mesbahi, M. (2005), "Agreement over random networks," *IEEE Transactions on Automatic Control* 50: 1867–1872.

Jadbabaie, A., Lin, J., and Morse, A. S. (2003), "Coordination of groups of mobile autonomous agents using nearest neighbor rules," *IEEE Transactions on Automatic Control* 48: 988–1001.

Johari, R. (2007), "The price of anarchy and the design of scalable resource allocation mechanisms," in N. Nisan, T. Roughgarden, E. Tardos, and V. Vazirani, eds. *Algorithmic Game Theory* (Cambridge: Cambridge University Press, pp. 543–568).

Johnson, K. E. and Thomas, N. (2009), "Wind farm control: Addressing the aerodynamic interaction among wind turbines," in ACC'09: Proceedings of the 2009 American Control Conference, 2104-2109.

Komali, R. S. and MacKenzie, A. B. (2006), "Distributed topology control in ad-hoc networks: A game theoretic perspective," CCNC 2006. 2006 3rd IEEE Consumer Communications and Networking Conference.

Krstic, M., Frihauf, P., Krieger, J., and Basar, T. (2010), "Nash equilibrium seeking with finitely—and infinitely—many players," in *Proceedings of the 8th IFAC Symposium on Nonlinear Control Systems.*

Li, N. and Marden, J. R. (2013), "Designing games for distributed optimization," *IEEE J. Selected Topics in Signal Processing* 7: 230–242.

Li, N. and Marden, J. R. (2014), "Decoupling coupled constraints through utility design, *IEEE Transactions on Automatic Control* 59: 2289–2294.

Lobel, I. and Ozdaglar, A. (2011), "Distributed subgradient methods for convex optimization over random networks," *IEEE Transactions on Automatic Control* 56: 1291–1306.

Marden, J. R. (2012), "State based potential games," *Automatica* 48: 3075–3088.

Marden, J. R. and Wierman, A. (2008), "Distributed welfare games," discussion paper, Department of ECEE, University of Colorado, Boulder.

Marden, J. R., Arslan, G., and Shamma, J. S. (2009a), "Connections between cooperative control and potential games," *IEEE Transactions on Systems, Man and Cybernetics. Part B: Cybernetics* 39: 1393–1407.

Marden, J. R., Arslan, G., and Shamma, J. S. (2009b), "Joint strategy fictitious play with inertia for potential games," *IEEE Transactions on Automatic Control* 54: 208–220.

Marden, J., Ruben, S. D., and Pao, L. Y. (2013), "A model-free approach to wind farm control using game-theoretic methods," *IEEE Transactions on Control Systems Technology* 21: 1207–1214.

Marden, J., Young, H. P., and Pao, L.Y. (2012), "Achieving Pareto optimality through distributed learning," in *Proceedings of the 51st IEEE Conference on Decision and Control.*

Marden, J. R., Young, H. P., Arslan, G., and Shamma, J. S. (2009), "Payoff based dynamics for multi-player weakly acyclic games," *SIAM Journal on Control and Optimization* 48: 373–396.

Monderer, D., and Shapley, L. S. (1996), "Potential games," *Games and Economic Behavior* 14: 124–143.

Nedic, A., Olshevsky, A., Ozdaglar, A., and Tsitsiklis, J. N. (2009), *IEEE Transactions on Automatic Control* 54: 2506–2517.

Nisan, N., Roughgarden, T., Tardos, E., and Vazirani, V. V. (2007), *Algorithmic Game Theory* (New York: Cambridge University Press).

Osborne, M. and Rubinstein, A. (1994), A Course in Game Theory (Cambridge, MA: MIT Press).

Pradelski, B. R. and Young, H. P. (2012), "Learning efficient Nash equilibria in distributed systems," *Games and Economic Behavior* 75: 882–897.

Solodov, M. V. (1998), "Incremental gradient algorithms with stepsizes bounded away from zero," *Computational Optimization and Applications* 11: 23–35.

Stankovic, M. S., Johansson, K. H., and Stipanovic, D. M. (2010), "Distributed seeking of Nash equilibria in mobile sensor networks," in *Proceedings of the 2010 IEEE Conference on Decision and Control.*

Tsitsiklis, J. and Athans, M. (1984), "Convergence and asymptotic agreement in distributed decision problems," *IEEE Transactions on Automatic Control* 29: 42–50.

Tsitsiklis, J., Bertsekas, D., and Athans, M. (1986), "Distributed asynchronous deterministic and stochastic gradient optimization algorithms," *IEEE Transactions on Automatic Control* 31: 803–812.

Wolpert, D. and Tumer, K. (1999), "An overview of collective intelligence," in J. M. Bradshaw, ed., *Handbook of Agent Technology* (Cambridge, MA: AAAI Press/ MIT Press).

Young, H. P. (2009), "Learning by trial and error," *Games and Economic Behavior* 65: 626–643.

Zhu, M. and Martinez, S. (2010a), "Distributed coverage games for energy-aware mobile sensor networks," *SIAM Journal on Control and Optimization* 51: 1–27.

Zhu, M. and Martinez, S. (2010b), "On distributed convex optimization under inequality and equality contraints via primal-dual subgradient methods," preprint *arXiv*: 2001.2612.

Zhu, M., Tembine, H., and Basar, T. (2010), "Heterogenous learning in zero-sum stochastic games with incomplete information," in *Proceedings of the 2010 IEEE Conference on Decision and Control.*

PART III
THE DIFFUSION OF INNOVATIONS

Preface to Chapters 11–13

How Do New Ways of Doing Things Become Generally Accepted?

New ways of doing things are rarely adopted all at once, but tend to spread gradually as people learn about the innovation from others. The adoption process is often decentralized, undirected, and bottom-up; in this sense it is similar to the dynamic models studied in previous chapters. Unlike those cases, however, the focus here will be on the *interim* rather than the long-run dynamics of the adoption process. Does the rate of adoption accelerate initially and then taper off? Or does it start up slowly and only accelerate later on? How long does it take, in expectation, for an innovation to spread to a large proportion of the population? As we shall see, the answers to these questions depend on the structure of the social network through which people interact, the gain in payoff realized by switching from the old to the new way of doing things, the factors motivating people to adopt, and the amount of behavioral heterogeneity in the population.

Chapter 11 is concerned with the question of what motivates people to adopt, and shows that different types of motivation leave a characteristic "footprint" on the shape of the adoption curve. For example, suppose that everyone has heard of a given innovation, but people delay in adopting purely out of inertia. Over time, the rate of adoption increases but at a decreasing rate (the adoption curve is concave); moreover, this remains true even when people have different degrees of inertia. An accelerating pattern of adoption is therefore incompatible with a pure inertia model. An alternative possibility is that people adopt through a kind of contagion process: the probability that a given individual adopts increases with the number of prior adopters with whom he comes in contact. Unlike the inertial model, a contagion process accelerates initially and then decelerates as the saturation level is approached. Less obviously, it has a characteristic footprint in which the hazard rate (the rate at which non-adopters become adopters) must be monotone non-increasing relative to the number of prior adopters; in fact, this result holds for any amount of heterogeneity in contagion rates among members of the population.

Another factor in the adoption decision is the information gleaned from the experiences of prior adopters. What matters in this case is the information accumulated by prior adopters over time, not just the *current* number of adopters. The resulting dynamics can be quite complex. Indeed, the early stages can exhibit a prolonged period in which the rate of adoption declines, followed by an extremely rapid "take-off' phase in which adoption accelerates at a super-exponential pace. Chapter 11 applies this framework to Ryan and Gross's landmark study of the adoption of hybrid corn by Iowa farmers in the 1930s and 1940s.[1] The analysis shows that the dynamic pattern of adoption is most consistent with learning from observation rather than contagion. This finding is backed up by Ryan and Gross's survey data, which suggests that farmers paid more attention to the experiences of their neighbors than to centralized information campaigns by the government (and commercial seed companies) touting the benefits of hybrid corn.

The adoption of hybrid corn was one of the earliest systematic studies of innovation diffusion; many subsequent studies have confirmed that adoption tends to be largely a bottom-up process in which individuals' decisions are strongly influenced by information gathered from neighbors about thebenefits of adopting the new technology.[2] However, the precise mechanisms driving the adoption decision can only be teased out by detailed event studies in which each individual's choice can be observed contingent on information they have about the benefits and costs of the innovation, whether garnered from their neighbors or more public sources. This is a challenging empirical exercise and remains one of the most important open topics of research on innovation diffusion.[3]

Chapters 12 and 13 focus on innovations with increasing returns: the more people who adopt the new technology or practice, the more valuable it becomes for each successive adopter. Contemporary examples include social media and computer software. Social and institutional innovations, such as forms of contracts and forms of organization, also exhibit increasing returns because they require a coordinated change in expectations and behaviors by multiple individuals to have any force. For example, the value of a legal contract depends on how widely it is used, because the meaning of its terms (and their enforcement) depends on legal precedents. More generally, norms and institutions exhibit increasing returns precisely because they function as coordination technologies.

[1] Bryce Ryan and Neal C. Gross (1943), "The Diffusion of Hybrid Corn in Two Iowa Communities," *Rural Sociology* 8(1): 15–24.

[2] For a landmark study of the effect of payoff gains on the rate of adoption, see Z. Griliches (1957), "Hybrid corn: an exploration in the economics of technological change," *Econometrica* 25: 501–522.

[3] For a particularly detailed study of learning from neighbors, see Timothy G. Conley and Christopher R. Udry (2010), "Learning about a new technology: pineapple in Ghana," *American Economic Review* 100: 35–69.

In Chapter 4 we studied the long-run dynamics of adoption in a setting where there are increasing returns and agents interact globally. Here we shall focus instead on the intermediate-run dynamics when agents are embedded in a social network and there are two competing technologies: an innovation (A) and the status quo (B). The returns from adoption are represented by a two-person coordination game. If both parties adopt the innovation A they each get $1 + \alpha$, if they both adopt the status quo B they each get 1, and if they miscoordinate they get zero. Thus the parameter α represents the "relative advance" of the innovation, compared to the status quo, from interacting with other adopters. Assume that agents are embedded in a social network with undirected edges, and that the payoff to a potential adopter is the expected payoff from interacting at random with someone with whom he is linked. Let us assume further that agents usually choose best responses given their neighbors' choices, but they sometimes deviate for idiosyncratic reasons.

Starting from the state where everyone chooses the status quo, how long does it take for the innovation to spread to a given proportion of the population? The answer depends on the probability distribution of idiosyncratic choices, the payoff advance of the innovation relative to the status quo, and the details of the network structure. Consider first the situation where everyone interacts with everyone else with equal probability. In other words, the network forms a complete, undirected graph with equal weights on all the edges. Suppose that agents update according to i.i.d. Poisson arrival processes, where each agent updates once in expectation per unit of time. Suppose further that, at any given update, the log probability of choosing a best reply minus the log probability of choosing a non-best reply is β times the difference in payoff for some constant $\beta > 0$. (This is equivalent to assuming that agents optimize subject to payoff shocks that are extreme-value distributed.) Wee can think of β as a measure of (myopic) rationality: when β is large an agent is nearly certain to choose a best response; when β is near zero the agent chooses an action more or less at random.

It can be shown that some combinations of α and β reach 50 percent penetration very quickly no matter how large the population is, whereas other combinations are extremely slow: in fact, the expected waiting time grows exponentially with the population size.[4] This "phase transition" between fast and slow convergence is illustrated in Figure P.1. Each value of β determines the probability of choosing A when everyone else chooses the status quo B; this is the *initial error rate* ε. Let us compute, as a function of α and ε, the

[4] G. E. Kreindler and H. P. Young (2013), "Fast convergence in evolutionary equilibrium selection," *Games and Economic Behavior* 80: 39–67.

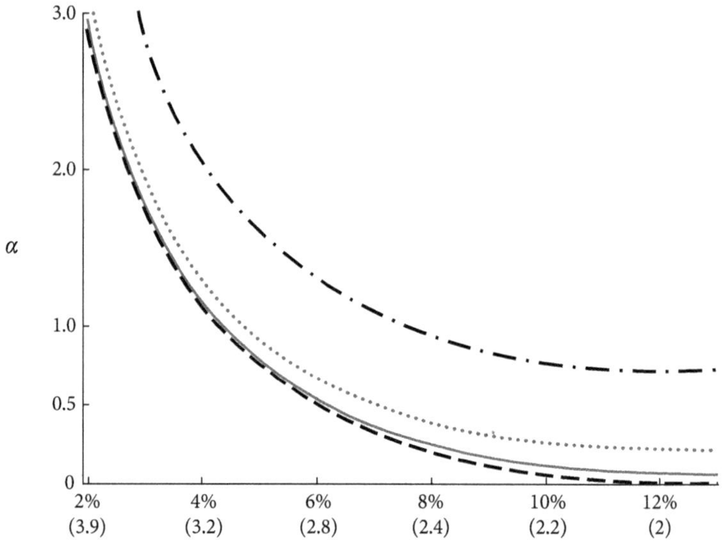

Figure P.1 Expected waiting times to reach 50% penetration as a function of α and ε for arbitrarily large population sizes. (The corresponding value of β is in brackets.) The dot-dash line corresponds to 10 periods, the dotted line to 20 periods, and the solid line to 40 periods. Below the dashed line the waiting time to reach 50% is unboundedly large as the population grows.

Source: G. E. Kreindler and H. P. Young (2013), "Fast convergence in evolutionary equilibrium selection," *Games and Economic Behavior* 80: 39–67.

expected waiting time until at least half of the population has adopted the innovation. Combinations of α and ε that lie below the dashed line have expected waiting times that increase exponentially with the size of the population. Combinations that lie above the dashed line have expected waiting times that are bounded above *independently* of the population size, and are often surprisingly small. For example, the dotted line corresponds to combinations that take only 20 periods in expectation to reach the 50 percent penetration level. (A specific example is $\alpha = 1$ combined with an initial error rate of about 5 percent.)

As Chapter 12 shows, the expected waiting time depends not only on the level of perturbations and the size of the innovation's advance, but also on the specific topology of the social network in which the subjects are embedded.[5] Unlike complete graphs, there are some network topologies for which the waiting time is bounded independently of the number of agents even when the

[5] The effect of local interactions on speed of convergence was first pointed out by G. Ellison (1993), "Learning, local interaction, and coordination," *Econometrica* 61 (5): 1047–1071; see also H. P. Young (1998), *Individual Strategy and Social Structure*, Chapter 6 (Princeton, NJ: Princeton University Press).

size of the advance is very small. This is the case, for example, in networks that consist of many small "villages" or "clusters" that are only loosely connected. It also holds for spatial networks in which agents are linked by an edge if and only if they are close to one another geographically.

Consider, for example, a two-dimensional grid embedded on a torus, where everyone is linked to their four closest neighbors. For given α, $\beta > 0$, the expected waiting time to reach at least 50 percent penetration is bounded independently of the population size. The reason is that the innovation can gain a foothold in many distinct areas of the network more or less simultaneously, and from there it spreads to the rest of the population. The smaller the advance α is, the longer it takes to reach the target; in fact the expected waiting time grows exponentially with respect to $1/\alpha$ for a given population size This result suggests that social innovations that represent a small advance in welfare may take so long to diffuse that they will tend to be overtaken by those (rarer) innovations that offer a substantial welfare gain.

Instead of focusing on expected waiting times for particular classes of networks, Chapter 13 shows that, under certain conditions, one can establish a *uniform upper bound* that holds for networks of arbitrary size and topological structure. More precisely, when the welfare advance from the innovation is sufficiently large and/or idiosyncratic departures from best response are sufficiently frequent, there exists an upper bound on the expected waiting time to reach a given level of penetration that holds for *all* social networks of arbitrary size and complexity, including networks with asymmetric degrees of influence.[6]

This estimated upper bound is "universal," in the sense that it holds for any given network, hence it holds for networks whose structure is evolving due to exogenous factors. The analysis does not cover the situation where the network is evolving *endogenously*, that is, links are being formed and severed depending on players' prior actions and realized payoffs. Estimating the expected waiting time in this case represents an important open question for future research. A second open question is how these theoretical results relate to the actual choice behavior of human subjects when they play a coordination game on a network. Experimental work suggests that subjects frequently do make non-best replies, and the time it takes to reach or come close to the stochastically stable state can

[6] Chapter 13 generalizes results of Kreindler and Young for regular networks (G. E. Kreindler and H. P. Young (2014), "Rapid innovation diffusion in social networks," *Proceedings of the National Academy of Sciences* 111: 10881–10888.

be quite short in practice.[7] To the best of my knowledge, however, no systematic attempt has been made to study empirical convergence times as a function of the structure of the game, or to estimate the frequency of deviations from best reply.

A general implication of the results in Chapters 12–13 is that, in 2×2 coordination games played on social networks, evolutionary learning processes can come close to the stochastically stable state surprisingly quickly. In this sense, stochastic stability constitutes a useful predictive tool about the medium-run behavior of the dynamics when deviations from best reply occur fairly frequently. A challenging open question is whether similar results hold for more general classes of games when the stochastic perturbations in players' behavior are sufficiently (but not arbitrarily) small.

[7] Empirical convergence times for a particular class of coordination games played on networks are studied by D. Centola and A. Baronchelli (2015), "The spontaneous emergence of conventions: an experimental study," *Proceedings of the National Academy of Sciences* 112 (7): 1989–1994. See also E. Bilancini, L. Boncinelli, and H. H. Nax (2021), "What Noise Matters?: Experimental Evidence for Stochastic Deviations in Social Norms," *Journal of Behavioral and Experimental Economics* 90: 101626.

11

Innovation Diffusion in Heterogeneous Populations

Contagion, Social Influence, and Social Learning

1 Models of Innovation Diffusion

A basic puzzle posed by innovation diffusion is why long lags occur between an innovation's first appearance and its general acceptance by a population. Among the factors that have been suggested are delays in acting on information, a desire to conform, learning from others, and changes in external factors such as prices. There is an extensive literature on this phenomenon that spans economics, sociology, and marketing.[1] Although the models are highly varied, a common theme is that delay results in part from heterogeneity among potential adopters. People may realize different benefits and costs from the innovation, hear about it at different times, or have different amounts of information, different predispositions to conform, and so forth. Nevertheless, most of the extant models incorporate heterogeneity in a very restricted fashion, say by considering two homogeneous classes of agents, or by assuming that the heterogeneity is described by a particular class of distributions.[2]

The purpose of this chapter is threefold. First, it will be shown that the benchmark models in economics, marketing, and sociology proceed from fundamentally different assumptions that lead to key differences in their dynamical structure. Second, each type of model can be formulated at a high level of generality that allows for essentially any distribution of heterogeneous characteristics without losing analytical tractability. Third, each model leaves a distinctive "footprint" in its pattern of acceleration that holds with few or no restrictions on the distribution of characteristics. The reason is that they have

[1] For general overviews, see Mahajan and Peterson (1985); Mahajan, Muller, and Bass (1990); Valente (1995, 1996, 2005; Geroski (2000); Stoneman (2002); and Rogers (2003)).

[2] See, in particular, Jeuland (1981); Jensen (1982); Karshenas and Stoneman (1992); and Geroski (2000).

Spontaneous Order. H. Peyton Young, Oxford University Press. © H. Peyton Young (2024).
DOI: 10.1093/oso/9780198892908.003.0011

fundamentally different structures that details in the underlying distributions cannot overcome.

We shall consider three broad classes of models—contagion, social influence, and social learning—that are drawn from the literature on marketing, sociology, and economics, respectively.[3] The focus of attention will be on situations where the dynamics are driven from within; that is, there are internal feedback effects from prior to future adopters. Diffusion models that are driven mainly by *external* factors, such as price and quality changes, obviously depend on the exogenous rates of change in these factors, so much less can be said about them from an a priori standpoint.[4]

1. *Contagion.* People adopt when they come in contact with others who have already adopted; that is, innovations spread much like epidemics.
2. *Social influence.* People adopt when enough other people in the group have adopted; that is, innovations spread by a conformity motive.
3. *Social learning.* People adopt once they see enough empirical evidence to convince them that the innovation is worth adopting, where the evidence is generated by the outcomes among prior adopters. Individuals may adopt at different times due to differences in their prior beliefs, amount of information gathered, and idiosyncratic costs.

Of the three categories of models, social learning is certainly the most plausible from an economic standpoint, because it has firm decision-theoretic foundations: agents are assumed to make rational use of information generated by prior adopters in order to reach a decision.[5] By contrast, the contagion and social influence models are based on the notion of *exposure* rather than on utility maximization. Why, then, should economists be concerned with them? One reason is that they are the most prominent models in the literature outside of economics, so it is worth understanding how they work in a general setting. A second reason is that we would like to know whether adoption curves

[3] Some of these ideas have also been applied in anthropology; see, in particular, Boyd and Richerson (1985) and Henrich (2001).

[4] These are sometimes called "moving equilibrium" models; see, in particular, David (1968, 1969, 1975, 2005); David and Olsen (1984, 1986); and Stoneman (2002). There are also hybrid models in which the dynamics are driven by a combination of external and internal factors. The most common example is the Bass model, which will be considered in Section 3.

[5] There is a large literature on social learning, but the specific assumptions are highly varied. See, among others, Banerjee (1992); Bikchandani, Hirshleifer, and Welsh (1992); Ellison and Fudenberg (1993, 1995); Kirman (1993); Kapur (1995); Bala and Goyal (1998); Smith and Sorensen (2000); Banerjee and Fudenberg (2004); Chatterjee and Xu (2004); and Manski (2004). However, relatively little prior work has been done on the implications of social learning for the *shape* of the adoption curve. Notable exceptions are Jensen (1982); Henrich (2001); and Lopez-Pintado and Watts (2008).

generated by social learning differ from those generated purely by exposure, that is, whether we can distinguish empirically between the two explanations.

We shall find that under certain conditions this is indeed the case. To make progress on this question, however, we need to formulate the aggregate dynamics of each class of model at a high level of generality without making restrictive assumptions about the distribution of individual characteristics. This is particularly challenging in the case of social learning models, where there are multiple sources of heterogeneity, including different costs of adoption, different prior beliefs, and different amounts of information-gathering. It turns out that all of these considerations can be folded together in a simple dynamical equation that is easy to analyze.

One limitation of the analysis is the use of a "mean-field" approach in which the population is assumed to be infinitely large and encounters between individuals are purely random. If the population is finite and agents interact through a fixed social network, the aggregate dynamics are substantially more complex and may depend on the topology of the network. The extent to which our results can be generalized to this case is an interesting question that will be left to future work.[6]

The main results on shape and acceleration patterns can be summarized as follows.

1. A process that is driven purely by inertia must decelerate from start to finish, that is, *the adoption curve is strictly concave.* Acceleration—in particular *S*-shaped curves—cannot be explained solely by delay no matter how the rates of inertia are distributed in the population.[7]
2. A process that is driven purely by contagion typically accelerates initially and then decelerates as it approaches saturation (the curve is *S*-shaped), but it cannot accelerate beyond the 50 percent adoption level. When everyone is alike, the hazard rate (the rate at which nonadopters become adopters) is linear and increasing in the current number of adopters. From this it follows that *the hazard rate must be nonincreasing relative*

[6] For models of diffusion in social networks, see Blume (1993, 1995); Bala and Goyal (1998); Young (2003); Chatterjee and Xu (2004); Cowan and Jonard (2004); Jackson and Rogers (2007); Jackson and Yariv (2007); Vega-Redondo (2007); Golub and Jackson (2008) as well as Chapters 12 and 13 in this volume.

[7] This feature of inertia models distinguishes them from contagion models, as was pointed out in the classic study of medical innovation by Coleman, Katz, and Menzel (1957, 1966). It is also a well-known feature of heterogeneous duration models (Lancaster and Nickell 1980; Heckman and Singer 1982; Heckman, Robb, and Walker 1990). We draw attention to it here because it shows in a particularly transparent way how restrictions on shape can hold irrespective of the distribution of individual characteristics (in this case the inertia rates).

to the number of adopters. Less obviously, this result holds for any joint distribution of "infection" rates from a combination of internal and external sources. This forms a significant and testable implication of the contagion model, as we shall see in Section 6.

3. A process that is driven purely by social influence can either decelerate or accelerate initially, but in the latter case it typically accelerates at a *superexponential rate* for some period of time.
4. A process that is driven purely by learning from the experiences of others (social learning with direct observation) *usually begins slowly and may even decelerate initially.* If such a process does eventually accelerate, it typically does so at a *superexponential rate* for some period of time. Moreover, the overall rate at which adoption proceeds is a monotone increasing function of the expected payoffs from adoption.

These results provide a general framework for assessing the relative plausibility of different mechanisms that could be driving a given diffusion process. It is not an identification strategy, which is virtually impossible with aggregate-level data and difficult enough with micro-level data. Rather, our results establish restrictions on the shape of adoption curves that are associated with particular explanatory mechanisms. The overall plausibility of a given explanation should be examined in conjunction with other information about the specific nature of the process, e.g., whether it was driven largely by internal communication among members of a population or by external sources such as advertising campaigns; whether payoffs played a prominent role (as in new production methods); or conformity was the main motive (as in new fashions). We outline how this framework can be applied empirically using Bryce Ryan and Neal C. Gross's classic study of the diffusion of hybrid corn in the 1920s and 1930s (Ryan and Gross 1943). While the data in this study are rather limited, it illustrates how the approach can be brought to bear on actual cases, and also highlights the kinds of data that one would need to conduct a full-fledged empirical analysis.

2 Inertia

Before launching into a discussion of the three main models, it will be useful to consider a much simpler reason why innovations take time to diffuse; namely, people sometimes *delay* in acting on new information. This hypothesis leads to a particularly tractable model that has been studied in other contexts, notably heterogeneous duration models (see, among others, Lancaster and Nickell 1980; Heckman and Singer 1982; Heckman, Robb, and Walker 1990). We include it here because it forms a useful foil for the models to come later; furthermore, all of the other models include inertia as one of the parameters, so this is by far the simplest case to analyze.

First, consider a model of delayed adoption in which there is no heterogeneity among the agents. Let $\lambda > 0$ be the instantaneous rate at which any given nonadopter first adopts. Let $p(t)$ be the proportion of adopters at time t, and let us set the clock so that $p(0) = 0$. The function $p(t)$ is called the *adoption curve*. Assume for simplicity that once agents have adopted the new technology, they do not switch back to the old technology within the time frame of the analysis. Then the expected motion is described by the ordinary differential equation $\dot{p}(t) = \lambda(1 - p(t))$, which has the solution $p(t) = 1 - e^{-\lambda t}$ given the initial condition $p(0) = 0$.

Notice that this curve is *concave* throughout; in particular, it is not S-shaped. It is a rather remarkable fact that this remains true when *any amount of heterogeneity* is introduced. To see this, suppose that $v(\lambda)$ is the distribution of λ in the population. Then the expected trajectory of the process is given by

$$p(t) = 1 - \int e^{-\lambda t}\, dv. \tag{1}$$

Differentiating (1) twice over, we see that $p(t) < 0$ *irrespective of the distribution* $v(\lambda)$. The intuition is straightforward: agents with high values of λ tend to adopt earlier than those with low values of λ. It follows that the average value of λ in the current population of nonadopters is nonincreasing, and the number of such individuals is strictly decreasing. Thus the flow of new adopters is strictly decreasing.

This simple example illustrates the kinds of results that hold in more complex situations: the structure of the model has implications for the *shape* of the curve that remain true even when an arbitrary amount of heterogeneity is introduced.

3 Contagion

Contagion refers to a process in which people adopt a new product or practice when they come in contact with others who have adopted it. An everyday example would be a new fashion that spreads because people imitate those who have already adopted. The resulting dynamics are similar to those of an epidemic; indeed, some of the models are borrowed directly from the epidemiology literature. In the context of innovation diffusion, it is common to use a two-parameter model that allows for contagion from within the group at one rate, and from sources outside the group at a different rate. This is known as the *Bass model* of new product diffusion (Bass 1969, 1980) or the *mixed-influence diffusion model* (Mahajan and Peterson 1985). In the context of a new fashion in clothing, for example, these two rates would correspond to seeing other people on the street who are wearing it, and seeing television ads that promote it.[8]

Let us begin by describing the homogeneous version of the model, later we will introduce heterogeneity. Let λ be the instantaneous rate at which a current nonadopter "hears about" the innovation from a previous adopter *within* the group, and let γ be the instantaneous rate at which he hears about it from sources *outside* of the group. We shall assume that λ and γ are nonnegative, and that not both are zero. In the absence of heterogeneity, such a process is described by the ordinary differential equation $\dot{p}(t) = (\lambda p(t) + \gamma)(1 - p(t))$, and the solution is

$$p(t) = \left[1 - \beta\gamma e^{-(\lambda+\gamma)t}\right] / \left[1 + \beta\lambda e^{-(\lambda+\gamma)t}\right], \quad \lambda > 0. \tag{2}$$

When contagion is generated purely from internal sources ($\gamma = 0$), this boils down to the ordinary logistic function, which is of course *S*-shaped.[9] When innovation is driven solely by an external source ($\gamma > 0$ and $\lambda = 0$), we obtain the pure inertia model discussed earlier. When both γ and λ are positive, we can choose β in expression (2) so that $p(0) = 0$; namely, with $\beta = 1/\gamma$ we obtain

$$p(t) = \left[1 - e^{-(\lambda+\gamma)t}\right] / \left[1 + (\lambda/\gamma)e^{-(\lambda+\gamma)t}\right]. \tag{3}$$

This model has spawned many variants, some of which assume a degree of heterogeneity, such as two groups with different contagion parameters (Karshenas and Stoneman 1992; Geroski 2000), while others employ distributions from a specific parametric family, such as gamma distributions (Jeuland 1981).

In fact, we can formulate a fully heterogeneous version that is analytically tractable and places virtually no restrictions on the joint distribution of the parameters. Specifically, let μ be the *joint distribution* of the contagion parameters λ and γ. Assume for analytical convenience that μ has bounded support,

[8] Some forms of Internet advertising attempt to exploit internal feedback effects by informing members of a target group how many other members of their group have already adopted.

[9] The logistic model was common in the early work on innovation diffusion; see, for example, Griliches (1957) and Mansfield (1961).

which we may take to be $\Omega = [0, 1]^2$. (Rescaling λ and γ by a common factor is equivalent to changing the time scale, so this involves no real loss of generality.) The only substantive restriction that we shall place on μ is that, averaged over the whole population, external influence is positive $\left(\bar{\gamma} = \int_\Omega \gamma d\mu > 0\right)$, for otherwise the process cannot get out of the initial state $p(0) = 0$.

Let $p_{\lambda,\gamma}(t)$ be the *proportion* of all type-(λ, γ) individuals who have adopted by time t. Then the proportion of all individuals who have adopted by time t is

$$p(t) = \int p_{\lambda,\gamma}(t) d\mu. \tag{4}$$

(Hereafter integration over Ω is understood.) Each subpopulation of adopters $p_{\lambda,\gamma}(t)$ evolves according to the differential equation

$$\dot{p}_{\lambda,\gamma}(t) = (\lambda p(t) + \gamma)\left(1 - p_{\lambda,\gamma}(t)\right). \tag{5}$$

This defines an infinite system of first-order differential equations coupled through the common term $p(t)$. We can reduce it to a single differential equation by the following device: let $x_{\lambda,\gamma}(t) = \ln(1 - p_{\lambda,\gamma}(t))$ and observe that (5) is equivalent to the system $\dot{x}_{\lambda,\gamma}(t) = -(\lambda p(t) + \gamma)$ for all (λ, γ). From this and the initial condition $x_{\lambda,\gamma}(0) = 0$ we obtain

$$x_{\lambda,\gamma}(t) = -\int_0^t (\lambda p(s) + \gamma) ds = -\lambda \int_0^t (p(s) ds - \gamma t. \tag{6}$$

From the definition of $x_{\lambda,\gamma}(t)$ it follows that

$$p(t) = 1 - \int e^{x_{\lambda,\gamma}(t)} d\mu, \tag{7}$$

that is, $p(t)$ satisfies the integral equation

$$p(t) = 1 - \int e^{-\gamma t - \lambda \int_0^t p(s) ds} d\mu. \tag{8}$$

In other words, the current number of nonadopters $1 - p(t)$ decays according to a weighted average of exponential functions, where the exponent depends on the elapsed time t and the average adoption level up to t. Starting from the initial condition $p(0) = 0$, one can solve for $p(t)$ by successive approximation.

In this model, the hazard rate $\dot{p}(t)/(1 - p(t))$ may be increasing or decreasing depending on the relative importance of the internal and external sources of contagion. It turns out, however, that the hazard rate cannot increase relative to the current number of adopters, that is, the ratio $\dot{p}(t)/[p(t)(1 - p(t))]$ must be nonincreasing. Furthermore, if contagion from external sources is present, the ratio must be strictly decreasing. We shall call this ratio the *relative hazard rate*.

When the parameters λ, γ are fixed, it is obvious that the relative hazard rate is nonincreasing; indeed, this follows at once from the equation $\dot{p}(t) = (\lambda p(t) + \gamma)(1 - p(t))$. The claim is not so obvious when the parameters are heterogeneously distributed, including the possibility that λ and γ are negatively correlated: people who are heavily influenced by outside sources (high γ) might be less susceptible to information from inside sources (low λ), and so forth.

Proposition 1 *Suppose that diffusion is driven by heterogeneous contagion with positive external influence. Then the hazard rate decreases relative to the number of adopters, which implies that the process cannot accelerate beyond* $p = 1/2$.

The last statement follows immediately from the prior one, because if $p(t)$ were to accelerate beyond ½, then at that point the numerator of $\dot{p}(t)/[p(t)(1 - p(t))]$ would be increasing and the denominator would be decreasing, contradicting the assertion that the hazard rate decreases relative to the number of adopters (this is proved in the Appendix).

We remark that there are perfectly reasonable *S*-shaped curves that do not satisfy the monotonicity condition in Proposition 1. Consider, for example, an adoption process described by the differential equation $\dot{p}(t) = p^a(t)(1 - p(t))$,

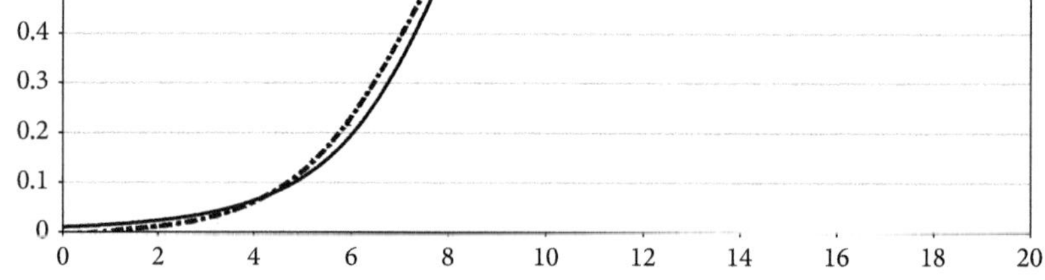

Figure 11.1 Two adoption curves. *(The solid line is generated by* $\dot{p}(t) = p^{1.1}(1 - p)$ *and* $p(0) = 0.01$*, the dashed line by a Bass model with* $\lambda = 0.75$ *and* $\gamma = 0.0025$.*)*

which was first proposed as a model of innovation diffusion by Easingwood, Mahajan, and Muller (1981, 1983). When $a > 1, \dot{p}(t)/[p(t)\,(1 - p(t)] = p^{a-1}(t)$ is strictly increasing. Hence Proposition 1 shows that such a process cannot arise from a contagion model with any amount of heterogeneity. Nevertheless, it is possible to generate an *S*-shaped curve from a contagion model—in fact from a homogeneous contagion model—whose overall appearance is very similar (see Figure 11.1).

4 Social Influence

The sociological literature on innovation stresses the importance of social pressure on individual decisions about whether to adopt. In the standard model individuals are assumed to have different "thresholds" that determine whether they will adopt as a function of the number (or proportion) of others in the population who have already adopted. (The dynamics of these models were first studied by Schelling 1971, 1978; Granovetter 1978; and Granovetter and Soong 1988; for more recent work in this vein, see Macy 1991; Valente 1995, 1996, 2005; Dodds and Watts 2004, 2005; Centola 2006; and Lopez-Pintado and Watts 2008.)

For each agent *i*, suppose that there exists a minimum proportion $r_i \geq 0$ such that *i* adopts as soon as r_i or more of the group has adopted. (If $r_i > 1$, the agent never adopts.) This is called the *social threshold* of agent *i*. The precise meaning of these thresholds varies from one context to another; broadly speaking we can think of them as *degrees of responsiveness to social influence*. A concrete example would be the transmission of rumors: some people would need to hear the rumor from many people to pass it on while others might need to hear it only once. Another is fashion: some people need to see only a few others wearing a new kind of clothing in order to try it, whereas others will wait until the fashion is widespread before jumping on the bandwagon. A key feature of such a model is that the adoption depends on the innovation's current *popularity* rather than on how good or desirable the innovation has proven to be.[10] The latter is the basis of *social learning models*, which we shall take up in the Section 5.

We wish to model the aggregate dynamics without assuming a specific parametric form for the distribution of thresholds. To this end, let $F(r)$ be the cumulative distribution function of thresholds in some given population. At

[10] Popularity is a factor in some of the economics models of diffusion; see, in particular, Ellison and Fudenberg (1993, 1995).

time t the proportion of people whose thresholds have been crossed is $F(p(t))$. Of these, $p(t)$ have already adopted, so the proportion whose thresholds have been crossed but have not yet adopted is $F(p(t)) - p(t)$. Let $\lambda > 0$ be the instantaneous rate at which these people convert. Then the adoption process is described by the differential equation

$$\dot{p}(t) = \lambda[F(p(t)) - p(t)], \quad \lambda > 0. \tag{9}$$

Assume that $F(0) > 0$ and let b be the *first fixed point* of F, that is, the smallest number in $(0, 1]$ such that $F(b) = b$, if any such exists; otherwise, let $b = 1$. We then have $F(p) > p$ for all $p \in [0, b)$. Since (9) is a separable ordinary differential equation, we obtain the following explicit solution for the inverse function $t = p^{-1}(x)$:

$$\forall x \in [0, b), t = p^{-1}(x) = (1/\lambda) \int_0^x dr/(F(r) - r). \tag{10}$$

Notice that the right-hand side is integrable because $F(r)$ is monotone nondecreasing and $F(r) - r$ is bounded away from zero for all r in the interval $[0, x]$ whenever $x < b$. (The constant of integration is zero because of the initial condition $p(0) = 0$.) As $x \to b$, the right-hand side of (10) goes to infinity, which implies that the adoption curve approaches b asymptotically. In particular, *the adoption process peters out as it approaches the first fixed point of F.*

This phenomenon is illustrated in Figure 11.2. Here we assume that the thresholds are normally distributed to the right of the origin, and there is a point mass at the origin corresponding to the subset of innovators—the people who are willing to adopt even when no one else adopted. Notice that the adoption curve asymptotes to $p = 0.50$, which is the first fixed point of F. There is nothing special about the normal distribution in this regard; similar results hold for any cumulative distribution function (CDF) where it first crosses the 45° line.

The adoption curve in Figure 11.2 is concave, but this is by no means necessary or even typical for this family of models. Figure 11.3 illustrates an entirely different adoption curve generated by a truncated normal with smaller mean and variance. An interesting feature of this curve is that it accelerates very sharply in the early stages; indeed, it can be shown that it grows at a faster-than-exponential rate up to $p = 0.10$.

We claim that whenever the process starts in the left tail of the distribution, the curve must have one of two shapes: either it decelerates initially, or it accelerates initially at a superexponential rate. To see why this is so, consider the basic dynamic equation in (9). Assume that $F(0) > 0$ and that F has

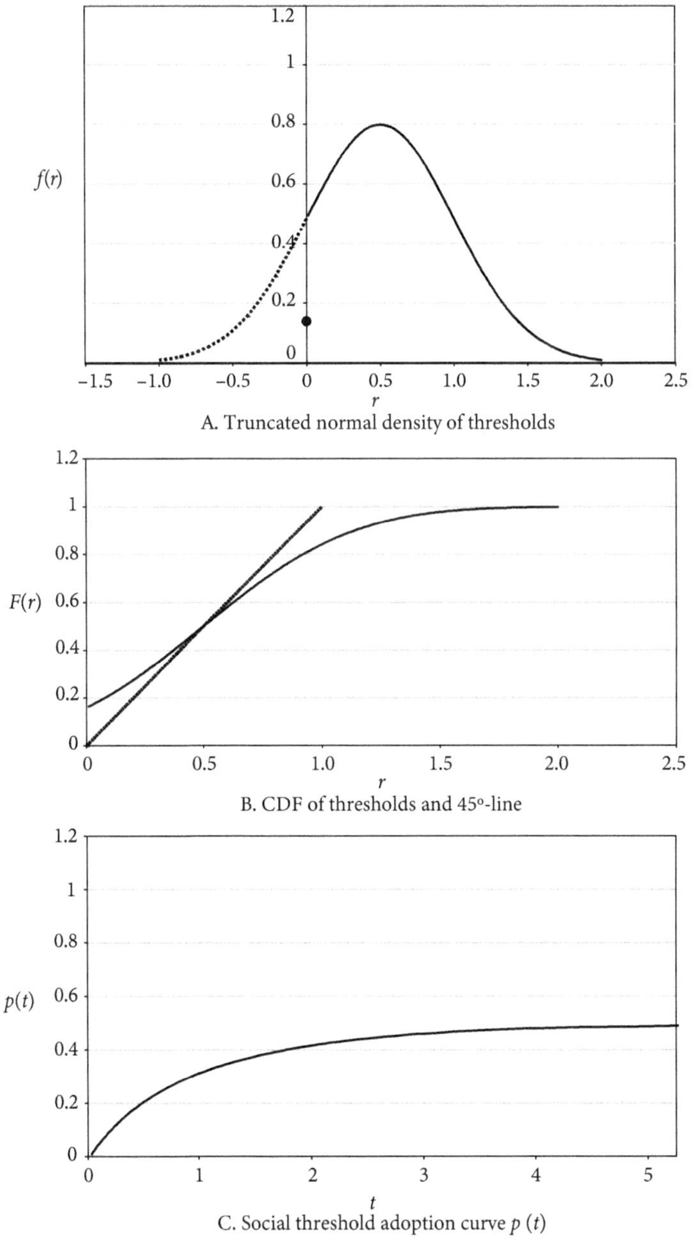

Figure 11.2 Density, CDF, and social threshold adoption curve generated by $N(0.50, 0.25)$ and $\lambda = 4$.

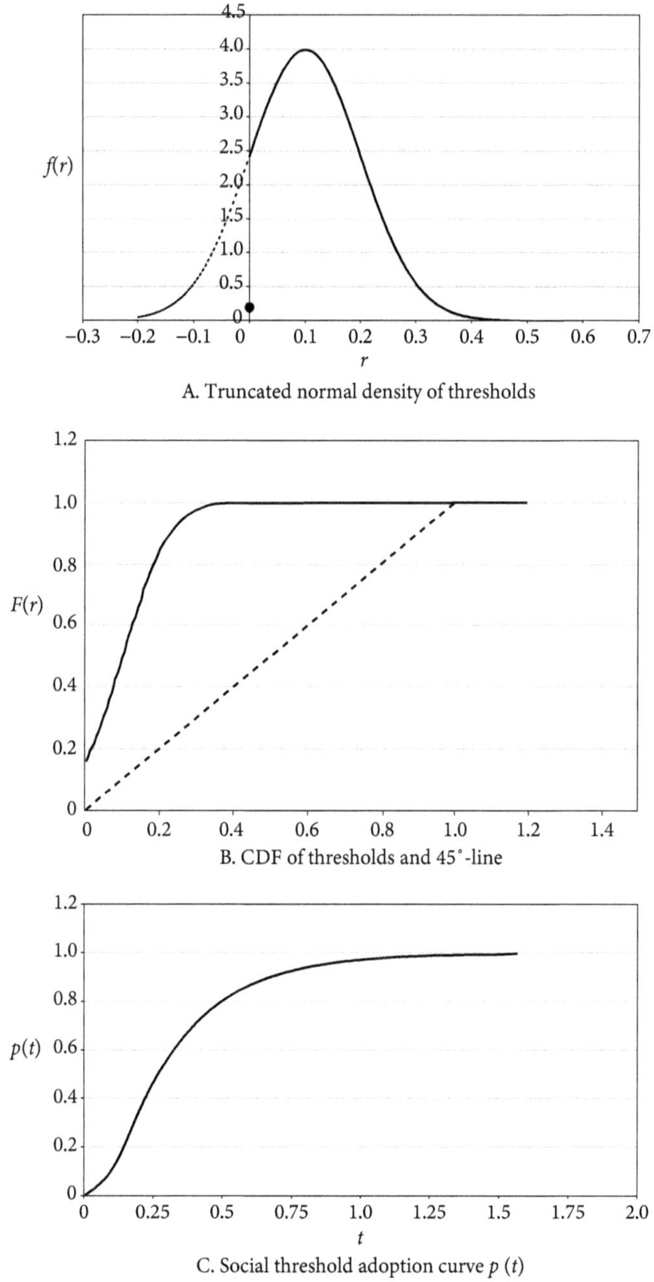

Figure 11.3 Density, CDF, and social threshold adoption curve generated by $N(0.10, 0.01)$ and $\lambda = 4$.

a continuous density $f(p)$ defined for all $p \in (0, 1]$. (Note that the density is not defined at the origin because there is a point mass there.) Differentiating (9) with respect to t and dividing through by $\dot{p}(t)$, which by assumption is positive, we obtain

$$\ddot{p}(t)/\dot{p}(t) = \lambda[f(p(t)) - 1]. \tag{11}$$

In other words, the relative acceleration rate traces out a positive linear transformation of the underlying density. It follows that the process accelerates initially if and only if the initial density is large enough, that is, $f(p) > 1$ in a neighborhood of the origin. Suppose further that $f(p)$ is increasing in a neighborhood of the origin, that is, the process starts in the left tail of the distribution of thresholds. Then (11) shows that the relative acceleration rate $\ddot{p}(t)/\dot{p}(t)$ is also strictly increasing; that is, the adoption curve exhibits *superexponential growth*.[11] This phenomenon results from the compounding of two effects. First, as more and more people adopt, the amount of information available to the remainder of the population increases. Second, the number of people persuaded by each additional bit of information increases as the process moves up the left tail of the distribution.

These conclusions continue to hold when λ is heterogeneously distributed according to some distribution function $\nu(\lambda)$. Since λ is a scaling parameter, there is no real loss of generality in assuming that the support of ν lies in (0, 1]. Let $F_\lambda(p)$ be the cumulative distribution of thresholds p conditional on λ, and assume that the conditional density $f_\lambda(p)$ exists for every $p \in (0, 1]$. The following proposition is proved in the Appendix.

Proposition 2 *Suppose that diffusion is driven by a heterogeneous social threshold model, where there is a positive mass of individuals who are prepared to adopt when no one else adopts, and the threshold densities $f_\lambda(p)$ are positive and increasing in a common neighborhood of zero for all λ. If the process accelerates initially, then it accelerates initially at a superexponential rate for some period of time.*

[11] Growth is *exponential* if $\ddot{p}(t)/\dot{p}(t)$ is constant, *superexponential* if $\ddot{p}(t)/\dot{p}(t)$ is strictly increasing, and *subexponential* $\ddot{p}(t)/\dot{p}(t)$ if is strictly decreasing.

5 Social Learning

A notable limitation of the contagion and social influence models is that they provide no clear reason *why* people would adopt an innovation given that others have adopted it. In this section we consider a class of learning models in which the adoption decision flows directly from the rational evaluation of evidence. Specifically, let us suppose that an agent adopts if he has reason to believe the innovation is better than what he is doing now, where the evidence comes from directly observing the outcomes among prior adopters. For example, when a new product becomes available—e.g., a new type of medication (antibiotics) or a new communication technology (cell phones)—people will want to see how it works for others *over a period of time* before trying it themselves. These are variously known as *social learning models* or *social learning models based on direct observation.*

There is a sizable theoretical literature on social learning, but it is difficult to summarize due to the great diversity in behavioral and informational assumptions that different authors use. Some assume that payoff outcomes among prior adopters are fully observable, while others assume only that the act of adoption is observable (the latter are usually called herding models). Some assume that agents can recognize others' types, while others assume that types cannot be identified. There are also significant differences in the relevant *characteristics* that authors choose to focus on, including heterogeneity in risk aversion, discount rates, and amount of information.[12] There is considerable empirical evidence, however, that learning from the experience of others does in fact occur.[13]

Here we shall make a number of simplifying assumptions in order to get a handle on an issue that has not received much previous attention in this literature, namely, *what do the short-run aggregate dynamics of a heterogeneous learning model look like, and do they differ qualitatively from the dynamics generated by other classes of models?*

To make some progress on this question let us make the following assumptions:

(i) payoffs are observable;

(ii) payoffs generated by different individuals and/or at different points in time are independent and equally informative;

[12] See, among others, Banerjee (1992); Bikchandani, Hirshleifer, and Welsh (1992); Ellison and Fudenberg (1993, 1995); Kirman (1993); Kapur (1995); Bala and Goyal (1998); Smith and Sorensen (2000); Banerjee and Fudenberg (2004); Chatterjee and Xu (2004);Manski (2004); Golub and Jackson (2008).

[13] Foster and Rosenzweig (1995); Munshi (2004); Conley and Udry (2005).

(iii) agents are risk-neutral and myopic;
(iv) there is no idiosyncratic component to payoffs due to differences in agents' types, but agents may have different costs (not necessarily observable);
(v) there are differences in agents' prior beliefs about how good the innovation is relative to the status quo;
(vi) there are differences in the average number of people they observe and hence in the amount of information they have;
(vii) the population is fully mixed.

Many other complicating factors could be introduced, such as discounting, one-time switching costs, risk aversion, imperfect observability, and restricted information acquisition (such as through social networks), but these extensions would distract from the main point, which is to identify the dynamic characteristics of a fairly general class of learning models without attempting to formulate the most general such model.

We claim that, under the assumptions above, the dynamical system has a surprisingly simple structure. One can reduce the various types of heterogeneity to a *composite index* that measures the *probability of a given agent adopting, conditional on the amount of information that has been generated thus far in the population.* The equation of motion of such a process turns out to be analogous to the one obtained for the social influence model, except that here the relevant state variable at time t turns out to be the *integral of the adoption curve* up through t, $\int_0^t p(s)ds$, rather than the *level of adoption* $p(t)$.[14] The reason is that $\int_0^t p(s)ds$ measures the cumulative information generated by all prior adopters *from the time they first adopted*, which is the relevant variable in the learning context.

To illustrate how such a model works, let us walk through a particular example using a standard normal–normal updating framework. This is chosen mainly for its computational transparency; similar results hold under alternative assumptions.[15] Assume that the payoff from the innovation is a normally distributed random variable X with mean $\mu > 0$ and variance σ^2, which is i.i.d. among agents and time periods. We shall interpret μ as the *mean payoff gain per period from using the innovation as compared to the status quo technology.* Each agent is assumed to have an idiosyncratic variable cost c_i of using

[14] Recall that the integral of the adoption curve also featured in the heterogeneous contagion model (see equation (8)).

[15] Jensen (1982) and Lopez-Pintado and Watts (2008) study the case where the outcome variable is binomial (payoffs are "high" or "low"). They assume that agents pay attention only to current outcomes, not the cumulative amount of information generated from earlier periods.

the innovation, so he adopts if and only if he believes that the mean payoff per period is at least c_i (thus, he may switch back if after receiving more information he believes that the mean payoff is less than c_i).

If everyone knew the true value of μ from the outset, then everyone would adopt for whom $c_i < \mu$. This is the *efficient outcome*. Ex ante, however, people do not know the true value of μ; furthermore, they may start with substantially different beliefs (based on their prior private information) about what the true value is. As more information comes in, they update their beliefs. If this information is sufficiently favorable, more people will adopt, which creates a still-larger base of information, which causes even more people to adopt, and so on. This is the essential logic driving the learning dynamics.

To continue with our example, suppose that each agent i has a prior belief about the unknown mean μ and unknown precision $\rho = 1/\sigma^2$ such that: (i) the marginal of ρ is gamma-distributed, and (ii) for each value of ρ the conditional distribution of μ is normal with mean μ_{i0} and precision $\rho\tau_i$. (This is a standard normal–normal updating model; see De Groot 1970.) Low values of τ_i reflect *flexibility* in beliefs, that is, relatively little evidence is needed to shift i's belief about the mean by a given amount. Low values of μ_{i0} reflect *pessimism* about the payoffs from the innovation. In particular, if i initially believes that the mean is less than his costs, $\mu_{i0} < c_i$, he will not want to adopt. As more information comes in, however, his posterior estimate of the mean, μ_{it}, may increase sufficiently that he changes his mind. The point at which this happens depends, among other things, on how much information i collects and how flexible his beliefs are.

At a given point in time t, agent i will have seen (or heard about) a random assortment of prior outcomes depending on the chance encounters he had with other members of the population. In a continuous-time setting, the total "number" or measure of prior outcomes up to time t is given by the integral under the adoption curve, namely, $r(t) = \int_0^t p(s)ds$. Assuming that i is equally likely to see any particular outcome, we can model i's flow of information by a Poisson arrival process. In particular, let N_{it} be a Poisson random variable representing the number of i's observations up to time t, where $E[N_{it}] = \beta_i\, r(t)$ and the parameter β_i is a measure of the extent to which i "gets around."[16]

Let n_{it} denote the realization of N_{it}, and let $\bar{x}_{it}$ denote the mean payoff among these n_{it} observations. Given our assumptions, $\bar{x}_{it}$ is normal with mean

[16] If agents were embedded in a fixed social network, the analogous parameter would be the number of other agents with whom a given agent is connected. In this case, however, the aggregate amount of information $r(t)$ will generally not be sufficient to describe the state of the system; the dynamics of the process will depend on the specific network topology. See Golub and Jackson (2008).

μ and standard deviation $\sigma/\sqrt{n_{it}}$. In the normal–normal updating framework, i's Bayesian posterior estimate of the mean, μ_{it}, can be expressed very simply as a convex combination of μ_{i0} and $\bar{x}_{it}$, namely,

$$\mu_{it} = \frac{n_{it}\bar{x}_{it} + \tau_i\mu_{i0}}{n_{it} + \tau_i}. \tag{12}$$

In other words, the posterior estimate is just a weighted average of the prior and the observed mean, where the weight on the mean is the number of independent observations that produced it.

Given our assumption that i is myopic, she is prepared to adopt once μ_{it} is at least equal to her variable cost c_i, which by (12) is equivalent to

$$(\bar{x}_{it} - c_i)\, n_{it} \geq \tau_i \left(c_i - \mu_{i0}\right). \tag{13}$$

By assumption, $(\bar{x}_{it} - \mu)\left(\sqrt{n_{it}}/\sigma\right) = z_{it}$ is $N(0,1)$; hence (13) can be rewritten as

$$\left(\frac{\sigma z_{it}}{\sqrt{n_{it}}} + \mu - c_i\right) n_{it} \geq \tau_i \left(c_i - \mu_{i0}\right). \tag{14}$$

To make sense of this expression, let us focus on the subpopulation of agents for whom adoption is worthwhile, $P^0 = \{i\colon \mu > c_i\}$, and substitute the expected value $E[n_{it}] = \beta_i\, r(t)$ into (14). After rearranging terms we obtain

$$r(t) \geq \frac{\tau_i \left(c_i - \mu_{i0}\right)}{\beta_i \left(\mu - c_i\right)} - \frac{\sigma\sqrt{r(t)}\,z_{it}}{\left(\mu - c_i\right)\sqrt{\beta_i}}. \tag{15}$$

Define i's *resistance level* (or *information threshold*) to be the expected value of the right-hand side of this inequality, namely,

$$r_i = \frac{\tau_i \left(c_i - \mu_{i0}\right)}{\beta_i \left(\mu - c_i\right)}. \tag{16}$$

We shall say that an agent is of *type i* if she has the vector of characteristics $(c_i, \tau_i, \beta_i, \mu_{i0})$. From (15) and (16) we see that an agent of type i is increasingly likely to adopt as $r(t)$ passes the threshold r_i. Moreover, this threshold has a natural interpretation: agents with high r_i are those who are initially *pessimistic* that the innovation will cover their costs ($c_i - \mu_{i0}$ is large), *inflexible* in their initial beliefs (τ_i is large), *marginally profitable* ($\mu - c_i$ is low), and *relatively uninformed* (β_i is small).

The precise form of expressions (15) and (16) is not essential for our purpose, however: the key point is that each agent i has a *response function* $\varphi_i\,(r)$, *which represents the probability that i believes the innovation is worth adopting, given that the total amount of information generated by the prior adopters equals r.*[17]

Let us now abstract from this particular example and take the notion of a response function as a primitive of the learning model. Suppose that each agent in the population has a "type" i that is characterized by a response function φ_i: $R_+ \to [0,1]$, where $\varphi_i(r)$ is the probability that i's information threshold has been crossed when the total amount of information generated by the prior adopters is r. For ease of interpretation we shall assume that the functions $\varphi_i(r)$ are monotone nondecreasing, though this is not actually necessary for some of the results to follow. Notice that a given individual will typically know only a small fraction of the prior outcomes, that is, r is a state variable that represents a common pool of information but it is not common knowledge.

Assume that there are countably many types in the population, and let p_i be the *proportion* of i-types.[18] When the total information generated by prior adopters equals r, the proportion of the population whose thresholds have been crossed is given by the function

$$F(r) = \sum_i p_i \varphi_i(r). \tag{18}$$

$F(r)$ is a monotone nondecreasing function which we can interpret as a *notional distribution function* of agents' information thresholds. We allow for the possibility that some agents have an infinite threshold; hence, $\lim_{r\to\infty} F(r)$ may be less than one.[19]

[17] In the present case, the response function has the following explicit representation. Given that $r(t) = r$, n_{it} is Poisson-distributed with mean $\beta_i\, r$. Given a realization $n_{it} = k > 0$, the mean observed payoff, $\bar{x}_{(it)}$, is normal with mean μ and variance σ^2/k. Let Φ denote the standard normal CDF. Then the probability that i's posterior estimate exceeds i's costs, i.e., the probability that (14) holds, is

$$\varphi_i(r) = \sum_{k=1}^{\infty} \frac{(\beta_i r)^k e^{-\beta_i r}}{k!} \Phi\left[\frac{(\mu - c_i)\sqrt{k}}{\sigma} - \frac{\tau_i\,(c_i - \mu_{i0})}{\sigma\sqrt{k}}\right]. \tag{17}$$

[18] Alternatively we could assume that the types are uncountable and distributed according to a density; this makes no material difference in the subsequent analysis.

[19] The closest model to this one in the literature is due to Dodds and Watts (2004, 2005). They consider a cumulative-dose model of infection with heterogeneity in the thresholds at which agents become infected (including social as well as biological interpretations of infection), though their analysis of the dynamics emphasizes different features from the ones considered here.

Recalling that $r(t) = \int_0^t p(s)ds$, it follows that the aggregate dynamics is described by the differential equation

$$\dot{p}(t) = \lambda\left[F\left(\int_0^t p(s)ds\right) - p(t)\right], \lambda > 0. \tag{19}$$

Observe that (19) is analogous to the dynamical equation (9) defining a social threshold model, except that in the present case the argument of F is the *integral of the adoption curve* rather than the adoption curve itself. This arises because agents use all past information generated by previous adopters rather than just the most recent information. Note that this formulation hinges on our assumption that payoff information is equally informative no matter when it was received. If this were not the case, for example, if payoffs received in earlier time periods were less valuable, the defining equation would have to be modified accordingly.[20]

The cumulative feature of the social learning model has some important implications for the shape of the adoption curve. Compared to the social influence model, acceleration is initially quite weak; in fact, in a neighborhood of the origin the process strictly *decelerates* irrespective of the distribution generating it. To see why this is so, differentiate equation (19) to obtain the acceleration equation

$$(1/\lambda)\ddot{p}(t) = p(t)f(\mathrm{r}(t)) - \dot{p}(t). \tag{20}$$

By assumption, $p(0) = 0$ and the solution $p(t)$ is continuous, so the first term is close to zero when t is close to zero. However, the second term is bounded away from zero when t is close to zero, because (19) implies that $\dot{p}(0) = \lambda F(0) > 0$. Hence,

$$\lim_{t\to 0^+}\ddot{p}(t) = -\lambda^2 F(0) + f(p(0))p(0) < 0. \tag{21}$$

More generally, when we compare (20) with the acceleration equation for the social influence model, namely, $(1/\lambda)\ddot{p}(t) = \dot{p}(t)f(p(t)) - \dot{p}(t)$, we see that the former has weaker initial acceleration than the latter because $\dot{p}(t)$ is initially positive, whereas $p(t)$ is initially zero. *The result may be an extended period of weak growth in the early phases of social learning.* The reason is that the initial

[20] If payoff information is discounted at some rate $\delta > 0$, the dynamical equation takes the form $\dot{p}(t) = \lambda\left[F\left(\int_0^t e^{-\delta(t-s)}p(s)ds\right) - p(t)\right]$. Note that when δ is large, this process is similar to the social influence model. Another modification is to allow for joint heterogeneity in λ and r; this extension is straightforward and is left to the reader.

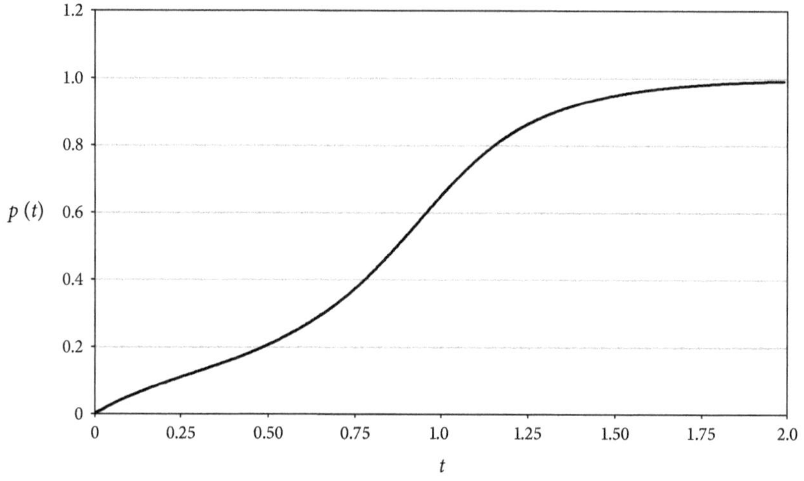

Figure 11.4 Adoption curve generated by social learning, a normal distribution of information thresholds. $N(0.10, 0.01)$, and $\lambda = 4$.

block of optimists $F(0)$ exerts a decelerative drag on the process: they contribute at a decreasing rate as their numbers diminish, while the information generated by the new adopters gathers steam slowly because there are so few of them to begin with.[21] Figure 11.4 illustrates this phenomenon for the same density that generated the social thresholds adoption curve in Figure 11.3.

Next we shall investigate the behavior of the relative acceleration rate $\ddot{p}(t)/\dot{p}(t)$. To avoid uninteresting technical complications let us assume that the density is well behaved near the origin, that is, $\bar{f}(0) = \lim_{r \to 0^+} f(r) > 0$, and that $f'(r)$ is continuous and bounded. Define

$$\alpha(t) = (1/\lambda)\ddot{p}(t)/\dot{p}(t). \tag{22}$$

From (20) we deduce that

$$\alpha(t) = f(r(t))p(t)/\dot{p}(t) - 1. \tag{23}$$

Differentiating (23), we obtain

$$\dot{\alpha}(t) = f'(r(t))p^2(t)/\dot{p}(t) + f(r(t)) - f(r(t))\,\ddot{p}\,(t)\,p(t)/\dot{p}^2(t). \tag{24}$$

[21] Initial deceleration does not necessarily occur, however, if observations are *bunched* at particular points in time. An example would be an agricultural innovation (e.g., a new type of crop) that is tried once in each growing season, and whose outcomes farmers observe at the end of the season. This case will be discussed further in Section 6.

As $t \to 0^+$, the first term in (24) goes to zero, because by assumption $f'(r)$ is bounded, $p(t) \to 0$, and $\dot{p}(0) > 0$. The third term also goes to zero. However, $f(r(t)) \to \bar{f}(0) > 0$, so the second term is positive in the limit. It follows from continuity that $\dot{\alpha}(t)$ is strictly positive on some initial interval $0 \leq t \leq T$. In the region near the origin where $\ddot{p}(t) < 0$, this says that the relative acceleration rate is becoming less negative.

Now, suppose that at some time t_0, the process begins to accelerate. If at this point the density is increasing, then $f'(r(t_0)) > 0$, and it follows from (24) and continuity that $\dot{\alpha}(t)$ is positive for some interval of time beginning at t_0. In other words, if the density is increasing when the process begins to accelerate (assuming it ever does accelerate), then the process undergoes a period of superexponential growth.

Proposition 3 *Suppose that diffusion is driven by social learning, where there is a positive mass of individuals who are willing to adopt even when no one else adopts* ($F(0) > 0$)*, and the distribution of resistances has a density* $f(r)$ *that is bounded away from zero near the origin and has a continuous bounded derivative. Then, initially the process decelerates whereas the relative acceleration rate strictly increases. If the process begins to accelerate, and if the density is increasing at that point, then the process undergoes a period of superexponential growth.*

Propositions 2 and 3 show that both the social influence model and the social learning model can exhibit periods of superexponential growth. This is consistent with observations by various authors that adoption curves often exhibit a sudden take-off phase after a slow start (Ryan and Gross 1943; Henrich 2001; Vendrik 2003; Rogers 2003). The propositions above are more specific in predicting superexponential growth, and they clarify why one would expect it theoretically. Theory also suggests why the curves generated by social influence and social learning will differ in other respects. One difference is that, in a social learning model, the speed of adoption at a given point in time depends on adoption levels at earlier points in time, whereas in a social influence model this is not the case. Given sufficiently detailed adoption data, this effect could be tested by regression analysis with lagged variables.

A second distinction between the models is the way in which the resistance thresholds are constructed. In a social influence model the thresholds reflect differences in a single scalar parameter (people's tendency to conform), whereas in a social learning model they reflect differences in a variety of factors, including prior beliefs, amount of information gathered, and the payoff gain from the innovation. Some of these factors are unobservable and would be difficult to estimate even with micro-level adoption data. However, the payoff gain—the parameter μ in the social learning model—can sometimes be estimated from the data, as we shall see in Section 6.

The key point to observe is that higher average payoffs from adoption shift the individual response functions upward: for a given "number" of prior outcomes r in the population, the probability $\varphi_i(r)$ that any given individual i adopts increases the higher the potential payoff gain μ. Thus, if we consider two distinct populations $\{1, 2\}$ such that one has a higher mean payoff than the other, say $\mu_2 > \mu_1$, then all else being equal the resistance distributions satisfy $F_2(r) \geq F_1(r)$ for all r. (F_1 first-order stochastically dominates F_2.) Under these circumstances, adoption occurs more rapidly in population 2 than in population 1; that is, the solutions satisfy $p_2(t) \geq p_1(t)$ for all t.[22] This is one way to evaluate the plausibility of the social learning model without knowing much about the distribution of the learning parameters.

6 The Diffusion of Hybrid Corn

Diffusion data that permit definitive tests of the preceding propositions are not easy to come by. One reason is that an adoption curve must be charted at frequent time intervals in order to conduct tests of statistical significance. Consider, for example, a product or practice that took ten years to go from introduction to completion. If the adoption level is measured annually, there will be only ten data points, only nine first differences, and only eight second differences; furthermore, the errors in successive differences will not be independent. Of course, the situation would be different if we had monthly or weekly data over a period of ten years, but there appear to be very few datasets of this sort in the literature. Moreover, the data that do exist tend to be for new products where mass advertising played a critical role. In these cases it is difficult to sort out the relative importance of internal and external sources of information, though the mixed-influence contagion model is certainly a step in that direction.

Fortunately, there exist some data that are sufficiently detailed to carry out a preliminary analysis of the ideas discussed above. The example we choose here

[22] Consider the dynamical equations $\dot{p}_i(t) = \lambda\,[F_i(p_i(t)) - p_i(t)]$, $i = 1,2$. Let $p_1(t), p_2(t)$ be the solutions, which we assume to be continuous. Note that $\dot{p}_2(0) \geq \dot{p}_1(0)$, because by assumption $F_2(0) \geq F_1(0)$ and $p_1(0) = p_2(0) = 0$. Suppose, by way of contradiction, that $p_2(t) < p_1(t)$ at some time $t > 0$. By continuity there is a time $\tilde{t} > 0$ such that $p_2(s) \geq p_1(s)$ for all $s \leq \tilde{t}$ and $\dot{p}_2(\tilde{t}) < \dot{p}_1(\tilde{t})$. Therefore, $\int_0^{\tilde{t}} p_2(s)ds \geq \int_0^{\tilde{t}} p_1(s)ds$ and $F_2\left(\int_0^{\tilde{t}} p_2(s)ds\right) \geq F_1\left(\int_0^{\tilde{t}} p_1(s)ds\right)$, from which it follows that $\dot{p}_2(\tilde{t}) \geq \dot{p}_1(\tilde{t})$, a contradiction.

is Ryan and Gross's classic study of the diffusion of hybrid corn in two Iowa communities (Ryan and Gross 1943). This work prompted a number of later studies on innovation diffusion, including Zvi Griliches's analysis of hybrid corn diffusion across the midwestern and southern United States (Griliches 1957). Although the latter is more comprehensive in its geographical coverage, the study by Ryan and Gross has several advantages for our purposes. First, it was carried out at the community level, whereas the adoption curves in Griliches's study were aggregated over several counties and involved scores of dispersed communities. Second, Ryan and Gross provided the proportion of hybrid corn adopters, whereas Griliches gave the proportion of *acreage* planted in hybrid corn, which (among other things) biases the results toward large landowners. Third, the Ryan and Gross data covered the critical start-up phase in the 1920s, which was absent in Griliches's analysis. Finally, Ryan and Gross's study included important related information, such as the year that farmers first learned about the new technology and the sources from which they learned about it. These data provide additional contextual details that, together with the shape of the curve itself, allow us to assess the relative plausibility of different diffusion mechanisms.

The adoption curve is shown in Figure 11.5. Particularly noteworthy features are the long choppy start, which lasted for some six years (1927–1932), followed by a sharp acceleration for about four years (1933–1936), and then a rapid deceleration as nearly 100 percent acceptance levels were reached in 1941.[23]

Ryan and Gross repeatedly emphasize that natural *conservatism* was one of the key reasons why farmers delayed so long in adopting a technology that could have substantially increased their profits. Indeed, their survey data reveal that, although at least two-thirds of the farmers had heard about the advantages of hybrid corn as early as 1931, only about 8 percent had adopted by then. Thus, pure inertia could well have been the explanation for the long delay. Nevertheless, the shape of the curve suggests that this was not the case. The reason is that, in a pure inertia model with heterogeneous levels of inertia, the adoption curve must be concave throughout. This hypothesis can be rejected with reasonable confidence, since there was deceleration in only one of the first ten time periods, and sharp acceleration in most of them. Thus, although inertia may have been a contributing factor, it was unlikely to have been the sole factor.

[23] These features have been noted by other authors, notably Henrich (2001), who explains them by a combination of learning from direct experience and biased cultural transmission.

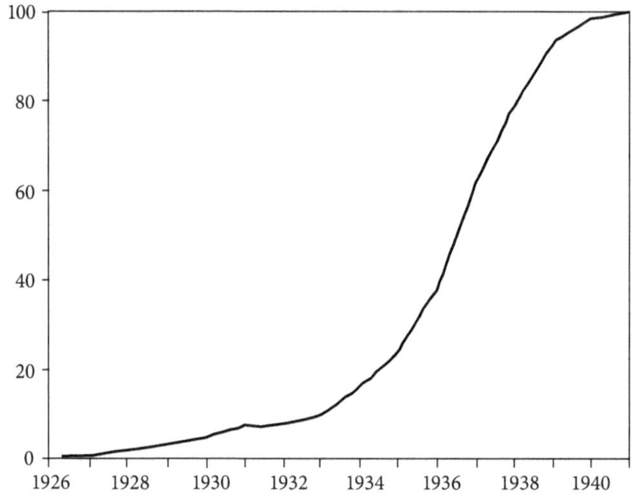

Percent	0.0	0.4	1.4	3.0	4.6	6.9	7.3	9.6	15.8	24.0	38.0	62.0	79.0	93.0	98.5	99.7
Year	1926	1927	1928	1929	1930	1931	1932	1933	1934	1935	1936	1937	1938	1939	1940	1941

Figure 11.5 Percent of adopters of hybrid corn in two Iowa communities, 1926–1941.

Source: From Ryan and Gross 1943, Figure 4.

Another interesting aspect of Ryan and Gross's survey is that farmers reported their *sources of information* about the new technology, and identified the sources they believed were most influential in their decision to adopt. Whereas seed salesmen were listed as the most common *original source* of knowledge, neighbors were listed as the *most influential in deciding whether to implement it.* This is broadly consistent with the three classes of models under consideration, all of which posit that adoption results from an internal feedback loop between earlier and later adopters.

Let us consider the contagion model first. This posits that adoption decisions are driven by a combination of internal and external influences, where external influence operates at a constant rate and internal influence operates at a rate proportional to the current number of adopters. We know that this model generally produces *S*-shaped curves, but we also know that the curve has a particular property: the hazard rate must be nonincreasing relative to the current number of adopters. (It can be verified that this property also holds in the discrete-time version.)

To check this, let p_t be the proportion of adopters at the end of period *t*, and let $\Delta_t = p_{t+1} - p_t$ be the rate of change at *t*. Then, $H_t = \Delta_t / p_t\,(1 - p_t)$ should be nonincreasing in *t*, and strictly decreasing if there is any outside influence. The piecewise linear curve shown in Figure 11.6 is represented by H_t.

To test whether it is monotone nonincreasing, let us fit a cubic polynomial using OLS (the smooth line in Figure 11.6). The overall fit is very good ($R^2 = 0.86$), and the quadratic and cubic coefficients *differ in sign* at a highly significant level. We therefore have reason to believe that H_t is not a monotone function, which is corroborating evidence against the contagion hypothesis. Of course this is not the same as saying that there is *no* contagion effect, merely that this effect *by itself* is unlikely to have produced the Ryan and Gross data.

Next let us consider the social influence and social learning models. Ryan and Gross argued that the balance of evidence from the survey suggested some form of learning by farmers: "In some sense the early acceptors provided a community laboratory from which neighbors could gain some vicarious experience with the new seed over a period of some years" (Ryan and Gross 1943: 16). This is precisely the effect posited in the social learning model. Is social learning consistent with the shape of the curve? The answer is affirmative. Indeed the curve has two features that are quite typical of social learning models: there is a long slow start-up followed by very rapid acceleration. We remind the reader why this is so: in the early phases, adoption is driven mainly by people who are already convinced that the new technology is a good idea but who

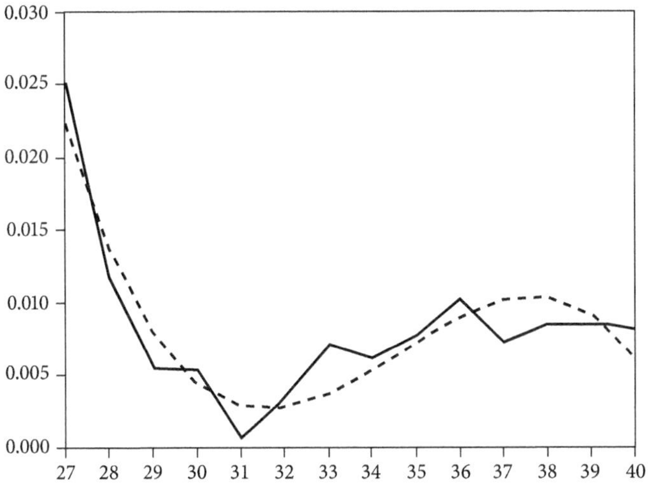

$$H_t = a + b_1 t + b_2 t^2 + b_3 t^3 + \varepsilon$$

Coefficient	*t-statistic*
$a = 0.034$ (0.0034)	10.11
$b_1 = -0.013$ (0.0019)	−7.15
$b_2 = 0.0018$ (0.00029)	6.23
$b_3 = -0.000069$ (0.000013)	−5.48

Figure 11.6 Cubic Polynomial (*Smooth Line*) fitted by OLS to the relative hazard rate $H_t = \Delta_t / p_t(1 - p_t)$ (*Jagged Line*).

may delay due to inertia; once this effect is overcome, there is rapid acceleration caused by the accumulation of information provided by prior adopters, combined with the rising number of people at the margin who are persuaded by the new information.

Unfortunately, the curve does not have enough data points to carry out tests of significance, but we can make a rough estimate of the relative acceleration rate in the following way. First, compute the average slope over successive three-year intervals in order to smooth out the data. Then, compute the relative acceleration rate as the ratio of successive slopes minus unity.[24] The results are shown in Table 11.1 for three different ways of grouping the data. In every case, the relative acceleration rate rises sharply from the second to the third period. (Recall that in an exponential growth model the relative acceleration rate would be constant.) While this is not proof of superexponential acceleration, it is certainly consistent with it.

These features provide corroborating evidence for social learning, but they certainly do not constitute identification. For example, we cannot rule out the possibility that in the mid-1930s there was an exogenous shock (e.g., a

Table 11.1 Relative acceleration rate measured over three-year intervals.

Interval	$\bar{\Delta}_i$	$\bar{\Delta}_{i+1}/\bar{\Delta}_i - 1$
1927–1929	1.0	—
1930–1932	1.4	1.4
1933–1935	5.6	4.0
1936–1938	18.3	3.3
1928–1930	1.4	—
1931–1933	1.7	1.2
1934–1936	9.5	5.6
1937–1939	18.3	1.9
1929–1931	1.8	—
1932–1934	3.0	1.7
1935–1937	15.4	5.1
1938–1940	12.2	0.8

Note: $\bar{\Delta}_i$ is the average annual change (in percent) over the *i*th interval, and relative acceleration is $\bar{\Delta}_{i+1}/\bar{\Delta}_i - 1$.

[24] If Δ_t is the slope in period t, then $(\Delta_{t+1} - \Delta_t)/\Delta_t = \Delta_{t+1}/\Delta_t - 1$ is an estimate of the relative acceleration rate at t.

price war among seed companies) that caused the sudden uptick.[25] Moreover, the sharp rise in acceleration is equally consistent with the social influence model. In this case, however, there is evidence from another source—namely Griliches's study—that social influence was probably not the sole explanation. The reason is that, in a social learning model, the rate of adoption depends on the potential gain from adoption, whereas in a social influence model this is not the case. One of Griliches's main findings was that in the midwestern states generally (including Iowa), the rate of adoption in each region was positively correlated with the estimated monetary gain from adoption, just as the social learning model predicts.

Putting all of this information together, we can say that social learning is consistent with the observed pattern of diffusion of hybrid corn, although we cannot say that it was the sole explanatory factor. We can also say with some confidence, however, that inertia and contagion were probably *not* the sole explanatory factors, and given Griliches's findings neither was social influence.

While the Ryan and Gross study is quite limited in scope, it highlights some of the issues that would need to be addressed in a more complete empirical analysis. First, one would need a curve—or set of curves—that are measured at frequent time intervals. The reason is that one requires an accurate estimate of the acceleration pattern at precisely the point where acceleration is most rapid; hence, frequent observations in this part of the curve are crucial for running tests of significance. Second, one needs data that cover the whole life cycle of the adoption process, including the very early phases when adoption is just getting started. Many of the datasets in the literature are quite weak in this respect; the reason, of course, is that researchers are not likely to focus their attention on an innovation until it is already well underway.

Third, the framework developed here applies to situations where there is informational feedback between members of a group who interact more or less at random. If information flows through a fixed network, say between individuals who are near neighbors (in a geographical or social sense), then a different analysis will be required. (Ryan and Gross did not say whether farmers learned more or less uniformly from others in the community, or mainly from their immediate neighbors.) Fourth, we have restricted our attention to social learning models in which people can directly observe outcomes generated by other adopters. In the case of hybrid corn, this assumption seems reasonable; indeed Ryan and Gross specifically say that farmers looked at their neighbors' fields

[25] It is worth noting, however, that Griliches investigated whether price changes in hybrid seed could explain the rates of adoption in various parts of the Midwest (including Iowa), and he found no significant effect (Griliches 1957: 503).

to assess the viability of the new technology (Ryan and Gross 1943: 16). In situations where direct observation is not possible, one would need an analog of the theory in which inferences are made indirectly.

Finally, the Ryan and Gross study highlights the importance of having supporting data about the agents' sources of information and the reasons they delayed in acting on it. These contextual details, combined with an analysis of the shape of the curve, have the potential to discriminate between alternative diffusion mechanisms even when detailed micro-level data are unavailable.

Appendix

Proof of Proposition 1 Define the function $H(t) = h(t)/p(t) = \dot{p}(t)/[p(t)(1 - p(t))]$; this is well defined for all $t > 0$ because, by assumption, $\dot{p}(0) = \bar{\gamma} > 0$; hence, $p(t) > 0$ when $t > 0$. We need to show that $\dot{H}(t) < 0$.

For each parameter pair (λ, γ), let $q_{\lambda,\gamma}(t) = (1 - p_{\lambda,\gamma}(t))$ denote the *proportion of the* (λ, γ)*-population that has not yet adopted by time t*. The proportion of the total population that has not adopted by t is therefore

$$q(t) = \int q_{\lambda,\gamma}(t)d\mu. \tag{A1}$$

For each (λ, γ), we have

$$\dot{p}_{\lambda,\gamma}(t) = (\lambda p(t) + \gamma)q_{\lambda,\gamma}(t)\ . \tag{A2}$$

Integration with respect to μ yields

$$\dot{p}(t) = [\lambda(t)p(t) + \gamma(t)]q(t)\ , \tag{A3}$$

where

$$\lambda(t) = \frac{\int \lambda q_{\lambda,\gamma}(t)d\mu}{q(t)} \qquad \gamma(t) = \frac{\int \gamma q_{\lambda,\gamma}(t)d\mu}{q(t)}. \tag{A4}$$

Note that $\lambda(t)$ and $\gamma(t)$ are the *expected* values of λ and γ in the population of nonadopters at time t. It follows that

$$H(t) = \dot{p}(t)/[p(t)q(t)] = \lambda(t) + \gamma(t)/p(t). \tag{A5}$$

Claim:

$$\textit{For every } t > 0, \ \dot{\lambda}(t)p(t) + \dot{\gamma}(t) \leq 0. \tag{A6}$$

Proof of claim For every $t > 0$, we have

$$\dot{\lambda}(t) = \frac{\int \lambda \dot{q}_{\lambda,\gamma}(t)d\mu}{\int q_{\lambda,\gamma}(t)d\mu} - \frac{\left[\int \lambda q_{\lambda,\gamma}(t)d\mu\right]\left[\int \dot{q}_{\lambda,\gamma}(t)d\mu\right]}{\left[\int q_{\lambda,\gamma}(t)d\mu\right]^2}. \tag{A7}$$

and

$$\dot{\gamma}(t) = \frac{\int \gamma \dot{q}_{\lambda,\gamma}(t)d\mu}{\int q_{\lambda,\gamma}(t)d\mu} - \frac{\left[\int \gamma q_{\lambda,\gamma}(t)d\mu\right]\left[\int \dot{q}_{\lambda,\gamma}(t)d\mu\right]}{\left[\int q_{\lambda,\gamma}(t)d\mu\right]^2}. \tag{A8}$$

To show that $\dot{\lambda}(t)p(t) + \dot{\gamma}(t) \leq 0$, multiply (A7) by $p(t)$ and add it to (A8); after simplifying, we obtain the equivalent condition

$$\left[\int (\lambda p(t) + \gamma)\dot{q}_{\lambda,\gamma}(t)d\mu\right]\left[\int q_{\lambda,\gamma}(t)d\mu\right] - \left[\int (\lambda p(t) + \gamma)q_{\lambda,\gamma}(t)d\mu\right]$$
$$\left[\int \dot{q}_{\lambda,\gamma}(t)d\mu\right] \leq 0 \tag{A9}$$

We need to show that (A9) holds for every $t > 0$. (Notice that t does not vary in this expression; t is fixed and integration is taken with respect to λ and γ.) We know from (A2) that $\dot{q}_{\lambda,\gamma}(t) = -(\lambda p(t) + \gamma)q_{\lambda,\gamma}(t)$ for every λ, γ, and t. Substituting this into (A9) we obtain

$$\left[\int (\lambda p(t) + \gamma)^2 q_{\lambda,\gamma}(t)d\mu\right]\left[\int q_{\lambda,\gamma}(t)d\mu\right] \geq \left[\int (\lambda p(t) + \gamma)q_{\lambda,\gamma}(t)d\mu\right]^2. \tag{A10}$$

Fix $t > 0$ and define the random variables

$$X = (\lambda p(t) + \gamma)\sqrt{q_{\lambda,\gamma}(t)} \text{ and } Y = \sqrt{q_{\lambda,\gamma}(t)}\ .$$

The realizations of X and Y are determined by random draws from μ. Thus, (A10) follows directly from Schwarz's inequality: $E[X^2]\,E[Y^2] \geq (E[XY])^2$. This establishes the claim.

We can now apply this result to show that $H(t)$ is strictly decreasing in t for all $t > 0$. Differentiating $H(t)$, we obtain

$$\dot{H}(t) = \dot{\lambda}(t) + \dot{\gamma}(t)/p(t) - \gamma(t)\dot{p}(t)/p^2(t). \tag{A12}$$

By the claim above, $\dot{\lambda}(t)p(t) + \dot{\gamma}(t) \leq 0$, so division by $p(t) > 0$ yields $\dot{\lambda}(t) + \dot{\gamma}(t)/p(t) \leq 0$. Thus, the sum of the first two terms on the right-hand side of (A12) is nonpositive. But the last term is strictly negative, because $\gamma(t) > 0$ for all $t > 0$, given the initial condition $\bar{\gamma} = \gamma(0) > 0$. Hence, $\dot{H}(t) < 0$ as was to be shown.

Proof of proposition 2 The equations of motion are

$$\dot{p}_\lambda(t) = \lambda\,[F_\lambda(p(t)) - p_\lambda(t)]\ , \tag{A13}$$

where the initial conditions are $p_\lambda(0) = 0$.

By assumption, $F_\lambda(0) > 0$; hence $\dot{p}_\lambda(t) > 0$ for all sufficiently small $t > 0$ and

$$\ddot{p}_\lambda(t)/\dot{p}_\lambda(t) = \lambda\,[f_\lambda(p(t)) - 1]\,. \tag{A14}$$

By hypothesis, the functions $f_\lambda(p(t))$ are strictly increasing on some common neighborhood of the origin $0 < p(t) < \bar{p}$. It follows that, for every λ and all t in a suitable interval $(0, T]$,

$$\dddot{p}_\lambda/\dot{p}_\lambda(t) - (\ddot{p}_\lambda(t)/\dot{p}_\lambda(t))^2 > 0\,. \tag{A15}$$

Hence, we have

$$\sqrt{\dddot{p}_\lambda(t)\dot{p}_\lambda(t)} > \ddot{p}_\lambda(t)\,, \tag{A16}$$

$$\int \sqrt{\dddot{p}_\lambda(t)\dot{p}_\lambda(t)}dv > \int p_\lambda(t)dv = \ddot{p}(t). \tag{A17}$$

By Schwarz's inequality,

$$\sqrt{\dddot{p}(t)\dot{p}(t)} = \left(\int \dddot{p}_\lambda(t)dv \int \dot{p}_\lambda(t)dv\right)^{1/2} \geq \int \sqrt{\dddot{p}_\lambda(t)\dot{p}_\lambda(t)}dv. \tag{A18}$$

Combining this with (A17), we conclude that $\sqrt{\dddot{p}(t)\dot{p}(t)} > \ddot{p}(t)$, which implies that $d(\ddot{p}(t)/\dot{p}(t))/dt > 0$, that is, $\ddot{p}(t)/\dot{p}(t)$ is strictly increasing on $(0, T]$. This concludes the proof of Proposition 2.

Acknowledgments

I would like to thank Paul David, Steven Durlauf, Joshua Epstein, Edoardo Gallo, James Heckman, Josef Hofbauer, Charles Manski, David Myatt, Thomas Norman, Thomas Valente, Duncan Watts, Tiemen Woutersen, and two anonymous referees for valuable comments on previous drafts. Pierangelo de Pace and Jon Parker assisted with the data analysis. This research was supported in part by the Brookings Institution and the Santa Fe Institute.

References

Bala, Venkatesh and Goyal, Sanjeev (1998), "Learning from neighbours," *Review of Economic Studies* 65 (3): 595–621.

Banerjee, Abhijit V. (1992), "A simple model of herd behavior," *Quarterly Journal of Economics* 107 (3): 797–817.

Banerjee, Abhijit V. and Fudenberg, Drew (2004), "Word-of-mouth learning," *Games and Economic Behavior* 46 (1): 1–22.

Bass, Frank M. (1969), "A new product growth model for consumer durables." *Management Science*,15 (5): 215–227.

Bass, Frank M. (1980), "The relationship between diffusion rates, experience curves, and demand elasticities for consumer durable technological innovations." *Journal of Business* 53 (3): S51–67.

Bikhchandani, Sushil, Hirshleifer, David, and Welch, Ivo (1992), "A theory of fads, fashion, custom, and cultural change in informational cascades," *Journal of Political Economy* 100 (5): 992–1026.

Blume, Lawrence E. (1993), "The statistical mechanics of strategic interaction," *Games and Economic Behavior* 5 (3): 387–424.
Blume, Lawrence E. (1995), "The statistical mechanics of best-response strategy revision," *Games and Economic Behavior* 11 (2): 111–145.
Boyd, Robert, and Richerson, Peter J. (1985), *Culture and the Evolutionary Process* (Chicago: University of Chicago Press).
Centola, Damon (2006), "Free riding and the weakness of strong incentives," Unpublished.
Chatterjee, Kalyan, and Xu, Susan H. (2004), "Technology diffusion by learning from neighbors," *Advances in Applied Probability* 36 (2): 355–376.
Coddington, E. A. and Levinson, N. (1955), *Theory of Ordinary Differential Equations* (New York: McGraw-Hill).
Coleman, James S., Katz, Elihu, and Menzel, Herbert (1957), "The diffusion of an innovation among physicians," *Sociometry* 20 (4): 253–270.
Coleman, James S., Katz, Elihu, and Menzel, Herbert (1966), *Medical Innovation: A Diffusion Study* (Indianapolis: Bobbs-Merrill).
Conley, Timothy G. and Udry, Christopher R. (2010), "Learning about a new technology: pineapple in Ghana," *American Economic Review, 100 (1):35-69.*
Cowan, Robin and Jonard, Nicolas (2004), "Network structure and the diffusion of knowledge," *Journal of Economic Dynamics and Control* 28 (8): 1557–1575.
David, Paul A. (1968), "The mechanization of reaping in the ante-bellum midwest," in Henry Rosovksy ed.,*Industrialization in Two Systems: Essays in Honor of Alexander Gershenkron* (New York: Wiley, 3–28).
David, Paul A. (1969), "A contribution to the theory of diffusion," Stanford University Research Center in Economic Growth Memorandum 71.
David, Paul A. (1975), *Technical Change, Innovation, and Economic Growth* (Cambridge: Cambridge University Press).
David, Paul A. (2005), "Zvi Griliches on diffusion, lags, and productivity growth…connecting the dots," *Labor and Demography* Working Paper 0502012.
David, Paul A. and Olsen, Trond E. (1984), "Anticipated automation: a rational expectations model of technological diffusion," Stanford University Center for Economic Policy Research Technological Innovation Project Working Paper 2.
David, Paul A. and Olsen, Trond E. (1986), "Equilibrium dynamics of diffusion when incremental technological innovations are foreseen," *Ricerche Economiche* 40 (4): 738–770.
De Groot, Morris (1970), *Optimal Statistical Decisions* (New York: McGraw Hill).
Dodds, Peter Sheridan and Watts, Duncan J. (2004), "Universal behavior in a generalized model of contagion," *Physical Review Letters* 92 (21): 218701–218704.
Dodds, Peter Sheridan and Watts, Duncan J. (2005), "A generalized model of social and biological contagion." *Journal of Theoretical Biology* 232: 587–604.
Easingwood, Christopher J., Mahajan, Vijay, and Muller, Eitan (1981), "A non-symmetric responding logistic model for technological substitution." *Technological Forecasting and Social Change* 20: 199–213.
Easingwood, Christopher J., Mahajan, Vijay, and Muller, Eitan (1983), "A nonuniform influence innovation diffusion model of new product acceptance," *Marketing Science* 2: 273–296.
Ellison, Glenn and Fudenberg, Drew (1993), "Rules of thumb for social learning," *Journal of Political Economy* 101 (4): 612–643.
Ellison, Glenn and Fudenberg, Drew (1995), "Word-of-mouth communication and social learning," *Quarterly Journal of Economics* 110 (1): 93–125.

Foster, Andrew D. and Rosenzweig, Mark R. (1995), "Learning by doing and learning from others: human capital and technical change in agriculture," *Journal of Political Economy* 103 (6): 1176–1209.

Geroski, Paul A. (2000), "Models of technology diffusion," *Research Policy* 29 (4–5): 603–625.

Golub, Benjamin and Jackson, Matthew O. (2008), "Naive learning in social networks: convergence, influence, and the wisdom of crowds," Unpublished.

Granovetter, Mark (1978), "Threshold models of collective behavior," *American Journal of Sociology* 83: 1420–1443.

Granovetter, Mark and Soong, Roland (1988), "Threshold models of diversity: chinese restaurants, residential segregation, and the spiral of silence," *Sociological Methodology* 18: 69–104.

Griliches, Zvi (1957), "Hybrid corn: an exploration of the economics of technological change," *Econometrica* 25: 501–522.

Heckman, James J. and Singer, Burt (1982), "The identification problem in econometric models for duration data," in Werner Hildenbrand ed.,*Advances in Econometrics* (Cambridge: Cambridge University Press, 39–77).

Heckman, James J., Robb, Richard, and Walker, James R. (1990), "Testing the mixture of exponentials hypothesis and estimating the mixing distribution by the method of moments," *Journal of the American Statistical Association* 85 (410): 582–589.

Henrich, Joseph (2001), "Cultural transmission and the diffusion of innovations: adoption dynamics indicate that biased cultural transmission is the predominant force in behavioral change," *American Anthropologist* 103: 992–1013.

Jackson, Matthew O. and Rogers, Brian W. (2007), "Relating network structure to diffusion properties through stochastic dominance," *BE Journal of Theoretical Economics: Advances in Theoretical Economics* 7 (1): 1–13.

Jackson, Matthew O. and Yariv, Leeat (2007), "Diffusion of behavior and equilibrium properties in network games," *American Economic Review* 97 (2): 92–98.

Jensen, Richard (1982), "Adoption and diffusion of an innovation of uncertain profitability," *Journal of Economic Theory* 27 (1): 182–193.

Jeuland, Abel P. (1981), "Incorporating heterogeneity into parsimonious models of diffusion of innovation," University of Chicago Center for Research on Marketing Working Paper 45.

Kapur, Sandeep (1995), "Technological diffusion with social learning," *Journal of Industrial Economics* 43 (2): 173–195.

Karshenas, Massoud and Stoneman, Paul (1992), "A flexible model of technological diffusion incorporating economic factors with an application to the spread of colour television ownership in the UK," *Journal of Forecasting* 11: 577–601.

Kirman, Alan (1993), "Ants, rationality, and recruitment," *Quarterly Journal of Economics* 108 (1): 137–156.

Lancaster, Tony and Nickell, Stephen (1980), "The analysis of re-employment probabilities for the unemployed," *Journal of the Royal Statistical Society, Series A* 143: 141–152.

Lopez-Pintado, Dunia and Watts, Duncan J. (2008), "Social influence, binary decisions, and collective dynamics," *Rationality and Society* 20 (4): 399–443.

Macy, Michael W. (1991), "Chains of cooperation: threshold effects in collective action," *American Sociological Review* 56: 730–747.

Mahajan, Vijay and Peterson, Robert A. (1985), *Models for Innovation Diffusion* (Beverly Hills, CA: Sage Publications).

Mahajan, Vijay, Muller, Eitan, and Bass, Frank M. (1990), "New product diffusion models in marketing: a review and directions for further research," *Journal of Marketing* 54: 1–26.

Mansfield, Edwin (1961), "Technical Change and the Rate of Innovation," *Econometrica* 29: 741–766.

Manski, Charles F. (2004), "Social learning from private experiences: the dynamics of the selection problem" *Review of Economic Studies* 71 (2): 443–458.

Munshi, Kaivan (2004), "Social learning in a heterogeneous population: technology diffusion in the Indian Green Revolution," *Journal of Development Economics* 73 (1): 185–213.

Rogers, Everett M. (2003), *Diffusion of Innovations*. 5th ed (New York: Free Press).

Ryan, Bryce and Gross, Neal C. (1943), "The diffusion of hybrid corn in two Iowa communities" *Rural Sociology* 8 (1): 15–24.

Schelling, Thomas C. (1971). "Dynamic Models of Segregation," *Journal of Mathematical Sociology* 1: 143–186.

Schelling, Thomas C. (1978), *Micromotives and Macrobehavior* (New York: Norton).

Smith, Lones and Sorensen, Peter (2000), "Pathological outcomes of observational learning," *Econometrica* 68 (2): 371–398.

Stoneman, Paul. (2002), *The Economics of Technological Diffusion*. Oxford: Blackwell.

Valente, Thomas W. (1995), *Network Models of the Diffusion of Innovations* (Cresskill, NJ: Hampton Press).

Valente, Thomas W. (1996), "Social network thresholds in the diffusion of innovations" *Social Networks* 18: 69–89.

Valente, Thomas W. (2005), "Network models and methods for studying the diffusion of innovations," in Peter J. Carrington, John Scott, and Stanley Wasserman, eds,*Models and Methods in Social Network Analysis* (Cambridge: Cambridge University Press, 98–116).

Vega-Redondo, Fernando (2007), *Complex Social Networks* (Cambridge: Cambridge University Press).

Vendrik, Maarten C. M. (2003), "Bandwagon effects on female labour force participation: an application to the Netherlands," in Robin Cowan and Nicholas Jonard, eds., *Heterogeneous Agents, Interactions and Market Performance* (Berlin: Springer Verlag, 205–225).

Young, H. Peyton (2003), "The diffusion of innovations in social networks," in Lawrence E. Blume and Steven N. Durlauf, eds., *The Economy as an Evolving Complex System, Vol. III* (Oxford: Oxford University Press, 267–282).

12
The Dynamics of Social Innovation

Social institutions are forms of capital that, together with physical and human capital, determine the rate of economic growth. Notable examples include private property rights and legal systems to protect them, accounting methods, forms of corporate governance, and the terms of economic contracts. However, in contrast with the literature on technological progress, relatively little is known about the ways in which new institutions are created and how they become established within a given social framework. In this chapter I discuss one approach to this problem using methods from evolutionary game theory.

In the abstract, an institution can be viewed as a set of rules that structure a given type of interaction between individuals (North 1990). These rules can be simple coordination devices, such as which hand to extend in greeting or who goes through the door first. Or they can be very elaborate, such as rituals of courtship and marriage, cycles of retribution, performance criteria in employment contracts, and litigation procedures in the courts. We want to know how a particular set of rules becomes established as common practice and what process describes the displacement of one set of rules by another.

The viewpoint we adopt here is that new norms and institutions are introduced through a process of experimentation by individuals and that the value of adoption by a given individual is an increasing function of the number of his neighbors who have adopted. Under certain conditions that we discuss below, this trial-and-error process will eventually trigger a general change in expectations and behaviors that establishes the new institution within society at large. However, it can take a very long time for this to happen even when the new way of doing things is superior to the status quo.

One reason for this inertia is lack of information: it may not be immediately evident that an innovation actually is superior to the status quo, due to the small number of prior instances and variability in their outcomes. Thus, it

Originally published as H. P. Young, "The dynamics of social innovation," Proceedings of the National Academy of Sciences USA. 2011 December 27; 108 (Suppl 4): 21285–21291. Reproduced with permission.

Spontaneous Order. H. Peyton Young, Oxford University Press. © H. Peyton Young (2024).
DOI: 10.1093/oso/9780198892908.003.0012

may take a long time for enough information to accumulate before it becomes clear that the innovation is superior. A second reason is that an innovation as initially conceived may not work very well in practice; it must be refined over time through a process of learning by doing (Arrow 1962). A third reason is that social innovations often exhibit increasing returns. Indeed this property is characteristic of social innovations, because they require *coordinated* change in expectations and behaviors by multiple individuals. For example, an individual who invents a new form of legal contract cannot simply institute it on his own: first, the other parties to the contract must enter into it, and second, the ability to enforce the contract will depend on its acceptance in society more generally. Institutions exhibit strongly increasing returns precisely because of their function as coordination technologies. This property is of course not limited to social innovations; it is also a feature of technological innovations with positive network externalities (David 1985; Katz and Shapiro 1985; Arthur 1994), and the results in this chapter apply to these situations as well.

The increasing returns feature of social innovation has important implications for the dynamics governing institutional change. Of particular relevance is the social network that governs people's interactions. The reason is that, when a social innovation first appears, it will typically gain a foothold in a relatively small subgroup of individuals that are closely linked by geography or social connections. Once the new way of doing things has become firmly established within a local social group, it propagates to the rest of society through the social network. Thus, a key determinant of the speed with which institutional change occurs is the *network topology* and in particular the extent to which interactions are "localized."

The relationship between speed of diffusion and network structure has been investigated in a variety of settings. One branch of the literature is concerned with dynamics in which each agent responds *deterministically* to the number of his neighbors who have adopted the innovation. From an initial seeding among a subset of agents, the question is how long it takes for the innovation to spread as a function of network structure (Watts 1999; Morris 2000; Centola and Macy 2007; Jackson and Yariv 2007; Golub and Jackson 2008). A second branch of the literature explores dynamics in which agents respond to their neighbors' choices *probabilistically*; that is, actions with higher expected payoffs are more likely to be chosen than actions with low expected payoffs. Here the question is how long it takes in expectation for the innovation to spread in the population at large (Ellison 1993; Blume 1995; Young 1998, 2006;

Montanari and Saberi 2010). The rate of spread hinges on topological properties of the network that are different from those that determine the speed of convergence in the nonstochastic case. A third branch of the literature examines the issue of how the network structure itself evolves as agents are matched and rematched to play a given game (Skyrms and Pemantle 2000; Jackson and Watts 2002; Goyal and Vega-Redondo 2005; Jackson 2008).

In this chapter we assume that agents adjust their behavior probabilistically and that the network structure is fixed. In contrast to earlier work in this vein, however, we emphasize that it is not only the network topology that determines how fast an innovation spreads. Rather, the speed depends on the interaction between three complementary factors: (i) the *payoff gain* represented by the innovation in relation to the status quo; (ii) the *responsiveness* or *rationality* of the agents, that is, the probability that they choose a (myopic) best response given their information; and (iii) the presence of *small autonomous enclaves* where the innovation can gain an initial foothold, combined with pathways that allow the innovation to spread by contagion once it has gained a sufficient number of distinct footholds. We also use a somewhat subtle notion of what it means for an innovation to "spread quickly." Namely, we ask how long it takes in expectation for a high proportion of agents to adopt the innovation and stick with it with high probability. This latter condition is needed because, in a stochastic learning process, it is quite possible for an innovation to spread initially, but then go into reverse and die out. We wish to know how long it takes for an innovation to reach the stage where it is in widespread use and continues to be widely used with high probability thereafter.

The main results of this chapter can be summarized as follows. First we distinguish between fast and slow rates of diffusion. Roughly speaking, an innovation *spreads quickly* in a given class of networks if the expected waiting time to reach a high level of penetration and stay at that level with high probability is bounded above independently of the number of agents; otherwise the innovation *spreads slowly*. Whether it spreads quickly or slowly depends on the particular learning rule used, the degree of rationality of the agents, the gain in payoff from the innovation, and the structure of the network. In the interest of keeping the model simple and transparent, we restrict our attention to situations that can be represented by a two-person coordination game with two actions—the status quo and the innovation. Although this assumption is somewhat restrictive, many coordination mechanisms are essentially dyadic, such as principal–agent contracts, the division of property rights between married couples, and rules governing bilateral trade. We also focus mainly on log-linear response processes by individuals, because this class of rules is particularly easy to work with analytically. Among the key findings are the following:

1. If the agents' level of rationality is too low, the waiting time to spread successfully is very long (in fact it may be infinite) because there is too much noise in the system for a substantial proportion of the agents to *stay coordinated* on the innovation once it has spread initially. However, if the level of rationality is too high, it takes an exponentially long time in expectation for the innovation to gain a foothold anywhere. (Here "rationality" refers to the probability of choosing a myopic best response to the actions of one's neighbors. Under some circumstances a forward-looking rational agent might try to *induce* a change in his neighbors' behavior by adopting first, a situation that is investigated in Ellison (1997).) Hence only for intermediate levels of rationality can one expect the waiting time to be fast in an absolute sense and to be bounded independently of the number of agents. In this respect the analysis differs substantially from that in Montanari and Saberi (2010).
2. Certain topological characteristics of the network promote fast learning. In particular, if the agents fall into small clusters or enclaves that are mainly connected with each other as opposed to outsiders, then learning will be fast, assuming that the level of rationality is neither too low nor too high. One reason why clustering may occur in practice is homophily—the tendency of like to associate with like (Lazarfeld and Merton 1954; Golub and Jackson 2008). Moreover, recent empirical work suggests that some behaviors actually do spread more quickly in clustered networks than in random ones (Centola 2010). However, not everyone has to be contained in a small cluster for learning to be fast: it suffices that the innovation be able gain a foothold in a *subset* of clusters from which it can spread by contagion.
3. For convergence to be fast, it is not necessary for the agents to be contained in enclaves that are small in an absolute sense; it suffices that everyone be contained in a subgroup of bounded (possibly large) size that has a sufficiently high proportion of its interactions with other members of the group as opposed to outsiders. Various natural networks have this property, including those in which agents are embedded more or less uniformly in a finite Euclidean space and are neighbors if and only if they are within some specified distance of one another. (This result follows from Proposition 2 below; a similar result is proved in Montanari and Saberi (2010).)

4. The payoff gain from the innovation relative to the status quo has an important bearing on the absolute speed with which it spreads. We call this the *advance* in welfare represented by the innovation. If the advance is sufficiently large, *no* special topological properties of the network are required for fast learning: it suffices that the maximum degree is bounded.
5. An innovation that leads to a small advance will tend to take much longer to spread than an innovation with a large advance. A consequence is that successful innovations will tend to occur in big bursts, because a major advance will tend to overtake prior innovations that represent only minor advances.

1 Model

Let Γ be a graph with vertex set V and edge set E, where the edges are assumed to be undirected. Thus, E is a collection of unordered pairs of vertices $\{i, j\}$, where $i \neq j$. Assume that there are n vertices or *nodes*. Each edge $\{i, j\}$ has a weight $w_{ij} = w_{ji} > 0$ that we interpret as a measure of the *mutual influence* that i and j have on one another. For example, w_{ij} may increase the closer that i and j are to each other geographically. Because we can always assume that $w_{ij} = w_{ji} = 0$ whenever $\{i, j\}$ is *not* an edge, the graph Γ is completely specified by a set V consisting of n vertices and a symmetric nonnegative $n \times n$ matrix of weights $W = (w_{ij})$, where $w_{ij} = 0$ for all i.

Assume that each agent has two available choices, A and B. We think of B as the *status quo behavior* and A as the *innovative behavior*. The *state* of the evolutionary process at time t is a vector $x^t \in \{A, B\}^n$, where $x^t{}_i$ is i's choice at time t. Let G be a symmetric two-person game with payoff function $u(x, y)$, which is the payoff to the player who chooses x against an opponent who chooses y. The game G induces an n-person *network game* on Γ as follows: The payoff to individual i in any given period results from playing the game against each of his neighbors once, where i's current strategy $x^t{}_i$ is *unconditional* on which neighbor he plays. The payoff from any given match is weighted according to the influence of that neighbor. For each agent i, let N_i denote the set of i's neighbors, that is, the set of all vertices j such that $\{i, j\}$ is an edge. Thus, the payoff to i in state x is

$$U_i(x) = \sum_{j \in N_i} w_{ij} u\left(x_i, x_j\right). \qquad [1]$$

To give this a concrete interpretation, suppose that B is a form of contractual negotiation that relies solely on a verbal understanding and a handshake, whereas A requires a written agreement and a witness to the parties' signatures. If one side insists on a written agreement (A) whereas the other side views a verbal understanding as appropriate (B), they will fail to coordinate because they disagree on the basic rules of the game. As discussed in Chapter 4, this metagame (agreeing on the rules of the game) can be viewed as a pure coordination game with zero off-diagonal payoffs: There is no inherent value in adopting A or B unless someone else also adopts it. This situation is shown below, where α is the added benefit from the superior action:

$$\begin{array}{ccc} & A & B \\ A & 1+\alpha, 1+\alpha & 0,0 \\ B & 0,0 & 1,1 \end{array} \quad [2]$$

[In the case of a symmetric 2×2 coordination game with nonzero off-diagonal payoffs, the adoption dynamics are qualitatively similar, but they tend to select the risk-dominant equilibrium, which may differ from the Pareto-dominant equilibrium, as shown in Chapter 2 (Young 1993); see also Kandori, Mailath, and Rob 1993 and Blume (1993).).]

2 Learning

How do agents adapt their expectations and behaviors in such an environment? Several models of the learning process have been proposed in the literature; here we discuss two particular examples. Assume that each agent updates at random times according to the realization of a Poisson process with unit expectation. The processes are assumed to be independent among agents. Thus, the probability is zero that two agents update simultaneously, and each agent updates once per unit time period in expectation. We think of each update by some agent as defining a "tick" of the clock (a *period*), and the periods are denoted by $t = 1, 2, 3, \ldots$.

Denote the state of the process at the start of period t by x^{t-1}. Let $\varepsilon > 0$ be a small error probability. Suppose that agent i is the unique agent who updates at time t. Assume that, with probability $1 - \varepsilon$, i chooses an action that maximizes his expected payoff, and with probability ε he chooses an action uniformly at random. This process is the *uniform error model* treated in Chapters 2–4 (Ellison 1993; Kandori, Mailath, and Rob 1993; Young 1993).

An alternative approach is the *log-linear model* (Blume 1993). Given a real number $\beta \geq 0$ and two actions A and B, suppose that agent i chooses A with probability

$$\frac{e^{\beta U_i(A, x_{-i}^{t-1})}}{e^{\beta U_i(A, x_{-i}^{t-1})} + e^{\beta U_i(B, x_{-i}^{t-1})}}. \quad [3]$$

In other words, the log probability of choosing A minus the log probability of choosing B is β times the difference in payoff; hence the term "log-linear learning." [This model is also standard in the discrete choice literature (McFadden 1973).] The parameter β measures the *responsiveness* or *rationality* of the agent: the larger β is, the more likely it is that he chooses a best reply given the actions of his neighbors, i.e., the less likely it is that he experiments with a suboptimal action.[1]

More generally, we may define a *perturbed best response learning process* as follows. Let G be a 2×2 game with payoff matrix as in eq. 2, and let Γ be a weighted undirected network with n nodes, each occupied by an agent. For simplicity we continue to assume that agents update asynchronously according to i.i.d. Poisson arrival processes. Let $\varepsilon > 0$ be a *noise parameter*, and let $q_i^{\varepsilon}(x^{t-1})$ be the probability that agent i chooses action A when i is chosen to update and the state is x^{t-1}. We assume that (i) $q_i^{\varepsilon}(x^{t-1})$ depends only on the choices of i's neighbors; (ii) $q_i^{0}(x^{t-1}) = 1$ if and only if A is a strict best response to the choices of i's neighbors; (iii) $q_i^{\varepsilon}(x^{t-1})$ does not decrease when one or more of i's neighbors switches to A; and (iv) $q_i^{\varepsilon}(x^{t-1}) \to q_i^{0}(x^{t-1})$ in an exponentially smooth manner; that is, there is a real number $r \geq 0$ such that $0 < \lim_{\varepsilon \to 0} q_i^{\varepsilon}(x^{t-1})/\varepsilon^r < \infty$. (In the case of log-linear learning these conditions are satisfied if we take $\varepsilon = e^{-\beta}$.) This process defines a Markov chain on the finite set of states $\{A, B\}^n$, whose probability transition matrix is denoted by P^{ε}.

3 Speed of Diffusion

Let $\vec{B}$ be the initial state in which everyone is following the status quo behavior B. How long does it take for the innovative behavior A to become widely adopted? One criterion would be the expected waiting time until the first time that everyone plays A. Unfortunately this definition is not satisfactory due to the stochastic nature of the learning process. To understand the nature of the

[1] See Chapter 5 for an empirical application of log linear learning to contract choice.

problem, consider a situation in which β is close to zero, so that the probability of playing A is only slightly larger than the probability of playing B. No matter how long we wait, the probability is high that a sizable proportion of a large population will be playing B at any given future time. Thus, the expected waiting time until everyone first plays A is not the relevant concept. A similar difficulty arises for any stochastic learning process: if the noise is too large, it is unlikely that everyone is playing A in any given period.

We are therefore led to the following definition. Given a stochastic learning process P^{ε} on a graph Γ (not necessarily log-linear learning), for each state x let $a(x)$ denote the *proportion* of agents playing A in state x. Given a target level of *penetration* $0 < p < 1$, define

$$T(P^{\varepsilon}, \Gamma, \alpha, p) = E[\min\{t : a(x^{t}) \geq p \,\&\, \forall t' \geq t, P(a(x^{t'}) \geq p) \geq p\}]. \quad [4]$$

In words, $T(P^{\varepsilon}, \Gamma, \alpha, p)$ is the expected waiting time until at least p of the agents are playing A, and the probability is at least p that at least this proportion plays A at all subsequent times. The waiting time depends on the learning process P^{ε} (including the specific level of noise ε), as well as on the graph Γ and the size of the advance (the payoff gain) α.

To distinguish between fast and slow learning as a function of the number of agents, we consider families of networks of different sizes, where the *size* of a network is the number of nodes (equivalently, the number of agents).

Fast Versus Slow Learning. Given a family of networks $\mathcal{G}$ and a size of advance $\alpha > 0$ (the payoff gain from the innovation), learning is *fast* for $\mathcal{G}$ and α if, for every $p < 1$ there exists $\varepsilon_p > 0$ such that for each $\varepsilon \in (0, \varepsilon_p)$,

$$T(P^{\varepsilon}, \Gamma, \alpha, p) \text{ is bounded above for all } \Gamma \in \mathcal{G}. \quad [5]$$

Otherwise learning is *slow*, that is, for every $\varepsilon > 0$ there is an infinite sequence of graphs $\Gamma_1, \Gamma_2, \ldots, \Gamma_n, \ldots \in \mathcal{G}$ such that $\lim_{n\to\infty} T(P^{\varepsilon}, \Gamma_n, \alpha, p) = \infty$.

4 Autonomy

In this section we describe a general condition on families of networks that guarantees fast learning. Fix a network $\Gamma = (V, W)$, a learning process P^{ε}, and an advance $\alpha > 0$. Given a subset of vertices $S \subseteq V$, define the *restricted learning process* $\tilde{P}^{\varepsilon}_{S}$ as follows: All nodes $i \notin S$ are *held fixed* at B whereas the nodes in S update according to the process P^{ε}. Let $(\vec{A}_S, \vec{B}_{V-S})$ denote the state in which every member of S plays A and every member of $V - S$ plays B.

The set S is *autonomous for* $(P^\varepsilon, \Gamma, \alpha)$ if and only if $\left(\vec{A}_S, \vec{B}_{V-S}\right)$ is strictly stochastically stable for the restricted process $\tilde{P}^\varepsilon_S$. [A state is *strictly stochastically stable* if it has probability one in the limit as the noise goes to zero (Foster and Young 1990).]

Proposition 1 *Given a learning process* P^ε*, a family of networks* $\mathcal{G}$*, and an innovation with advance* $\alpha > 0$*, suppose that there exists a positive integer* s *such that for every* $\Gamma \in \mathcal{G}$*, every node of* Γ *is contained in a subset of size at most* s *that is autonomous for* $(P^\varepsilon, \Gamma, \alpha)$*. Then learning is fast.*

Concretely this means that given any target level of penetration $p < 1$ and advance $\alpha > 0$, there is an upper bound on the noise, $\varepsilon_{p,\alpha}$, such that for any given ε in the range $0 < \varepsilon < \varepsilon_{p,\alpha}$, the expected waiting time until at least p of the agents play A (and continue to do so with probability at least p in each subsequent period) is bounded above independently of the number of agents n in the network.

This result differs in a key respect from those in Montanari and Saberi (2010), who show that for log-linear learning the expected waiting time to reach all-A with high probability is bounded as a function of n *provided that the noise level is arbitrarily small.* Unfortunately this result implies that the *absolute* waiting time to reach all-A is arbitrarily large, because the waiting time until any given agent first switches to A grows exponentially as the noise level goes to zero.

The proof of Proposition 1 runs as follows. For each agent i let S_i be an autonomous set containing i such that $|S_i| \leq s$. By assumption the state $\left(\vec{A}_{S_i}, \vec{B}_{V-S_i}\right)$ in which all members of S_i play A is stochastically stable. Given a target level of penetration $p < 1$, we can choose ε so small that the probability of being in this state after some finite time $t \geq T_{S_i}$ is at least $1 - (1 - p)^2$. Because this result holds for the restricted process, and the probability that i chooses A does not decrease when someone else switches to A, it follows that in the *unrestricted* process $P\left(x_i^t = A\right) \geq 1 - (1 - p)^2$ for all $t \geq T_{S_i}$. (This result uses a coupling argument similar to that in Young (1998).) Because $|S_i| \leq s$ for all i, we can choose ε and T so that $P\left(x_i^t = A\right) \geq 1 - (1 - p)^2$ for all i and all $t \geq T$. It follows that the *expected proportion* of agents playing A is at least $1 - (1 - p)^2$ at all times $t \geq T$.

Now suppose, by way of contradiction, that the probability is less than p that at least p of the agents are playing A at some time $t \geq T$. Then the probability is greater than $1 - p$ that at least $1 - p$ of the agents are playing B. Hence the expected number playing A is less than $1 - (1 - p)^2$, which is a contradiction.

5 Autonomy and Close-Knittedness

Autonomy is defined in terms of a given learning process, but the underlying idea is topological: if agents are contained in small subgroups that interact to a large degree with each other as opposed to outsiders, then autonomy will hold under a variety of learning rules. The precise formulation of this topological condition depends on the particular learning rule in question; here we examine the situation for log-linear learning.

Let $\Gamma = (V, W)$ be a graph and $\alpha > 0$ the size of the advance. The *degree* of vertex i is $d_i = \sum_{j \in V} w_{ij}$. For every subset of vertices $S \subseteq V$ let

$$d(S) = \sum_{i \in S} d_i. \qquad [6]$$

Further, for every nonempty subset $S' \subseteq S$ let

$$d(S', S) = \sum_{\{i,j\}: i \in S', j \in S} w_{ij}. \qquad [7]$$

Thus, $d(S)$ is the sum of the degrees of the vertices in S, and $d(S', S)$ is the weighted sum of edges with one end in S' and the other end in S. Given any real number $r \in (0, 1/2)$, we say that the set S is *r-close-knit* if

$$\forall S' \subseteq S, S' \neq \emptyset, d(S', S)/d(S') \geq r. \qquad [8]$$

S is r-close-knit if no subset has "too many" (i.e., more than $1 - r$) interactions with outsiders. This condition implies in particular that no individual $i \in S$ has more than $1 - r$ of its interactions with outsiders, a property known as *r-cohesiveness* (Morris 2000).

One consequence of close-knittedness is that the "perimeter" of S must not be too large relative to its "area." Specifically, let us define the *perimeter* and *area* of any nonempty set of vertices $S \subseteq V$ as follows:

$$peri(S) = d(S, V - S), \text{area}(S) = d(S, S). \qquad [9]$$

Next observe that $d(S) = 2\,\text{area}(S) + \text{peri}(S)$. It follows from eq. 8 with $S' = S$ that

$$peri(S)/area(S) \leq (1/r) - 2. \qquad [10]$$

If S is autonomous under log-linear learning, then by definition the potential function of the restricted process is maximized when everyone in S chooses A. Straightforward computations show that this result implies the following.

Proposition 2 *Given a graph Γ and innovation advance $\alpha > 0$, S is autonomous for α under log-linear learning if and only if S is r-close-knit for some $r > 1/(\alpha + 2)$.*

Corollary 2.1 *If S is autonomous for α under log-linear learning, then $peri(S)/area(S) < \alpha$.*

A family of graphs $\mathcal{G}$ is *close-knit* if for *every* $r \in (0, 1/2)$ there exists a positive integer $s(r)$ such that, for every $\Gamma \in \mathcal{G}$, every node of Γ is in an r-close-knit set of cardinality at most $s(r)$ (Young 1998).

Corollary 2.2 *Given any close-knit family of graphs $\mathcal{G}$, log-linear learning is fast for all $\alpha > 0$.*

6 Examples

Consider n nodes located around a circle, where each node is linked by an edge to its two immediate neighbors and the edge weights are one. [This is the situation originally studied by Ellison (1993).] Any set of s consecutive nodes is $(1/2 - 1/2s)$-*close-knit.* It follows that for any $\alpha > 0$, log-linear learning is fast.

Next consider a 2D regular lattice (a square grid) in which every vertex has degree 4 (Figure 12.1). Assume that each edge has weight 1. The shaded region in Figure 12.1 is a subset of nine nodes that is 1/3-close-knit. Hence it is autonomous whenever $\alpha > 1$.

More generally, observe that any square S of side m has $2m(m - 1)$ internal edges and m^2 vertices, each of degree 4; hence

$$d(S, S)/d(S) = 2m(m - 1)/4m^2 = 1/2 - 1/2m. \qquad [11]$$

Furthermore it is easily checked that for every nonempty subset $S' \subseteq S$,

$$d(S', S)/d(S') \geq 1/2 - 1/2m. \qquad [12]$$

Therefore, every square of side m is $(1/2 - 1/2m)$-close-knit and hence is autonomous for all $\alpha > 2/(m - 1)$. A similar argument holds for any

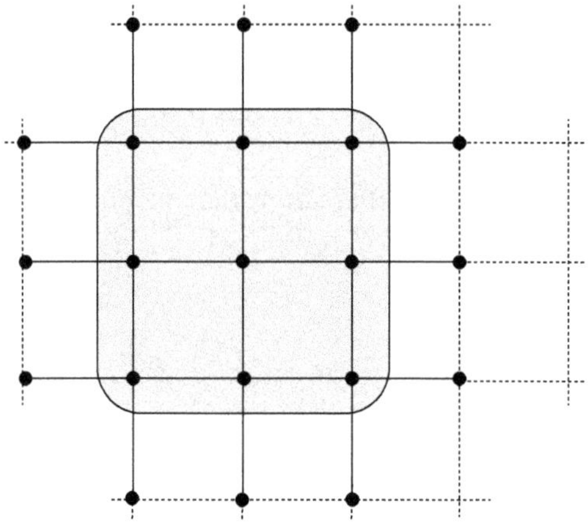

Figure 12.1 A 2D lattice with a subset of nine vertices (circled) that is autonomous for any $\alpha > 1$ under log-linear learning.

d-dimensional regular lattice: given any $\alpha > 0$, every sufficiently large sublattice is autonomous for α, and this holds independently of the number of vertices in the full lattice.

Note that in these examples fast learning does not arise because neighbors of neighbors tend to be neighbors of one another. In fact, a d-dimensional lattice has the property that *none* of the neighbors of a given agent are themselves neighbors. Rather, fast learning arises from a basic fact of Euclidean geometry: the ratio of "surface" to "volume" of a d-dimensional cube goes to zero as the cube becomes arbitrarily large.

A d-dimensional lattice illustrates the concept of autonomy in a very transparent way, but it applies in many other situations as well. Indeed one could argue that many real-world networks are composed of relatively small autonomous groups, because people tend to cluster geographically, or because they tend to interact with people of their own kind (homophily), or for both reasons.

To understand the difference between a network with small autonomous groups and one without, consider the pair of networks in Figure 12.2. Figure 12.2, *Upper* shows a tree in which every node other than the end nodes has degree 4, and there is a "hub" (not shown) that is connected to all of the end nodes. Figure 12.2, *Lower* shows a graph with a similar overall structure in which every node other than the hub has degree 4; however, in this case everyone (except the hub) is contained in a clique of size 4. In both networks all edges are assumed to have weight 1.

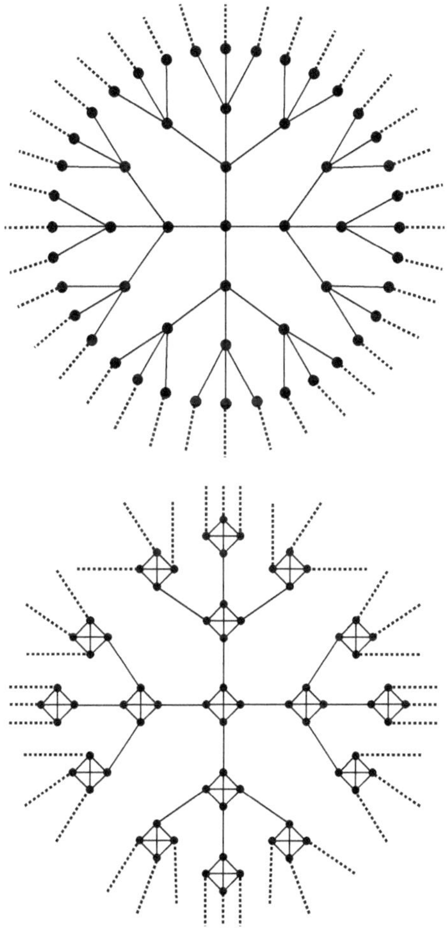

Figure 12.2 Two networks of degree 4, except for a hub (not shown) that is connected to every end node (dashed lines). All edge weights equal 1.

Suppose that we begin in the all-B state in both networks, that agents use log-linear learning with $\beta = 1$, and that the size of the advance is $\alpha > 2/3$. Let each network have n vertices. It can be shown that the waiting time to reach at least 99 percent A (and for this to hold in each subsequent period with probability at least 99 percent) is unbounded in n for the network on the top, whereas it is bounded independently of n for the network on the bottom. In the latter case, simulations show that it takes less than 25 periods (on average) for A to penetrate to the 99 percent level independently of n. The key difference between the two situations is that, in the network with cliques, the innovation can establish a toehold in the cliques relatively quickly, which then causes the hub to switch to the innovation also.

Note, however, that fast learning in the network with cliques does *not* follow from Proposition 2, because not every node is contained in a clique. In particular, the hub is connected to all of the leaves, the number of which grows with the size of the tree, so it is not in an r-close-knit set of bounded size for any given $r < 1/2$. Nevertheless learning is fast: Any given clique adopts A with high probability in bounded time; hence a sizable proportion of the cliques linked to the hub switch to A in bounded time, and then the hub switches to A also. In other words, fast learning occurs through a combination of autonomy and contagion, a topic that we explore in the next section.

7 Autonomy and Contagion

Contagion expresses the idea that once an innovation has become established for some core group, it spreads throughout the network via the best reply dynamic. Morris (2000) was the first to study the properties of contagion in the setting of local interaction games and to formulate graph-theoretic conditions under which contagion causes the innovation to spread throughout the network (for more recent work along these lines see Jackson and Yariv 2007; Centola and Macy 2007; Golub and Jackson 2008). Whereas contagion by itself may not guarantee fast learning in a stochastic environment, a combination of autonomy and contagion does suffice. The idea is that autonomy allows the innovation to gain a foothold quickly on certain key subsets of the network, after which contagion completes the job.

Consider a subset S of nodes, all of which are playing A, and choose some $i \notin S$. Let α be the size of the advance of A relative to B. Then A is a strict best response by i provided that

$$(1+\alpha)\sum_{j\in S} w_{ij} > \sum_{j\notin S} w_{ij}. \qquad [13]$$

Letting $r = 1/(\alpha+2)$, we can write this as follows:

$$\sum_{j\in S} w_{ij} > r\sum_{j\in V} w_{ij} = rd_i. \qquad [14]$$

Recall that for any vertex i and subset of vertices S, $d(i, S) = \sum_{j\in S} w_{ij}$ is the total weight on the edges linking i to a member of S. Given a graph $\Gamma = (V, W)$, a real number $r \in (0, 1/2)$, and a subset of vertices S, define the *first r-orbit* of S as follows:

$$O_r^1(S) = S \cup \{i \notin S : d(i, S) > rd_i\}. \quad [15]$$

Similarly, for each integer $k > 1$ recursively define the kth r-orbit by

$$O_r^k(S) = O_r^{k-1} \cup \left\{i \notin O_r^{k-1} : d\left(i, O_r^{k-1}\right) > rd_i\right\}. \quad [16]$$

Now suppose that $r = 1/(\alpha + 2)$. Suppose also that everyone in the set S is playing A. If the learning process is deterministic best response dynamics, and if everyone updates once per time period, then after k periods everyone in the kth r-orbit of S will be playing A. Of course, this argument does not show that log-linear learning with asynchronous updating will produce the same result. The key difficulty is that the core set S of A-players might unravel before contagion converts the other players to A. However, this problem can be avoided if (i) the core set S reaches the all-A state within a bounded period and (ii) S is autonomous and hence its members *continue to play A with high probability.* These conditions are satisfied if the core set is the union of autonomous sets of bounded size. We therefore have the following result.

Proposition 3 *Let $\mathcal{G}$ be a family of graphs. Suppose that there exist positive integers s, k, and a real number $r \in (0, 1/2)$ such that, for every $\Gamma = (V, W) \in \mathcal{G}$, there is a subset of vertices S such that (i) S is the union of r-close-knit sets of size at most s and (ii) $O^k{}_r(S) = V$. Then log-linear learning is fast on $\mathcal{G}$ whenever $\alpha > (1/r) - 2$.*

We illustrate with a simple example. Let the network consist of a circle of n agents (the rim) plus a central agent (the hub). Each agent on the rim is adjacent to the hub and to its two immediate neighbors on the rim. Note that the hub is not contained in an r-close-knit set of bounded size for any $r < 1/2$. However, for every $r < 1/2$, the hub is in the first r-orbit of the rim. Moreover, for every $r < 1/3$, there is an r-close-knit set of bounded size that consists of rim nodes; namely, choose any sequence of k adjacent rim nodes where $k > 1/(1 - 3r)$. It follows from Proposition 3 that learning is fast for this family of graphs whenever $\alpha > 1$.

Proposition 3 shows that for some families of networks, log learning is fast provided that is α large enough, where the meaning of "large enough" depends on the network topology. When the degrees of the vertices are bounded above, however, there is a simple lower bound on α that guarantees fast learning independently of the particular topology.

Proposition 4 *Let $\mathcal{G}$ be a family of graphs with no isolated vertices and bounded degree $d \geq 3$. If $\alpha > d - 2$, then log-linear learning is fast.*

Proof sketch Given $\alpha > d - 2$, let k be the smallest integer that is strictly larger than $(\alpha + 2)/(\alpha + 2 - d)$. I claim that every *connected* set S of size k is autonomous. By Proposition 2 it suffices to show that any such set S is r-close-knit for some $r > 1/(\alpha + 2)$. In particular it suffices to show that

$$\forall S' \subseteq S, S' \neq \emptyset, \frac{d(S', S)}{d(S')} = \frac{d(S', S)}{\sum_{i \in S'} d_i} > 1/(\alpha + 2). \quad [17]$$

Since S is connected, it contains a spanning tree, hence $d(S, S) \geq |S| - 1$. Furthermore, $d \geq d_i$ for all i. It follows that

$$\frac{d(S, S)}{d(S)} \geq \frac{|S| - 1}{d|S|} > 1/(\alpha + 2), \quad [18]$$

where the strict inequality follows from the assumption that $|S| = k > (\alpha + 2)/(\alpha + 2 - d)$. This verifies eq. 17 for the case $S' = S$. It is straightforward to show that eq. 17 also holds for all proper subsets S' of S. We have therefore established that every connected set of size k is autonomous. Clearly every connected component of the network with size less than k is also autonomous. Therefore every vertex is contained in an autonomous set of size at most k. It follows that log-linear learning is fast on all networks in $\mathcal{G}$ when $\alpha > d - 2$.

Fast learning says that the expected waiting time is bounded for a given family of networks, but it does not specify the size of the bound concretely. Actual examples show that the expected waiting time can be surprisingly short in an absolute sense. Consider an innovation with advance $\alpha = 1$, and suppose that all agents use log-linear learning with $\beta = 1$. Figure 12.3 shows the expected waiting time to reach the 99 percent penetration level for two families of networks: circles where agents are adjacent to their nearest four neighbors and 2D lattices. (Thus, in both cases the networks are regular of degree 4.) The expected waiting time is less than 25 periods in both situations. In other words, almost everyone will be playing A after about 25 revision opportunities per individual.

Note that this waiting time is substantially shorter than it takes for a given individual to switch to A when his neighbors are playing B. Indeed, the probability of such a switch is $e^0/(e^0 + e^{4\beta}) \approx e^{-4\beta} \approx 0.018$. Hence, in expectation, it takes about $1/0.018 = 54$ periods for any given agent to adopt A when none

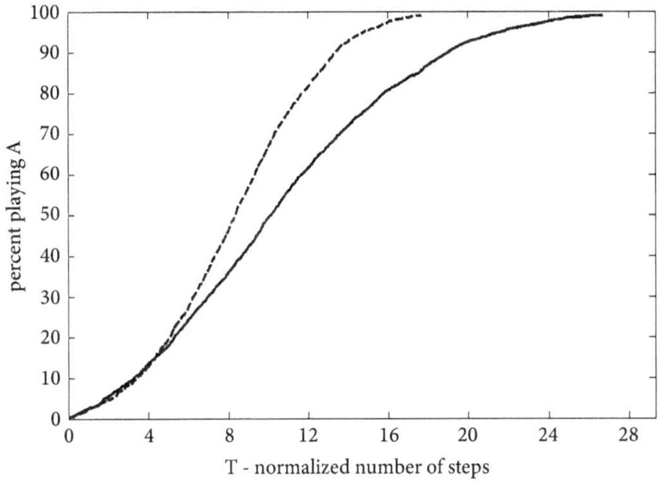

Figure 12.3 Simulated waiting times to reach 99% A starting from all B for circles (solid curve) and 2D lattices (dashed curve). Time periods represent the average number of updates per individual.

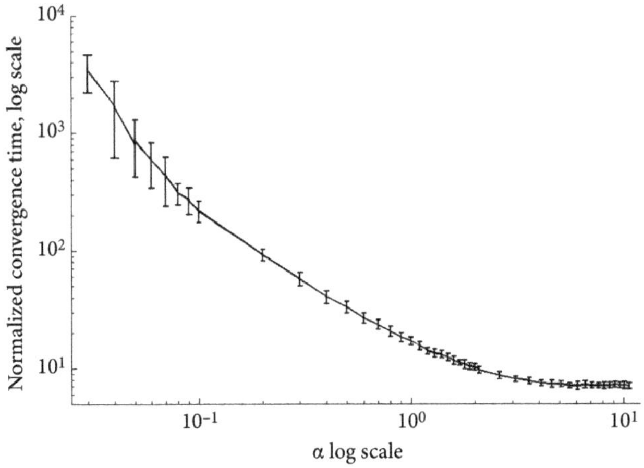

Figure 12.4 Expected number of periods to reach 99% playing A as a function of the size of advance α when $\beta = 1$, estimated from 10,000 simulations starting from the all-B state.

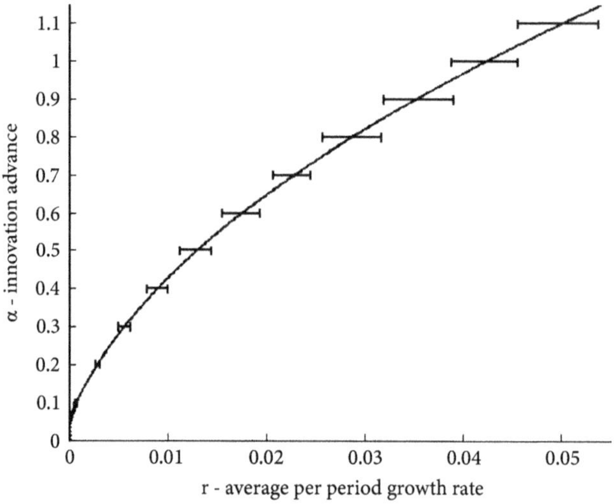

Figure 12.5 The level of innovative advance $\alpha = f(r)$ required to achieve an average growth rate of r per period. Simulation results for log-linear learning with $\beta = 1$ on a 30×30 2D lattice are shown.

of his neighbors has adopted, yet it takes only about half as much time for nearly everyone to adopt. The reason, of course, is that the process is speeded up by contagion. The rate at which the innovation spreads results from a combination of autonomy and contagion.

8 Bursts of Innovation

We have seen that the speed with which innovations spread in a social network depends crucially on the interaction between three features: the size of the advance α, the degree of rationality β, and the existence of autonomous groups that allow the innovation to gain a secure foothold. The greater the advance is from the innovation relative to the status quo, the more rapidly it spreads for any given topology, and the more that people are clustered in small autonomous groups the more rapidly the innovation spreads for any given size of advance. Furthermore the degree of rationality must be at an intermediate level for the rate of spread to be reasonably fast: If β is too high it will take a very long time before anyone even tries out the innovation, whereas if β is too low there will be so much random behavior that even approximate convergence will not take place.

In this section we examine how the size of the advance α affects the rate of diffusion for a given family of networks. The results suggest that social innovations tend to occur in large bursts rather than through small incremental improvements. The reason is that a small improvement takes a much longer time to gain an initial foothold than does an innovation that results in a major gain.

To illustrate this point concretely, let $\mathcal{G}$ be the family of 2D lattices where each agent has degree 4. Figure 12.4 shows the simulated waiting time to reach the target $p = 0.99$ as a function of α. For small values of α the waiting time grows as a power of α and is many thousands of periods long, whereas for large values of α (e.g., $\alpha > 1$) the waiting time is less than 20 periods.

To understand the implications of this relationship, suppose that each successive innovation leads to an advance of size α. Let $T(\alpha)$ be the waiting time for such an innovation to spread (for a given level of penetration p and level of rationality β). Assume that each new advance starts as soon as the previous one has reached the target p. Then the *rate of advance* per period is $r(\alpha)$, where $(1 + r(\alpha))^{T(\alpha)} = 1 + \alpha$; that is,

$$r(\alpha) = (1 + \alpha)^{1/T(\alpha)} - 1. \qquad [19]$$

The inverse function $f(r)$ mapping r to α is shown in Figure 12.5 for a 2D lattice and log-linear learning with $\beta = 1$. Figure 12.5 shows that to achieve a 1 percent average growth rate per period requires innovative bursts of size at least 40 percent. A 3 percent growth rate requires innovative bursts of at least 80 percent, and so forth.

Of course, these numbers depend on the topological properties of the grid and on our assumption that agents update using a log- linear model. Different network structures and different learning rules may yield different results. Nevertheless, this example illustrates a general phenomenon that we conjecture holds across a range of situations. Institutional change that involves a series of small step-by-step advances may take a very long time to spread compared with a change of comparable total magnitude that occurs all at once. The basic reason is that it takes much longer for a small advance to gain a secure foothold in an autonomous group: the group must be quite large and/or it must be quite interconnected to prevent a small advance from unraveling. Furthermore, under a small advance there are fewer pathways through the network that allow contagion to complete the diffusion process. The key point is that social innovations are technologies that facilitate—and require—coordination with others to be successful. It is this feature that makes social change so difficult and that favors large advances when change does occur.

Acknowledgments

I am indebted to Gabriel Kreindler for suggesting a number of improvements and for running the simulations reported here. The chapter also benefited from comments by Tomas Rodriguez Barraquer and participants at the Sackler Colloquium held at the Beckman Institute, December 3–4, 2010. This research was supported by Grant FA 9550-09-10538 from the Air Force Office of Scientific Research.

This chapter results from the Arthur M. Sackler Colloquium of the National Academy of Sciences, "Dynamics of Social, Political, and Economic Institutions," held December 3–4, 2010, at the Arnold and Mabel Beckman Center of the National Academies of Sciences and Engineering in Irvine, California.

References

Arrow, K. J. (1962), "The economic implications of learning by doing," *Review of Economic Studies* 29: 155–173.

Arthur, W. B. (1994), *Increasing Returns and Path Dependence in the Economy* (Ann Arbor, MI: University of Michigan Press).

Blume, L. E. (1993), "The statistical mechanics of strategic interaction," *Games and Economic Behavior* 5: 387–424.

Centola, D. (2010), "The spread of behavior in an online social network experiment," *Science* 329: 1194–1197.

Centola, D. and Macy, M. (2007), "Complex contagions and the weakness of long ties," *American Journal of Sociology* 113: 702–734.

David, P. A. (1985), "Clio and the economics of QWERTY," *American Economic Review Papers and Proceedings* 75: 332–337.

Ellison, G. (1993), "Learning, local interaction, and coordination," *Econometrica* 61: 1046–1071.

Ellison, G. (1997), "Learning from personal experience: One rational guy and the justification of myopia," *Games and Economic Behavior* 19: 180–210.

Foster, D. P. and Young, H. P. (1990), "Stochastic evolutionary game dynamics," *Theoretical Population Biology* 38: 219–232.

Golub, B. and Jackson, M. O. (2008), "How homophily affects the speed of learning and diffusion in networks," Stanford University arXiv:0811.4013v2 (physics.soc-ph).

Goyal, S. and Vega-Redondo, F. (2005), "Learning, network formation, and coordination," *Games and Economic Behavior* 50: 178–207.

Jackson, M. O. (2008), *Social and Economic Networks* (Princeton, NJ: Princeton University Press).

Jackson, M. O. and Watts, A. (2002), "The evolution of social and economic networks," *Journal of Economic Theory* 106: 265–295.

Jackson, M. O. and Yariv, L. (2007), "The diffusion of behaviour and equilibrium structure properties on social networks," *American Economic Review Papers and Proceedings* 97: 92–98.

Kandori, M., Mailath, G., and Rob, R. (1993), "Learning, mutation, and long run equilibria in games," *Econometrica* 61: 29–56.
Katz, M. and Shapiro, C. (1985), "Network externalities, competition, and compatibility," *American Economic Review* 75: 424–440.
Lazarsfeld, P. F. and Merton, RK (1954), *Freedom and Control in Modern Society*, M. Berger ed. (New York: Van Nostrand).
McFadden, D. (1973), *Frontiers in Econometrics*, P. Zarembka ed. (New York: Academic, 105–142).
Montanari, A. and Saberi, A. (2010), "The spread of innovations in social networks," *Proceedings of the National Academy of Science USA* 107: 20196–20201.
Morris, S. (2000), "Contagion," *Review of Economic Studies* 67: 57–78.
North, D. (1990), *Institutions, Institutional Change and Economic Performance* (Cambridge: Cambridge University Press).
Skyrms, B. and Pemantle, R. (2000), "A dynamic model of social network formation," *Proceedings of the National Academy of Science USA* 97: 9340–9346.
Watts, D. (1999), *Small Worlds: The Dynamics of Networks Between Order and Randomness* (Princeton, NJ: Princeton University Press).
Young, H. P. (1993), "The evolution of conventions," *Econometrica* 61: 57–84.
Young, H. P. (1998), *Individual Strategy and Social Structure: An Evolutionary Theory of Institutions* (Princeton, NJ: Princeton University Press).
Young, H. P. (2006), *The Economy as a Complex Evolving System III*, L. E. Blume and S. N. Durlauf eds. (Oxford: Oxford University Press, 267–282).

13

The Speed of Innovation Diffusion in Social Networks

With Itai Arieli, Yakov Babichenko, and Ron Peretz

1 Introduction

Social and technological advances are essential to economic development, but the mere existence of new and better ways of doing things does not guarantee that they will be widely used. The time it takes to dislodge inferior practices is another crucial factor in explaining how rapidly development can occur. This lag time depends on several crucial factors. One is lack of information: it may not be immediately evident that the innovation is in fact superior to the status quo. A second factor is network externalities: the desirability of an innovation depends not only on its inherent payoff but on how many others in one's social network have also adopted. Dislodging an inferior practice or technology requires a coordinated shift in expectations and behaviors among members of the group, which may take a long time even if it is already evident that everyone would be better off if they were to do so.

There is a substantial theoretical and empirical literature on these issues that we shall discuss below. In contrast to much of this literature, which is concerned with learning about payoffs from the actions of prior adopters, our focus here will be on the time it takes to dislodge an inferior practice when there are increasing returns from adoption and the payoffs are already known. This allows us to separate the effects of pure network externalities from the problem of learning about the payoffs by observing the behavior of others. Moreover, unlike much of the literature, we shall focus on the question of how long it takes to dislodge an inferior practice or technology when little or nothing is known about the topology of social interactions. Although this would appear to omit the main variable of interest, this is not the case. In particular,

Originally published in I. Arieli, Y. Babichenko, R. Peretz, and H. P. Young, (2020), "The speed of innovation diffusion in social networks," *Econometrica*, 88: 569–594. Reproduced with permission.

Spontaneous Order. H. Peyton Young with Itai Arieli, Yakov Babichenko, and Ron Peretz, Oxford University Press.
© H. Peyton Young (2024). DOI: 10.1093/oso/9780198892908.003.0013

Kreindler and Young (2014) demonstrate that the expected waiting time to overturn an inferior equilibrium can be usefully bounded from above for all undirected regular networks.

The theoretical contributions of this chapter are three-fold. First, we establish an upper bound on the expected waiting time that holds for networks of any *size* and *degree distribution, whether directed or undirected*. Second, we show how to extend the analysis to multiple competing innovations, instead of a single innovation versus the status quo, which is the usual assumption in the literature. Third, we show how the network topology affects the results by deriving waiting time bounds that hold for specific classes of networks, including regular networks and star networks. The mathematical techniques are novel and rely on results characterizing the distribution of arbitrarily large sums of independent random variables.

1.1 Related literature

The importance of social interactions in spreading new ideas and practices has been documented in a wide variety of situations. Ryan and Gross (1943) demonstrated that farmers' decisions to adopt an agricultural innovation—hybrid corn—were strongly influenced by the adoption decisions of their neighbors. Subsequently, Griliches (1957) showed that the decision to adopt also depended on the expected gains in payoff from the innovation relative to the status quo. At about the same time, sociologists Coleman, Katz, and Menzel (1957) analyzed the role of social networks among doctors in the adoption of a new medical treatment (in this case tetracycline), and showed that adoption was driven to a significant extent by peer effects.[1]

Since then a substantial theoretical literature has developed on the rate of innovation diffusion as a function of the network topology. A recurrent theme is that interaction among small close-knit groups can speed up the adoption process. The logic is that the innovation can gain a local foothold relatively quickly, and from different local footholds it then spreads throughout the network (Ellison (1993), and Young (1998, 2011); Montanari and Saberi (2010)); see also Chapter 12. Experimental studies of games played on networks are consistent with these predictions (Centola and Baronchelli (2015)).[2]

[1] For subsequent empirical work on innovation diffusion in networks, see Valente (1995, 2005); Foster and Rosenzweig (1995, 2010), Kohler (1997), Kohler, Behrman, and Watkins (2001), Udry and Conley (2001), Rogers (2003), and Munshi and Myaux (2006).

[2] There is also a recent literature on the speed of convergence in games where agents interact globally instead of with a fixed set of neighbors; see, in particular, Ellison, Fudenberg, and Imhof (2016) and Arieli and Young (2016).

A related line of work is concerned with the time it takes for a new idea to spread when it is seeded at one or more locations. Here, the key features are the centrality of the nodes where the new idea is seeded, and the degree of connectivity of the network (Morris (2000); Banerjee, Chandrasekhar, Duflo, and Jackson (2013)). These are instances of "threshold" models, in which a given individual adopts once a sufficient number of his neighbors have adopted (Watts (2002)). In this literature, adoption decisions are typically treated as irreversible, whereas in the present chapter we treat adoption (and disadoption) as random variables that depend on the distribution of choices by one's neighbors and stochastic shocks to their perceived payoffs; they are not deterministic.

Yet another branch of the literature investigates how network structure affects the rate at which agents update their priors about the desirability of the innovation, based on observation of their neighbors' choices. At issue here is the identification of conditions on the network topology and updating rules under which the process converges to correct beliefs and optimal actions.[3] The actual learning behavior of subjects who are embedded in different types of networks has been investigated experimentally by Gale and Kariv (2003) and Mäs and Nax (2016).

In this chapter, we focus instead on the situation where the payoffs are known in advance or have already been learned. In particular, we assume that people know that the new practice or technology would be inherently better than current practice provided that a sufficient number of people in one's network adopted it. The source of such network externalities depends very much on the situation. A social website is more valuable the more people who use it. A market is more valuable the more traders it attracts. A similar logic holds for communication technologies, technological standards, and many other innovations with increasing returns.[4]

A different type of network externality arises when people are sanctioned for not conforming to a current norm or practice. Demographers have found, for example, that social norms are a significant factor in explaining the pace and pattern of contraceptive use in developing countries.[5] In the United States,

[3] See Banerjee (1992), Bikhchandani, Hirschleifer, and Welch (1992), Ellison and Fudenberg (1993), Bala and Goyal (1998), Jackson (2008), Solan, Rosenberg, and Vieille (2009), Golub and Jackson (2010, 2012), Acemoglu, Dahleh, Lobel, and Ozdaglar (2011), Mueller-Frank (2013), Mueller-Frank and Pai (2016), and Mossel, Sly, and Tamuz (2015).

[4] David (1985) and Arthur (1994).

[5] See Bongaarts and Watkins (1996), Montgomery and Casterline (1996), Kohler (1997), Kohler, Behrman, and Watkins (2001), and Munshi and Myaux (2006).

norms of medical treatment for a given medical condition differ widely among states and even among counties within the same state. These differences appear to be the product of two types of network externality: peer effects and information sharing. Physicians tend to conform to the choices of local opinion leaders, and once a given practice becomes established its benefits are enhanced by local knowledge sharing within the group.[6]

These and other sources of network externalities can be modeled as a network game in which individuals repeatedly play a coordination game against their neighbors. Individuals periodically update their choices according to a random arrivals process. When individuals update they choose an optimal response given the inherent payoff from the choice as modified by the current choices of their neighbors (the coordination payoff) plus an idiosyncratic utility shock.[7]

The contribution of the current chapter is to establish an upper bound on the expected waiting time until a given proportion of the population has adopted, where the bound is independent of the size and topology of the network itself. This result is similar in spirit to the framework of Kreindler and Young (2013, 2014), who establish an upper bound on the waiting time for all regular undirected networks and a logit model of errors. Here, we employ different mathematical methods to establish a more general bound on the waiting time that holds for *all directed and undirected networks of arbitrary size*, and for a very broad class of error distributions. These bounds are especially useful when the network is difficult to observe or fluctuating due to exogenous influences.[8] The mathematical techniques rely on recent advances in estimating the distribution of arbitrarily large sums of independent random variables (Feige (2006), Garnett (2020)).

A common technique for the analysis of innovation diffusion in large networks is the mean field approach (see, e.g., Young (2009)). We use this approach in Proposition 6.1 to derive a tight bound on the expected waiting time in large regular networks. The mean field approach, however, is not suitable for proving our main result (Theorem 3.1) which holds even when agents are arbitrarily heterogeneous with respect to their degree and the influence they exert on one another.

[6] See Wennberg and Gittelsohn (1973), Phelps and Mooney (1993), Chandra and Staiger (2007), and Burke, Fournier, and Prasad (2007, 2010).

[7] See among others Blume (1993), Jackson and Yariv (2007), Jackson (2008), Vega-Redondo (2007), and Golub and Jackson (2010, 2012).

[8] Our bounds do not apply, however, when links are formed and severed endogenously as a consequence of agents' choices. For endogenous models of network formation see, among others, Skyrms and Pemantle (2000), Jackson and Watts (2002), Goyal and Vega-Redondo (2007), and Jackson (2008).

The plan of the chapter is as follows. In Section 2, we formulate the stochastic updating model. In Sections 3–4, we establish a general bound on the expected waiting time until a target proportion of the population has adopted, starting from the state where everyone is playing the status quo. The bound depends on the shape of the error distribution, on the payoff gap between the innovation and the status quo, and on the magnitude of the coordination payoffs relative to the inherent payoffs from different choices, but it does not depend on the topology of the network per se. Section 5 extends the analysis to multiple competing innovations. In Section 6, we show how the waiting time depends on the topology of the network by considering two contrasting cases: regular networks and star networks. In particular, we show that the bound established by Kreindler and Young (2014) for large regular networks can be substantially improved.

2 Model

A weighted directed network with m nodes can be represented by an $m \times m$ row stochastic matrix $P(t) = \{P_{ij}(t)\}$. We interpret $P_{ij}(t)$ as the probability that agent i interacts with j during the current period.[9] Alternatively, we can view $P_{ij}(t)$ as the relative weight that i attaches to interactions with j in the current period. For expositional simplicity, we shall begin by considering the situation where each agent chooses one of two actions: the innovation (action 1) or the status quo (action 0). In Section 5, we shall extend the analysis to multiple innovations with different payoffs.

The payoff from choosing an action consists of two parts: (i) an inherent payoff that is independent of how many others choose it, and (ii) a coordination payoff that results from coordinating with others. For an interaction with a single neighbor, we represent the payoffs by the following 2×2 matrix:

	0	1
0	c	0
1	a	$a+c$

Here, $a > 0$ is the payoff difference between the innovation 1 and the status quo 0, and $c > 0$ is the coordination payoff that results from making the same choice as someone with whom one interacts. Note that when $c < a$ the innovation is a dominant strategy and there is a unique equilibrium whereas if $c > a$ there are three equilibria: two pure and one mixed. The subsequent analysis

[9] We allow for the possibility that $P_{ii}(t) > 0$, that is, i's own action in the previous period increases the probability that he chooses it again next period. This can be interpreted as a form of inertia.

holds in either case but we shall typically assume that there are multiple equilibria.

The *state* of the process at the end of each period t is an m-vector $s(t) \in \{0,1\}^m$, where $s_i(t) = 1$ if agent i chooses the innovation at t, and $s_i(t) = 0$ otherwise. The updating process works as follows. Time is continuous and the initial state is $s(0) = (0, \ldots, 0)$. We suppose that every agent receives updating opportunities according to a Poisson arrival process with rate one per time period, where these processes are independent among the individuals. Suppose that i receives such an opportunity at time t. Given the current network structure $P = P(t)$ and the current state $s(t)$, let

$$x_i(t) = \sum_{j \in [m]} P_{ij} s_j(t).$$

Thus i's expected payoff from interacting with a randomly drawn neighbor (drawn according the distribution $P(t)$) is $u_i(1) = a + cx_i(t)$ if he chooses action 1, and $u_i(0) = c(1 - x_i(t))$ if he chooses action 0.

Let us assume that the difference between the payoff from 1 and the payoff from 0 is perturbed by a random payoff shock $\epsilon_i(t)$ with c.d.f. $F(\epsilon)$. For ease of interpretation, we shall assume that F has a density $f(\epsilon)$ and that all shocks are i.i.d. among agents and among time periods. If the perturbed payoff difference is positive, agent i chooses action 1; otherwise he chooses action 0. This class of perturbed best reply dynamics is very general and includes such standard dynamics as the logit response, where the payoff shocks are extreme-valued distributed. Experimental evidence on learning in networks shows that subjects do deviate from best reply with a probability that is decreasing in the resulting payoff loss, which is consistent with this class of models (Mäs and Nax (2016)).

Conditional on receiving an updating opportunity at time t, the probability that i chooses 1 is

$$\begin{aligned} \mathbf{P}\left[s_i(t) = 1 \mid x_i(t)\right] &= \mathbf{P}\left[a + \epsilon_i(t) + cx_i(t) > c\left(1 - x_i(t)\right)\right] \\ &= \mathbf{P}\left[\epsilon_i(t) > c - a - 2cx_i(t)\right] = 1 - F\left(c - a - 2cx_i(t)\right). \end{aligned} \tag{1}$$

We shall call this the *response function* and write

$$r\left(x_i(t)\right) = 1 - F\left(c - a - 2cx_i(t)\right). \tag{2}$$

Note that $r(\cdot)$ is monotone, nondecreasing, continuous, and $0 \leq r(0) \leq r(1) \leq 1$. (Continuity follows from our assumption that F has a density.) In what follows, we shall assume that $r(0) > 0$, that is, $F(c - a) < 1$. This amounts to saying that even when none of i's neighbors has adopted the innovation, there is a positive probability that i will switch from 0 to 1 due to receiving a sufficiently large payoff shock.

Figure 13.1 shows an example of a response function based on the normal distribution.

Although we have derived the response function from a specific shock distribution F, in what follows we can view the response function as a primitive of the model: $r(x)$ specifies the probability that an agent chooses action 1 whenever a (weighted) proportion x of the agent's neighbors choose action 1. Note that the same response function holds for all agents and is independent of the particular network through which they interact.

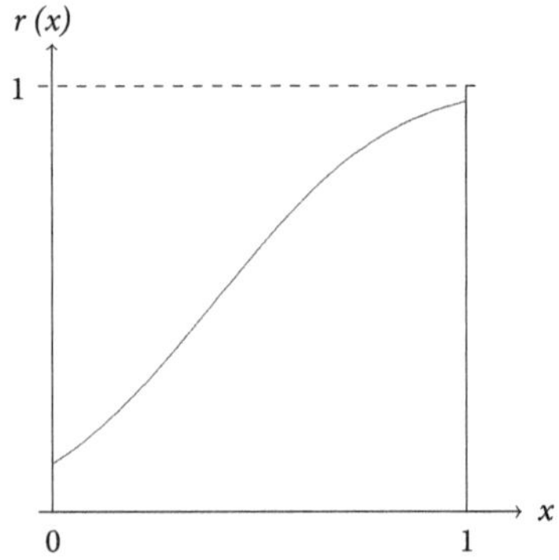

Figure 13.1 A response function based on the normal distribution.

Given a response function $r(x)$, define *the supporting line* $L(x)$ to be the unique line that solves the following maximization problem:

$$\max L(0)$$
$$\text{subject to } L(1) = r(1) \text{ and } L(x) \leq r(x) \text{ for } x \in [0, 1].$$

The *slope* of L will be denoted by $1 - \alpha := L(1) - L(0) < 1$, and its *fixed point* will be denoted by $p = \frac{L(0)}{\alpha}$, as shown in Figure 13.2.[10]

In what follows, we shall focus on the case $L(0) > 0$. The conditions under which this assumption holds are discussed in Section 3.1. In particular, it fails to hold if the response function is too "bowed," and crosses the 45° line at a

[10] We note that the fixed point p of the supporting line is always smaller than the fixed point of the response function.

low value of x as shown in Figure 13.3. In this case, the adoption process can become trapped in a low-level equilibrium for a very long time; indeed the expected waiting time until a low proportion q of the agents have adopted is unbounded in the network size for some classes of networks, as we show in Section 6.3.

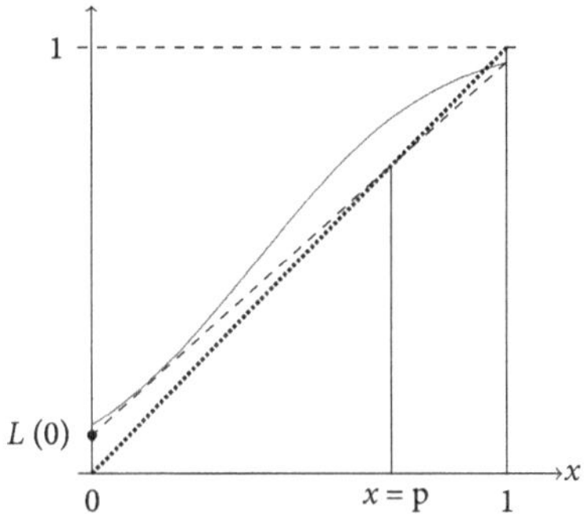

Figure 13.2 Supporting line $L(x)$ (dashed) and response function $r(x)$ (solid).

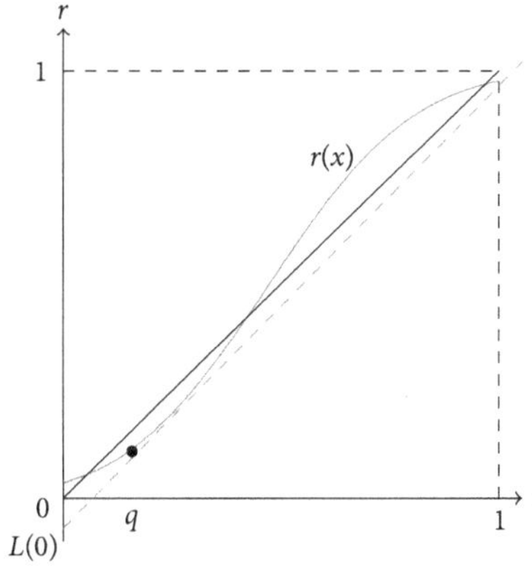

Figure 13.3 Response function with $L(0) < 0$, and a low value q such that $r(q) < q$.

3 A General Result on Waiting Times

The question we wish to address is the following: starting from the state $s(0) = (0, ..., 0)$ and given a target proportion $q \in [0, 1]$, how long does it take in expectation to reach a state in which at least qm agents have adopted the innovation? More precisely, we wish to find an upper bound on the expected waiting time defined as follows:

$$T_q(F, P(t)) = \mathbb{E}\left[\min\left\{t : \frac{1}{m}\sum_{i\in[m]} s_i(t) \geq q, s(0) = (0, \ldots, 0)\right\}\right].$$

We can now state our main result.

Theorem 3.1 *Let F be an error distribution function with density f. Suppose that the supporting line L(x) satisfies L(0) > 0, has slope 1 − α, fixed point p, and q < p. Then for every dynamic network P(t) we have*

$$T_q(F, P(t)) \leq \frac{7.2}{p\alpha}\left(1 + \ln\left(\frac{p}{p-q}\right)\right) = \frac{7.2}{L(0)}\left(1 + \ln\left(\frac{p}{p-q}\right)\right). \tag{3}$$

The crucial point is that the right-hand side does not depend on the network structure.

As we mentioned earlier, the closest result in the literature is due to Kreindler and Young (2014). Using martingale methods, they derive the following upper bound on the expected waiting time until at least qm agents have adopted and $P(t)$ is any regular, undirected network, namely,[11]

$$T_q(F, P(t)) \leq \frac{1}{\alpha(p-q)}. \tag{4}$$

The same bound holds for irregular networks provided that the target q is expressed as the degree-weighted proportion of adopting agents. This criterion is different from insisting that a given *proportion* of agents adopt. In a star network, for example, a degree-weighted majority ($q = \frac{1}{2}$) is achieved when just the central node has adopted. A comparison of (3) and (4) shows that the latter is a better (i. e., lower) bound unless q is very close to p. Indeed (4) is less than (3) provided that

$$\frac{p}{p-q} > 7.2\left(1 + \ln\left(\frac{p}{p-q}\right)\right),$$

which holds whenever $q < 0.969p$. However, our bound is much more general because it holds for directed and undirected networks of any size and degree distribution. In Section 6.1, we show that for large regular networks the waiting

[11] To illustrate their approach, let G be a complete undirected network. Given $q < p$, let $\mu = L(q) - q$. Assuming that $L(0) > 0$, we have $L(x) - x > \mu \geq 0$, for all $x \in [0, q]$. Given a proportion $x(t) \leq q$, the expected increase next period is $E[x(t+1) - x(t)|x(t)] \geq \mu > 0$. It follows that the expected waiting time until $x(t) \geq q$ is at most $\frac{1}{\mu}$. Since $L(x) = (1-\alpha)x + \alpha p$, $\mu = \alpha(p-q)$, and hence $T \leq \frac{1}{\alpha(p-q)}$. In fact, the same bound holds for all regular networks; moreover, if q = 1/2 the bound can be improved to $T \leq \frac{0.42}{\alpha(p-q)}$ (Kreindler and Young (2014), Lemma 2).

time is precisely $\frac{1}{\alpha}\ln\left(\frac{p}{p-q}\right)$, which is considerably tighter than the Kreindler–Young bound. Note, however, that for a fixed p our general bound (3) is just a linear transformation of the special bound $\frac{1}{\alpha}\ln\left(\frac{p}{p-q}\right)$, which is tight for regular networks. This result shows that our bound is of the right order with respect to q for a given p.

3.1 Discussion of the Condition L(0) > 0

A key assumption underlying Theorem 3.1 is that the tangent line $L(x)$ must have a strictly positive intercept $L(0)$. If this is not the case, the theorem does not apply. When will $L(0)$ be negative? The answer is when the payoff advantage a is small relative to the coordination payoff c, and/or the variance of the distribution F is small. Recall that $r(0) = 1 - F(c - a)$ and $r(1) = 1 - F(-a - c)$. For ease of exposition suppose also that f is symmetric about 0, so that the inflection point of F occurs at 0. Then $r(0) = F(a - c)$ and $r(1) = F(a + c)$. Increasing a to $a + b$ shifts the origin as shown in Figure 13.4. When b is large enough, the intercept of the supporting line becomes positive (see Figure 13.4, bottom panel).

The second case is illustrated in Figure 13.5. Rescaling the standard deviation by an amount σ is equivalent to rescaling the horizontal axis by $1/\sigma$, which leads to a positive intercept if σ is large enough, as shown in the bottom panel of Figure 13.5.

More generally, given a response function $r(x)$ we have

$$L(0) = \inf_{x\in[0,1)} \frac{r(x) - xr(1)}{1 - x}. \tag{5}$$

Therefore, $L(0) > 0$ iff $r(x) > xr(1)$ for every $x \in [0, 1)$. If r is induced by a c.d.f. F, then $r(x) = 1 - F(c - a - 2cx)$. In this case, the condition becomes $1 - F(c - a - 2cx) > x(1 - F(-a - c))$ for every $x \in [0, 1)$.[12]

3.2 Proof Sketch of Theorem 3.1

Conditional on receiving a revision opportunity, an agent chooses action 1 with probability $r(x)$, where x is the weighted average proportion of his neighbors who are currently choosing 1. The idea is to analyze the *linear*

[12] In case $a \geq c$ and the density f is symmetric and unimodal, the response function is concave, and hence $L(0) > 0$.

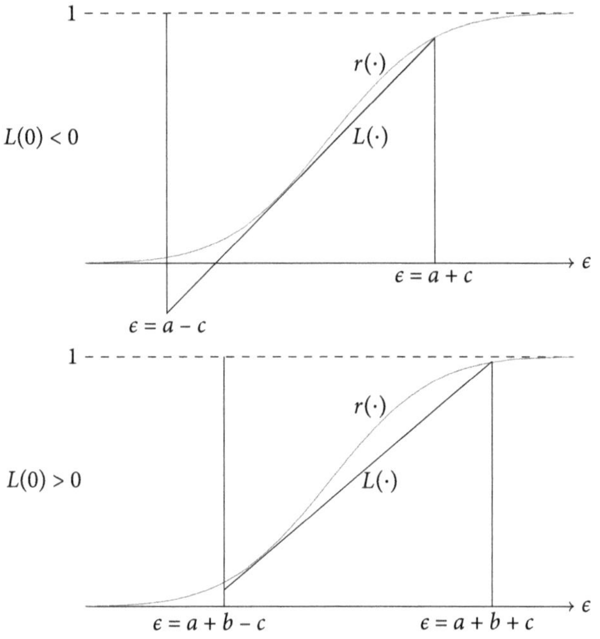

Figure 13.4 For a given c.d.f. $F(\epsilon)$, increasing the payoff advantage of the innovation by an amount b increases the intercept $L(0)$. Here, $F(a+b-c) = r(0)$ and $F(a+b+c) = r(1)$.

dynamic where an agent that faces the proportion x updates his action to 1 with probability $L(x)$ instead of $r(x)$. Since $L(x) \leq r(x)$, for every time t we have

$$r(x_i(t)) \geq p\alpha + (1-\alpha)\sum_j P_{ij}(t)s_j(t).$$

In particular, the expected time it takes for the linear dynamic to reach the threshold q is at least as long as the expected time under the actual dynamic.

An advantage of analyzing the linear dynamic is that it can be viewed as an *imitation dynamic*. Namely, at every revision opportunity, the revising agent i acts as follows: with probability α he chooses his action according to a coin toss with probability of success p. With probability $(1-\alpha)$, he imitates a neighbor that is drawn according to the distribution $P_i(t)$.[13] Thus the probability that i chooses 1 is

[13] This set-up bears some resemblance to the "voter model" in which a randomly drawn agent imitates a randomly drawn neighbor Liggett (1999). In our model, by contrast, imitation only occurs with probability $1-\alpha$; otherwise the agent chooses action 1 with probability p. This leads to a fully ergodic process whereas the voter model eventually absorbs into the all-0 or all-1 state.

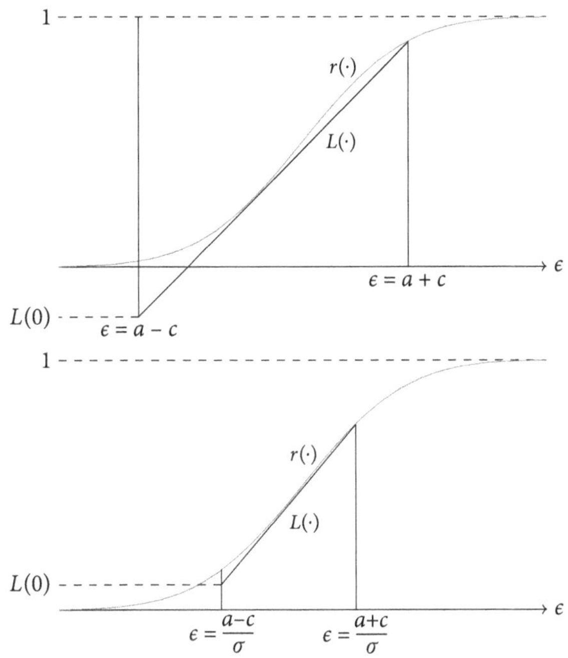

Figure 13.5 Increasing the standard deviation of the error distribution by a factor σ increases the intercept $L(0)$. Here, $F\left(\frac{a-c}{\sigma}\right) = r(0)$ and $F\left(\frac{a+c}{\sigma}\right) = r(1)$.

$$\alpha p + (1 - \alpha)\, x_i(t) = L(x_i(t)). \tag{6}$$

In what follows, it will be notationally convenient to analyze the corresponding *discrete time* linear dynamic such that at each time t one agent is chosen uniformly at random to update his action. (This discrete time dynamic is m times slower than the corresponding continuous time dynamic, where m is the number of agents.)

A history of the imitation dynamic up to time T induces an *imitation forest*, which is defined as follows. Each vertex of the imitation forest is a pair $(i, t) \in [m] \times [T]$. There is an edge from (i, t) to $(j, t-1)$ if at time t agent i copied agent j's action at time $t-1$. In addition, there is an edge from (i, t) to $(i, t-1)$ if i was not chosen to update at time t. This construction is illustrated in Figure 13.6.

The imitation forest admits two types of roots. A *coin-toss root* is a vertex (i, t) such that i tossed a coin at time t. An *origin root* is a vertex $(i, 0)$ for $i \in [m]$. This induces a partition of the population at time t into two groups. A

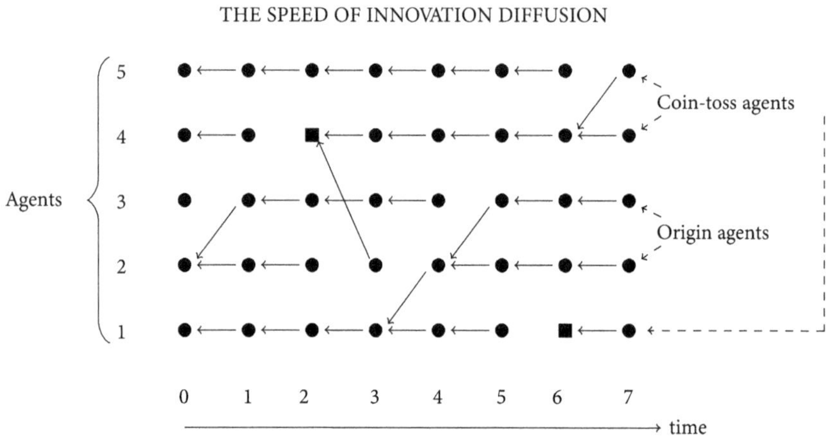

Figure 13.6 The imitation forest of the following history. At time t = 1, agent 3 imitates agent 2. At time t = 2, agent 4 tosses a coin. At time t = 3, agent 2 imitates agent 4. At time t = 4, agent 2 imitates agent 1. At time t = 5, agent 3 imitates agent 2. At time t = 6, agent 1 tosses a coin. At time t = 7, agent 5 imitates agent 4. Square nodes are points where an agent tossed a coin.

coin-toss agent is one that belongs to a tree with a coin-toss root. Similarly, an *origin agent* is one that belongs to a tree with an origin root.

The proof of Theorem 3.1 consists of two key steps.

Step 1. First, we show that the proportion of coin-toss agents increases at a rate that is bounded below by a simple formula that is independent of the network. This provides an explicit formula for the *expected* number of coin toss agents at each time *t*. We then apply Markov's inequality to show that *with high probability* the proportion of coin-toss agents is large at all sufficiently large times *t*.

Step 2. At each point in time, the proportion of agents that play 1 is a weighted average of a sequence of 0–1 random variables that correspond to prior coin tosses. Since we allow the network to evolve over time, analyzing these weights is extremely difficult. Nevertheless, we are able to overcome this hurdle by the following lemma, which follows readily from Feige's inequality (Feige (2006)) and its subsequent improvement by Garnett (2020). (See the Appendix for further details.)[14]

Lemma 1 *Let $c_1,\ldots, c_k$ be i.i d.* Bernoulli*(p) random variables. For all k and every sequence of weights $\beta_1,\ldots, \beta_k \geq 0$,*

[14] Billingsley ((2008), Theorem 9.2) has a different bound that depends on higher moments of the Bernoulli random variable, but the bound is not as good for values of *p* close to 1. See Appendix A for further details.

$$\mathbf{P}\left(\sum_{i=1}^{k}\beta_i c_i \geq \left[\sum_i \beta_i\right] p\right) \geq 0.14p$$

This corollary allows us to place a nonzero lower bound on the probability that a given proportion of the agents are choosing action 1 by time t, which leads to an upper bound on the expected waiting time to reach a given target proportion q for any interaction structure.

4 Proof of Theorem 3.1

As was mentioned in the proof sketch (Section 3.2), we shall analyze the discrete-time imitation dynamic in which one randomly drawn agent updates per period. Since each agent's Poisson clock has rate 1, the discrete process is m times slower than the actual continuous process. We shall prove that for the discrete dynamic the expected waiting time satisfies the bound

$$T_q(F, P(t)) \leq \frac{7.2m}{p\alpha}\left(1 + \ln\left(\frac{p}{p-q}\right)\right).$$

First, we analyze the diffusion of coin-toss agents. Consider the following process: the initial state is $y(0) = (y_i(0))_{i\in[m]} = (0, \ldots, 0)$, and in each period one player $i \in [m]$ is drawn at random to update. With probability α, $y_i(t) = 1$, and with probability $1 - \alpha$, $y_i(t) = y_j(t-1)$, where j is drawn according to $P_i(t)$.

Lemma 2 *The probability that an agent is coin-toss at time t is independent of the network structure and is equal to* $p(t) = \mathbf{P}\left(y_i(t) = 1\right) = 1 - \left(1 - \frac{\alpha}{m}\right)^t$.

Proof This claim is established by induction on t. For $t = 0$, the claim is trivial. Assume that the equality holds for t. Then it holds for $t + 1$, because

$$\begin{aligned}\mathbf{P}\left(y_i(t+1) = 1\right) &= \frac{1}{m}(\alpha + (1-\alpha)p(t)) + \left(1 - \frac{1}{m}\right)p(t)\\ &= \frac{\alpha}{m} + \left(1 - \frac{\alpha}{m}\right)p(t)\\ &= \frac{\alpha}{m} + \left(1 - \frac{\alpha}{m}\right)\left(1 - \left(1 - \frac{\alpha}{m}\right)^t\right)\\ &= 1 - \left(1 - \frac{\alpha}{m}\right)^{t+1} = p(t+1).\end{aligned}$$

This concludes the proof of Lemma 2.

Proof of theorem 3.1 Let $\tilde{T}_q$ be the first time at which at least qm agents have adopted. We seek to bound $E\left[\tilde{T}_q\right] = T_q(F, P(t))$ from above independently of $P(t)$. Lemma 2 shows that the expected number of coin-toss agents at time t is given by $\mathbb{E}\left[\sum_i y_i(t)\right] = m - m\left(1 - \frac{\alpha}{m}\right)^t$. Let us set a threshold of $\frac{q}{p}m$ for the number of coin-toss agents. We use Markov's inequality to bound the probability that the number of coin-toss agents is less than $\frac{q}{p}m$, namely,

$$\mathbf{P}\left[\sum_i y_i(t) < \frac{q}{p}m\right] = \mathbf{P}\left[\sum_i (1 - y_i(t)) > \left(1 - \frac{q}{p}\right)m\right]$$

$$\leq \frac{\mathbb{E}\left[\sum_i (1 - y_i(t))\right]}{\left(1 - \frac{q}{p}\right)m} = \frac{\left(1 - \frac{\alpha}{m}\right)^t}{1 - \frac{q}{p}}. \tag{7}$$

At each point in time t, let $R(t) = \{r_1(t), \ldots, r_{k(t)}(t)\}$, be the relevant coin-toss roots at t, and denote by $c_1(t), \ldots, c_{k(t)}(t) \in \{0, 1\}$ their realized actions. Note that for any fixed t, $c_1(t), \ldots, c_{k(t)}(t)$ are i.i.d. Bernoulli(p) random variables. For notational convenience, we let $c_0 = 0$. Let $\varphi(i, t)$ be the function that associates each coin-toss agent with the corresponding root. Formally, if i is a coin-toss agent at time t and (i, t) belongs to a tree with root $r_j(t)$ then $\varphi(i, t) = j$. If i is an origin agent at time t, we let $\varphi(i, t) = 0$. Finally, for every $j \in [k(t)]$ we let $\beta_j(t) = |\{i: \varphi(i, t) = j\}|$ be the number of appearances of the coin-toss root j, with the corresponding action $c_j(t)$. Note that $\sum_{i\in[m]} y_i(t) = \sum_{j\in[k(t)]} \beta_j(t)$. The state of the dynamic $s(t)$ can be expressed as $s_i(t) = c_{\varphi(i,t)}$, hence $\sum_i s_i(t) = \sum_{j\in[k(t)]} \beta_j(t)c_j(t)$.

In order to bound $\tilde{T}_q$ from above, we define the stopping time

$$\tau := \min\left\{t : \sum_i y_i(t) > \frac{q}{p}m\right\},$$

and the event

$$\mathcal{E} := \left\{\omega : \sum_j \beta_j(\tau(\omega))c_j(\tau(\omega)) \geq \sum_j \beta_j(\tau(\omega))p\right\}.$$

Note that $\omega \in \mathcal{E}$ implies that $\tilde{T}_q(\omega) \leq \tau(\omega)$. Since τ is measurable with respect to the imitation process and $\{c_1(\tau), \ldots, c_{k(\tau)}(\tau)\}$ are i.i.d. Bernoulli(p) random variables conditioned on the imitation process, Lemma 1 implies that

$$\mathbf{P}(\mathcal{E} \mid \tau) \geq 0.14p. \tag{8}$$

We next show that

$$\mathbb{E}[\tau] \leq \frac{m}{\alpha}\left(1 + \ln\left(\frac{p}{p-q}\right)\right). \tag{9}$$

Indeed,

$$\mathbb{E}[\tau] = \sum_{t=0}^{\infty} \mathbf{P}(\tau > t) \leq \sum_{t=0}^{\infty} \min\left\{1, \frac{\left(1-\frac{\alpha}{m}\right)^t}{1-\frac{q}{p}}\right\} \leq \frac{\ln\left(1-\frac{q}{p}\right)}{\ln\left(1-\frac{\alpha}{m}\right)} + \sum_{t=0}^{\infty}\left(1-\frac{\alpha}{m}\right)^t$$
$$= -\frac{\ln\left(\frac{p}{p-q}\right)}{\ln\left(1-\frac{\alpha}{m}\right)} + \frac{m}{\alpha} \leq \frac{m}{\alpha}\ln\left(\frac{p}{p-q}\right) + \frac{m}{\alpha}.$$

The first inequality follows from (7), and the last inequality from the inequality $\ln(1+x) \leq x$.

By restarting a new imitation process after τ, we can define a sequence of stopping times $\tau = \tau_1 < \tau_2 < \cdots$, and corresponding sequence of events $\mathcal{E}_1, \mathcal{E}_2, \ldots$ in the above fashion. Namely, τ_{i+1} is the first time at which there are at least $\frac{q}{p}m$ coin-toss agents in the imitation process that starts at time τ_i, and $\mathcal{E}_{i+1}$ is the event that the corresponding weighted sum of coin-toss realizations reaches or exceeds its expectation. Letting i^* be the first success of one of the events $\mathcal{E}_i$, we have $\tilde{T}_q \leq \tau_{i^*}$. Since the law of $(\tau_{i+1} - \tau_i, \mathcal{E}_{i+1})$ (conditioned on the history up to time τ_i) obeys the same conditions as those of $(\tau, \mathcal{E})$, the uniform bounds (8) and (9) apply also to the pairs $(\tau_{i+1} - \tau_i, \mathcal{E}_{i+1})$, $i = 1, 2, \ldots$, conditioned on the history up to time τ_i. Therefore,

$$T_q(F, P(t)) = \mathbb{E}\left[\tilde{T}_q\right] \leq \mathbb{E}\left[\tau_{i^*}\right] \leq \frac{1}{0.14p}\frac{m}{\alpha}\left(1 + \ln\left(\frac{p}{p-q}\right)\right)$$
$$= \frac{7.2m}{p\alpha}\left(1 + \ln\left(\frac{p}{p-q}\right)\right).$$

Recalling that the actual adoption dynamic with i.i.d. Poisson updating is m times faster than the discrete time dynamic, the statement of Theorem 3.1 follows. *Q.E.D.*

5 Multiple Technologies

So far we have analyzed the case where there are two competing technologies: the status quo and the innovation. In practice, multiple innovations may be competing for acceptance at any given point in time. In this section, we explain how our results can be extended to this more general case.

Consider the case where the action set $A = \{1, \dots, k\}$ consists of k alternatives, including the status quo. We shall assume that the payoff matrix U has the following form:

$$U = \begin{array}{c|c|c|c|c|} & 1 & 2 & \cdots & k \\ \hline 1 & a_1 + c & a_1 & \cdots & a_1 \\ \hline 2 & a_2 & a_2 + c & \cdots & a_2 \\ \hline \vdots & & & & \\ \hline k & a_k & a_k & \cdots & a_k + c \\ \hline \end{array}$$

Here, a_i is the utility from choosing technology i and c is the utility from coordination, which for simplicity we assume is the same for all technologies. (The model can also accommodate heterogeneous coordination payoffs but the notation is more cumbersome.) We shall assume, without loss of generality, that $a_1 > a_2 > \dots > a_k$. Let ϵ_j be the idiosyncratic payoff shock from playing alternative j. We assume that the shocks $(\epsilon_j)_{j=1,\dots,k}$ are nonatomic, i.i.d., and have mean zero.

Given a distribution $\vec{x} = (x_1, \dots, x_k) \in \Delta(A)$, let $r_i(\vec{x})$ be the probability that technology i is the best-reply alternative for an agent facing this distribution:

$$r_i(\vec{x}) = \mathbf{P}\left((U\vec{x})_i \geq (U\vec{x})_j \text{ for all } j \in A\right).$$

Let $\vec{r}(\vec{x}) = (r_1(\vec{x}), \dots, r_k(\vec{x}))$ denote the *multidimensional response function.*

For every value $x \in [0, 1]$, let $\rho(x)$ be the minimum probability that an agent adopts technology 1 when the proportion x of his neighbors have adopted technology 1:

$$\rho(x) = \min_{\vec{x} \in \Delta^k \text{ s.t. } x_1 = x} r_1(\vec{x}). \tag{10}$$

We replace the function r of the two technologies case with the function ρ. We now approximate ρ using a linear function L as described above. Note that for every vector of proportions $\vec{x}$,

$$r_1(\vec{x}) \geq \rho(x_1) \geq L(x_1).$$

We can therefore use Theorem 3.1 to approximate the expected waiting time until a proportion $q < p$ have adopted the superior technology 1.

Deriving the function ρ explicitly from the multidimensional response function $\vec{r}$ is not always straightforward. However, for a wide class of shock distributions used in empirical applications, ρ can easily be derived from $\vec{r}$, as the following result shows.

Lemma 3 *If the payoff shock distribution F is log concave, then for every* $x \in [0, 1]$,

$$\rho(x) = r_1(x, 1-x, 0, \ldots, 0).$$

In other words, if F is log concave, then the technology distribution that minimizes the switching probability to technology 1 across all distributions with $x_1 = x$ occurs when the proportion playing technology 2 equals $1 - x$.[15]

Before providing the proof, note that

$$U\vec{x} = (a_1 + cx_1, a_2 + cx_2, \ldots, a_n + cx_n).$$

Proof Consider the case where the realized payoff shock for technology 1 equals ε_1. In this case the payoff from using technology 1 is $u_1 = a_1 + cx + \varepsilon_1$. It is sufficient to show that across all $\vec{x} \in \Delta^k$ with $x_1 = x$, the vector $(x, 1 - x, 0, \ldots, 0)$ minimizes the probability

$$\mathbf{P}\left(\left(U\vec{x}\right)_j + \epsilon_j \leq u_j \text{ for } j = 2, \ldots, k\right). \tag{11}$$

Note that the probability in (11) equals

$$F(u_1 - a_2 - cx_2) \cdots F(u_1 - a_k - cx_k). \tag{12}$$

We wish to find $\vec{x} \in \Delta^k$ that minimizes (12) subject to $x_1 = x$. It follows from the log concavity of F that the product in (12) is also log concave. Since the domain, $\{\vec{x} \in \Delta^k : x_1 = x\}$, is convex and compact it follows that the minimum is attained at an extreme point.

It remains to show that the extreme point at which the minimum is attained is $(x, 1 - x, 0, \ldots, 0)$. We contend that the log concavity of F implies that for every $j \in A \backslash \{1, 2\}$ the function

$$g(z) = \ln\left(F\left(u_1 - a_j + z\right)\right) - \ln(F(u_1 - a_2 + z)),$$

is decreasing in z. To see this, note that $u_1 - a_j > u_1 - a_2$ for all $j \geq 3$. Since the function $h(z) = \ln(F(z))$ is concave, the slope

$$\frac{\ln\left(F\left(u_1 - a_j + z\right)\right) - \ln\left(F\left(u_1 - a_2 + z\right)\right)}{a_2 - a_j}$$

[15] For example, the extreme value distribution, which generates the logit response, is log concave, and so is the normal distribution, which generates the probit response. In contrast, the Cauchy and the Pareto distributions are not log concave.

is decreasing in z. In particular, $g(0) \le g(-(1-x)c)$. This implies that for every $j \in A \setminus \{1,2\}$,

$$F(u_1 - a_2 - (1-x)c)\,F(u_1 - a_j) \le F(u_1 - a_2)\,F(u_1 - a_j - (1-x)c).$$

Therefore, for every $j \in A \setminus \{1,2\}$

$$\begin{aligned} &F(u_1 - a_2 - (1-x)c)\,F(u_1 - a_3)\cdots F(u_1 - a_k) \\ &\le F(u_1 - a_2)\cdots F(u_1 - a_j - (1-x)c)\cdots F(u_1 - a_k). \end{aligned}$$

This concludes the proof of the lemma.

6 The Role of the Network

Our main result shows that, under certain conditions on the payoff shock distribution, the expected waiting time until the innovation is adopted by a large fraction of the population is uniformly bounded *for all networks of arbitrary size.* In this section, we discuss the role that the network topology plays in the diffusion process. We focus on simple classes of networks such as regular networks and star networks and demonstrate *network-dependent* behavior in these cases.

We address three closely related questions. First, for regular networks and star networks, we derive bounds on the speed of convergence that improve on the bound in Theorem 3.1. Furthermore, we show that for star networks, rapid convergence holds even for values of q that are greater than the fixed point p.

Second, we study the question of persistence. Once the adoption process reaches a high proportion q of adopters, how likely is it that the proportion remains high for an extended period of time? As we shall see, the answer depends on both the size of the network and its topological structure. When the network is small, it only takes a handful of agents to revert to playing action 0 for the proportion to drop substantially below q. Indeed, reversion for any particular agent will occur with probability at least $1 - r(1)$, which will typically be positive. Therefore, for small networks we cannot expect that a high proportion of adopters will be maintained for very long. Even if the network is large, the degree of persistence is sensitive to the network topology. We shall show in particular that for large regular networks persistence is high, whereas for large star networks persistence is low (see Section 6.2).

Finally we ask, given a response function r, for which values of q is the expected waiting time unbounded? The answer depends on the topology of the network. In star networks, the expected waiting time to reach a target $q < r(1)$ is bounded above for all monotone increasing response functions (Proposition 6.2). In contrast, for regular networks, the expected waiting time to reach a target $q < r(q)$ increases exponentially in the degree of the network (Proposition 6.5).

6.1 Convergence rates for specific networks

We begin by providing an exact formula for the convergence time of the linear dynamic in large regular networks. An undirected network G is *regular* if *all* agents have the same degree $d \geq 0$. As before, for every $q \in [0, 1]$, we let $T_q(G)$ be the expected waiting time until at least the proportion q of the agents have adopted.

Proposition 6.1 *Let $r(x) = L(x) = (1 - \alpha)x + \alpha p$ be a linear dynamic and let $q \in [0, p)$. For every $m \geq 1$, let R_m be an undirected, regular network with m nodes. Then* $\lim_{m \to \infty} T_q(R_m) = \frac{1}{\alpha} \ln\left(\frac{p}{p-q}\right)$.

Proof We rely on stochastic approximation results due to Benaim (1998) and Roth and Sandholm (2013). Let $\{x_{m,t}\}$ be the discrete time process representing the proportion of adopters in R_m at time t. Let $\mathcal{F}_t$ be the sigma algebra that is generated by the sequence of actions of the agents in R_m up to and including time t. For any proportion $x \in [0, 1]$, let $V(x) = \alpha(p - x)$. We contend that

$$E\left(x_{m,t+1} - x_{m,t} \mid \mathcal{F}_t\right) = \frac{V(x_{m,t})}{m}.$$

In particular, the expected change in the next period's proportion depends only on the current proportion and not on the network configuration.

To see this, consider the case where the current proportion $x_{m,t} = x$ and an agent is selected uniformly at random to revise his strategy choice. With probability α, the selected agent chooses action 1 with conditional probability p, and chooses action 0 with conditional probability $1 - p$. With probability $1 - \alpha$, the agent imitates the action of a randomly chosen neighbor. Let $d(t)$ be the number of edges i, j where the actions of agents i and j are distinct. We can write

$$\mathbf{P}\left[x_{m,t+1} - x_{m,t} = \frac{1}{m} \mid \mathcal{F}_t\right] = (1 - x_{m,t})\,\alpha p + (1 - \alpha)\frac{d(t)}{2|E|}\,. \tag{13}$$

Similarly,

$$\mathbf{P}\left[x_{m,t+1} - x_{m,t} = -\frac{1}{m} \mid \mathcal{F}_t\right] = \alpha x_{m,t}(1 - p) + (1 - \alpha)\frac{d(t)}{2|E|}\,. \tag{14}$$

It follows from (13) and (14) that

$$\begin{aligned} E\left(x_{m,t+1} - x_{m,t} \mid \mathcal{F}_t\right) &= \frac{\mathbf{P}\left[x_{m,t+1} - x_{m,t} = \frac{1}{m} \mid \mathcal{F}_t\right] - \mathbf{P}\left[x_{m,t+1} - x_{m,t} = -\frac{1}{m} \mid \mathcal{F}_t\right]}{m} \\ &= \frac{V(x_{m,t})}{m}. \end{aligned} \tag{15}$$

Hence for any m and t we can write

$$x_{m,t+1} - x_{m,t} - \frac{U_{m,t+1}}{m} = \frac{V(x_{m,t})}{m}, \tag{16}$$

where $U_{m,t+1} = m(x_{m,t+1} - x_{m,t}) - V(x_{m,t})$ is a bounded $\mathcal{F}_{t+1}$-measurable random variable. Furthermore, equations (15) and (16) imply that $E(U_{m,t+1} \mid \mathcal{F}_t) = 0$. Therefore, the process $\{(\mathrm{U}_{m,t})_t\}_m$ satisfies the conditions of Proposition 2.3 in Roth and Sandholm (2013) and we can use Theorem 3.2 in Roth and Sandholm (2013) to approximate our process $\{(x_{m,t})_t\}_m$ by a differential equation as illustrated below.

Let $\{x_m(s)\}_{s\geq 0}$ be the continuous-time process defined by

$$x_m(s) = x_{m,\lfloor ms \rfloor}.$$

For any $x \in [0, 1]$, let z_x be the solution of the following differential equation:

$$\dot{z} = L(z) - z = V(z),\; z(0) = x.$$

Theorem 3.2 in Roth and Sandholm (2013) implies that, for every $\delta > 0$, $x \in [0, 1]$ and $T > 0$,

$$\lim_{m\to\infty} \mathbf{P}\left(\sup_{s\in[0,T]} |x_m(s + t) - z_x(s)| \geq \delta \mid x_m(t) = x \right) = 0\,,$$

where the limit holds uniformly in x. Note that $z_0(s) = p(1 - \exp(-\alpha s))$ and for $s' = \frac{1}{\alpha}\ln\left(\frac{p}{p-q}\right)$ we have $z_0(s') = q$. In addition, for every $x \in [0, 1]$ and every

$\epsilon > 0$ we have $z_x(s' + \epsilon) > q$. Therefore, for every $\epsilon > 0$, $x \in [0, 1]$ and $t > 0$, $\lim_{m\to\infty} \mathbf{P}(x_m(s' + t + \epsilon < q | x_m(t) = x) = 0$, and hence

$$\lim_{m\to\infty} T_q(R_m) = \frac{1}{\alpha} \ln\left(\frac{p}{p-q}\right),$$

as was to be shown.

Next, we turn to estimating the expected convergence time for large star networks. Let S_m consist of m vertices with central vertex 1 and edges $\{1, j\}$ for all $j \in [m]$ such that $j > 1$. We shall show that for every monotone increasing response function, rapid convergence can hold for values $q > p$ beyond the fixed point.

Let $T_q(S_m)$ be the expected waiting time to reach at least qm adopters in the star network with m agents.

Proposition 6.2 *Consider any star network S_m and an adoption dynamic that is governed by a monotone increasing response function r: $[0, 1] \to [0, 1]$. For every $q < r(1)$,*

$$\limsup_{m\to\infty} T_q(S_m) \le \frac{r(1)}{(r(1)-q)}\left[\frac{1}{r(0)} + \ln\left(\frac{r(1)}{r(1)-q}\right)\right].$$

Proof Consider the following event: agent 1 updates his action to 1 and thereafter $\frac{q}{r(1)}m$ agents receive a revision opportunity. We claim that the expected waiting time for this event is bounded above by

$$\frac{1}{r(0)} + \ln\left(\frac{r(1)}{r(1)-q}\right).$$

Indeed, $\frac{1}{r(0)}$ bounds the expected waiting time for agent 1 to switch to action 1, and $\ln\left(\frac{r(1)}{r(1)-q}\right)$ bounds the expected waiting time for at least $\frac{mq}{r(1)}$ agents to have a revision opportunity (this is an instance of the coupon collector problem; see, e.g., Section 3.6 in Motwani and Raghavan (1995)). Note that with probability $\left(\frac{r(1)-q}{r(1)}\right)$ agent 1 is not part of the fraction $\frac{q}{r(1)}$ of agents who receive a revision opportunity.

Now consider the following event: agent 1 updates his action to 1 and thereafter at least $\frac{qm}{r(1)}$ of the agents, not including 1, receive a revision opportunity.

By the preceding, the expected waiting time for this event is at most

$$\frac{r(1)}{(r(1)-q)}\left[\frac{1}{r(0)}+\ln\left(\frac{r(1)}{r(1)-q}\right)\right].$$

Each of the $\frac{qm}{r(1)}$ agents who receives an updating opportunity chooses action 1 with probability $r(1)$ because during their revision time agent 1 is playing action 1. Therefore, conditional on this event, the probability of reaching a threshold of $q - \varepsilon$ adopters approaches one as m goes to infinity, for every $\epsilon > 0$. This establishes Proposition 6.2.

We note, in particular, that for a linear dynamic on star networks, the determinant of the fast adoption regime is $L(1)$ and not the fixed point $p < L(1)$. Hence even when q is close to p, the expected waiting time to reach at least qm adopters is much faster in large star networks than in large regular networks.

6.2 The persistence of innovation

In this section, we show that innovation is persistent for large regular networks, whereas it is not persistent for star networks.

Recall that p denotes the fixed point of the linear dynamic. Fix $q', q \in (0, p)$ with $q' < q$ and consider the expected waiting time for the population to reach a state with at most $q'm$ adopters from a state with at least qm adopters. Say that innovation is *persistent* for a class of networks if this expected waiting time grows to infinity with the population size m *for any* such pair q, q'.

Formally given a pair $q' < q < p$ and a network G, let $I_{q,q'}(G)$ be the minimal expected waiting time to reach a state in which the proportion of adopters is q' or less, where the minimum is taken across all initial conditions with proportion at least q. We next show that innovation is persistent for the class of regular networks.

Proposition 6.3 *Let R_m be a regular network with m agents. For every $0 < q' < q < p$, there exists a constant $c = c(q, q') > 1$ (independent of m) such that $I_{q,q'}(R_m) \geq c^m$.*

It follows that the expected waiting time until the fraction of adopters is lower than q', starting from a state where the fraction is at least $q > q'$, grows exponentially with the number of agents m. (The proof is contained in the Appendix).

In contrast, for the class of star networks, persistence fails and the adoption dynamic escapes from any targeted proportion relatively quickly, as the following proposition shows.

Proposition 6.4 *Let S_m be a star network with m agents and let r(x) be a response function with* $r(1) < 1$ *and* $r(0) < \frac{1}{2}$. *There exists a constant* $c > 0$ *(independent of m) such that* $I_{1,\frac{1}{2}}(S_m) \leq c$.

Thus, even if we start in the state where all agents have adopted, the expected waiting time to reach a state with a proportion of adopters below $\frac{1}{2}$ is bounded above.

Proof The proof is similar to the proof of Proposition 6.2. Let $\gamma = \frac{3-2r(0)}{4(1-r(0))}$. Note that since $r(0) < 1/2$ we have $\frac{\gamma-1/2}{\gamma} > r(0)$ and $\gamma < 1$. Consider the following event: agent 1 updates his action to 0 and thereafter a fraction γ of agents receive a revision opportunity. As above, the expected waiting time for this event is bounded above by

$$\frac{1}{1-r(1)} + \ln\left(\frac{1}{1-\gamma}\right).$$

Note that with probability $1-\gamma$ agent 1 is not part of the fraction $\frac{q}{r(1)}$ of agents who receive a revision opportunity.

Therefore, the expected waiting time of the following event: agent 1 updates his action to 0 and thereafter a fraction γ of the agents, *which does not include agent* 1, receive a revision opportunity is

$$\frac{1}{(1-\gamma)}\left[\frac{1}{1-r(1)} + \ln\left(\frac{1}{1-\gamma}\right)\right].$$

Each of the γm agents who receives an updating opportunity chooses action 1 with probability $r(0)$ because during their revision time agent 1 is playing action 0. By the choice of γ, there exists a positive probability $w > 0$ that at most $\frac{\gamma-1/2}{\gamma} > r(0)$ of the γm agents have chosen action 1, independently of m. If this holds, the number of adopters is at most $\gamma\frac{\gamma-1/2}{\gamma} + (1-\gamma) \leq 1/2$.

This shows that

$$I_{1,\frac{1}{2}} \leq \frac{1}{w(1-\gamma)}\left[\frac{1}{1-r(1)} + \ln\left(\frac{1}{1-\gamma}\right)\right],$$

which establishes Proposition 6.4.

6.3 Slow convergence

A crucial assumption for our main result (Theorem 3.1) is that the supporting line $L(x)$ of the response function $r(x)$ must intersect the y-axis at a positive value. If this is not the case, the corresponding imitation dynamic is not well-defined. In this section, we ask, given a monotone increasing response function $r(\cdot)$, when does fast convergence fail to hold for a given target q? A natural candidate would be any value of q such that $r(q) < q$, for then the adoption process gets caught in a bottleneck. In particular, this situation arises when r is convex–concave and is too sharply bowed in the convex part (see Figure 13.3 for an example). Proposition 6.2 shows, however, that this is not a sufficient condition: star networks exhibit fast convergence for any value $q < r(1)$. In this section, we show that the condition $r(q) < q$ does lead to slow convergence for large regular networks; in particular, the expected waiting time to reach such a target q grows exponentially with the degree of the network.

Proposition 6.5 *Let R_m be a sequence of undirected d_m-regular networks, and let the response function be monotone increasing with $r(q) < q$. There exists a constant $c > 0$ (independent of m) such that $T_q\,(R_m) \geq \frac{e^{cd_m}}{m}$.*

Proof Instead of considering the first time τ when the fraction of adopting agents reaches a threshold q, we study the first time $\tilde{\tau}$ where there exists an agent in the network who has a fraction of at least q adopting neighbors. (The regularity of the network implies that $\tilde{\tau} \leq \tau$.)

We rely on the fact that prior to time $\tilde{\tau}$, all updating agents in all periods revise their action to 1 with probability at most $r(q)$. (This follows from the monotonicity of r.) From the perspective of a single player, his neighbors update their actions in an i.i.d. manner with probability of success $c \leq r(q)$. We can therefore deduce (using Lemma 1 in Benaim and Weibull (2003)) that the expected waiting time to reach a fraction q of adopters in a d_m-regular network with m agents is bounded by $\frac{e^{cd_m}}{m}$ for some constant $c > 0$, as was to be shown.

Proposition 6.5 does not hold for regular networks of *fixed* degree d. To see this, note that for every constant d, the union of $\frac{m}{d+1}$ cliques of size $d + 1$ has the fast-convergence property. Since the mixing time T of a single clique of size $d + 1$ is constant, after time T we will obtain a large number of cliques whose configurations are distributed according to the invariant distribution, and this fraction will be very close to the expected fraction under the invariant distribution (i. e., a high fraction of adopters).

7 Conclusion

In this chapter, we have established an upper bound on the expected waiting time until an innovation is adopted by a large fraction of a given population. The formula for the bound is universal in the sense that it holds for all directed and undirected networks of arbitrary size and degree distribution. Previous results on this topic rely on the existence of a potential function and hold only for regular networks with symmetric interactions, as in Kreindler and Young (2014). We have also established waiting time bounds for the diffusion of multiple innovations instead of just a single innovation, which is the usual case treated in the literature.

Our bound holds for a wide variety of perturbed best response processes where agents choose optimal responses under random payoff shocks. Particular cases include errors that are normally or extreme-value distributed, for example. These models are consistent with empirical evidence on subjects' learning behavior in network games (Mäs and Nax (2016)). The formula for the bound is expressed in terms of the slope and intercept of a suitably chosen "linearization" of the perturbed response process. In particular, the expected waiting time to reach a given proportion of adopters is inversely proportional to the initial amount of "noise" in the linear process when no one has yet adopted ($L(0)$ in expression (3)).

Given more information on the topological structure of the network, one can obtain significantly tighter bounds on the expected waiting time, as we showed in Section 6. The usefulness of our main result arises from the fact that in practice it is very difficult to observe the topology of interactions and the degree of influence that different actors exert on one another. By contrast, it may be possible to estimate the response probabilities of agents conditional on the choices of their neighbors using longitudinal data. From this, one can derive the linearized process and the upper bound on the expected waiting time to reach a given adoption threshold.

Appendix A: Feige's Inequality and Lemma 1

The following theorem is due to Feige (2006).

Theorem A.1 *Let $X_1, \ldots, X_n$ be nonnegative independent random variables, with $E[X_i] \leq 1$ for all i,* and let $X = \sum_{i=1}^{n} X_i$. *Then for any n,*

$$\mathbf{P}[X < E[X] + 1] \geq \gamma > 0,$$

for some $\gamma \geq 1/13$.

Garnett (2020) improved upon the constant γ and show that $\gamma \geq 0.14$. We next prove Lemma 1 from the above theorem.[16]

Proof of Lemma 1 Let $c_1, \ldots, c_k$ be i.i.d. Bernoulli(p), and let $\beta_1, \ldots, \beta_k$ be positive real numbers. We can assume without loss of generality that $\beta_i \leq \frac{1}{1-p}$ for all $i \in [k]$, and that $\beta_1 = 1/(1-p)$. Let $X_i = \beta_i(1-c_i)$, $m_i = \beta_i(1-p)$, $X = X_2 + \ldots + X_k$, and $m = m_2 + \ldots + m_k$. We must show that $\mathbf{P}\left(\sum_{i \in [k]} \beta_i c_i \geq \sum_{i \in [k]} \beta_i p\right) = \mathbf{P}(X_1 + X \leq m + 1) \geq \gamma p$.

Indeed, by the above theorem,

$$\mathbf{P}(X_1 + X \leq m + 1) \geq \mathbf{P}(X_1 = 0)\,\mathbf{P}(X < m + 1) \geq \gamma p.$$

A related result due to Billingsley (2008, Theorem 9.2) says the following.

Theorem *If $E(Z) = 0$, $E(Z^2) = s^2$, and $E(Z^4) = \xi^4$, then* $\mathbf{P}(Z \geq 0) \geq s^4/4\xi^4$.

To apply this in our setting, let $c_1, \ldots, c_k$ be i.i.d. Bernoulli(p) and let $\beta_1, \ldots, \beta_k \geq 0$. Let $Z = \sum_{i=1}^{k} \beta_i (c_i - p)$. Note that $\mathbf{P}(Z \geq 0) = \mathbf{P}\left(\sum_{i=1}^{k} \beta_i c_i \geq \left(\sum_{i=1}^{k} \beta_i\right) p\right)$. The preceding theorem shows that

$$\mathbf{P}(Z \geq 0) \geq \frac{p(1-p)\left(\sum_{i=1}^{k} \beta_i^2\right)^2}{4\left[(1 - 3p + 3p^2)\sum_{i=1}^{k} \beta_i^4 + 3p(1-p)\sum_{i,j:\, i \neq j} \beta_i^2 \beta_j^2\right]}.$$

[16] The proof of Lemm 1 was first communicated to us by Terence Tao (see https://mathoverflow.net/questions/278117/convex-combination-iid-bernoulli-random-variables). We thank Yuval Peres for pointing out to us the result by Feige (2006).

For values of p that are close to $\frac{1}{2}$, Billingsley's inequality provides a better bound than the one by Feige and Garnett. However, for p close to 1 the right-hand side is close to zero whereas the corollary to Feige's lemma shows that $\mathbf{P}(Z \geq 0) \geq 0.14p$. For certain values of p, the bound of Theorem 3.1 can be further improved. Concretely, a result by Garnett (2020, Theorem 1.2), can be applied to establish the bound $\min\left\{\frac{1}{6}, \frac{p(1-p)}{2(1-3p+3p^2)}\right\}$ for $p \geq \frac{1}{2}$. This bound is slightly better than 0.14 for $p < 0.835$. To see this let $X = \sum_i \beta_i(p - c_i)$. By Garnett's theorem $\mathbf{P}(X < 0) \geq \frac{1}{2c}$ provided $E(X^4) \leq c[E(X^2)]^2$ and $E(X^3) \geq 0$. The latter holds for $p \geq \frac{1}{2}$. Now $\frac{E(X^4)}{[E(X^2)]^2} = \lambda \frac{1-3p+3p^2}{p(1-p)} + (1-\lambda)3$ where $\lambda = \frac{\sum_i \beta_i^4}{\left(\sum_i \beta_i^2\right)^2}$. Hence we may take $c = \max\left\{3, \frac{(1-3p+3p^2)}{p(1-p)}\right\}$ from which the bound follows.[17] Since we prefer not to limit the range of p, however, the Feige–Garnett inequality is more useful for our purposes.

Appendix B: Proofs of Propositions 6. 3 and 6.5

Proposition 6.3 *Let R_m be a regular network with m agents. For every $0 < q' < q < p$, there exists a constant $c = c(q, q') > 1$ (independent of m) such that $I_{q,\, q'}(R_{m'}) \geq c^m$.*

Proof of Proposition 6.3 As in the proof of Theorem 3.1, we analyze the discrete-time linear dynamic. Recall that $1 - \alpha = L(1) - L(0)$, and that $s(t)$ represents the imitation dynamic: with probability α the updating agent tosses a coin with success probability p, and with probability $1 - \alpha$ he imitates a random neighbor.

If we prove that the expected time to reach the fraction q' under the discrete-time dynamic is at least $\hat{c}^m$ for some $\hat{c} > 1$, then the original dynamic satisfies the exponential bound $\hat{c}^m/m$. (The factor m is a result of changing from discrete to continuous time.) Note that for all sufficiently small $1 < c < \hat{c}$ we have $\hat{c}^m/m \geq c^m$, hence the exponential lower bound on the discrete-time dynamic provides an exponential lower bound on the original dynamic.

Given $s(t) \in \{0, 1\}^m$, let $z(t) = \sum_i s_i(t) \in \mathbb{N}$ be the number of adopters. We denote by $d(t) = |\{[i, j] \in E\colon s_i(t) = 0, s_j(t) = 1\}|/|E| \in [0, 1]$ the fraction of "disagreement" edges in the graph, where the agents are playing opposite actions.

[17] We thank an anonymous referee for pointing this out.

The process $z(t)$ is a random walk with $z(t+1) - z(t) \in \{-1, 0, 1\}$. We shall provide an explicit formula for the probability that $z(t)$ moves to the left, to the right, or stays put. Given $s(t)$ one agent is drawn at random in period $t + 1$. The probability is α that this agent tosses a coin. Conditional on the coin toss, the agent switches from 0 to 1 with probability $(1 - z(t)/m)p)$. (The first term is the probability that he was initially choosing 0, and the second term is the probability that he updates to 1.) With probability $1 - \alpha$, the agent imitates. Let $d(t)$ be the number of edges $\{i, j\}$ where the actions of agents i and j are different.

Then

$$\mathbf{P}\left[z(t+1) - z(t) = 1 | s(t)\right] = \alpha\left(1 - z(t)/m\right)p + (1 - \alpha)d(t)/2. \tag{17}$$

Similarly, we deduce that

$$\mathbf{P}\left[z(t+1) - z(t) = -1 | s(t)\right] = \alpha\left(z(t)/m\right)(1 - p) + (1 - \alpha)\, d(t)/2. \tag{18}$$

For z(t) $\in$ [mq$'$, mq], the ratio between moving left and right can be bounded by

$$\begin{aligned}
&\frac{\mathbf{P}[z(t+1) - z(t) = -1 \mid s(t)]}{\mathbf{P}[z(t+1) - z(t) = 1 \mid s(t)]} \\
&= \frac{\alpha(z(t)/m)(1 - p) + (1 - \alpha)d(t)/2}{\alpha(1 - z(t)/m)p + (1 - \alpha)d(t)/2} \\
&\leq \frac{\alpha q(1 - p) + (1 - \alpha)d(t)/2}{\alpha(1 - q)p + (1 - \alpha)d(t)/2} \leq \frac{\alpha q + 1}{\alpha p + 1}.
\end{aligned}$$

We denote this constant by $\beta = \frac{\alpha q+1}{\alpha p+1} < 1$. Given $s(t)$ with $z(t) = mq - 1$, we want to estimate the probability that the process will reach the low value $z = nq'$ before it visits $z = nq$. This question has a precise answer for any biased random walk with bias $\beta < 1$ (see Feller 2008, Chapter XIV.2). Namely, if we let $n = (q - q')m$, the probability of reaching $z = mq'$ is $\frac{\beta^{n-1} - \beta^n}{1 - \beta^n}$, which is bounded above by β^{n-1}. Therefore, the expected time until our process reaches $z(t) = q'm$ is at least $\left(\frac{1}{\beta}\right)^{n-1} \geq \hat{c}^m$ for some sufficiently small constant $\hat{c} > 1$. This concludes the proof of Proposition 6.4.

Proposition 6.5 *Let R_m be a sequence of undirected d_m-regular networks, and let the response function be monotone increasing with $r(q) < q$. There exists a constant $c > 0$ (independent of m) such that $T_q(R_m) \geq e^{cd_m}/m$.*

Proof of Proposition 6.5 Consider a d_m-regular graph with m agents. Let $\tau := \min\left\{\frac{1}{m}\sum_i s_i(t) \geq q\right\}$ be the first time such that the fraction of adopters is at least q. Let $\tilde{\tau}$ be the first time such that there exists an agent i with at least $d_m q$ neighbors who have adopted. It follows from the regularity of the network that $\tilde{\tau} \leq \tau$. We shall bound $E(\tau)$ from below by providing a lower bound for $E(\tilde{\tau})$.

It follows from the monotonicity of r that, prior to time $\tilde{\tau}$, any agent that updates his strategy chooses action 1 with probability at most $r(q)$. Thus in order to bound $E(\tilde{\tau})$ from below we can assume that, prior to $\tilde{\tau}$, all agents choose action 1 with probability $r(q)$. Assume that at time $t < \tilde{\tau}$ the fraction of i's neighbors that have adopted lies weakly below $\frac{r(q)+q}{2}$. Consider the following two events: (a) at time $t+1$ the fraction of i's neighbors who have adopted lies above $\frac{r(q)+q}{2}$; (b) there exists a time $s \in [t, t+1]$ where the fraction of i's neighbors who have adopted exceeds q. It follows from Lemma 1 in Benaim and Weibull (2003) that if all neighbors of i choose action 1 with probability $r(q)$ at every revision opportunity, then the probability that either (a) or (b) holds is at most e^{-cd_m} for some constant $c > 0$.

Now assume that the fraction of adopting neighbors of *every* agent i at time $t < \tilde{\tau}$ lies below $\frac{r(q)+q}{2}$. Then the probability that there exists an agent i for which either (a) or (b) holds is at most me^{-cd_m}. If neither (a) nor (b) holds for every agent at every time period $t = 0, \ldots, n$, then $\tilde{\tau} > n + 1$. It follows that $E(\tilde{\tau}) \geq e^{-cd_m}/m$. This concludes the proof of Proposition 6.5.

Acknowledgments

The authors thank Uriel Feige, Abraham Neyman, Yuval Peres, Terence Tao, Amnon Schreiber, and several anonymous referees. Itai acknowledges support from the MOST (Grant 19400214). Yakov acknowledges support from the ISF (Grant 2021296).

References

Acemoglu, D., Dahleh, M. Z., Lobel, I., and Ozdaglar, A. (2011), "Bayesian learning in social networks," *Review of Economic Studies* 78: 1201–1236.

Arieli, I. and Young, H. P. (2016), "Stochastic Learning dynamics and speed of convergence in population games," *Econometrica* 84: 627–676.

Arthur, W. B. (1994), *Increasing Returns and Path Dependence in the Economy* (Ann Arbor: University of Michigan Press).

Bala, V. and Goyal, S. (1998), "Learning from neighbours," *Review of Economic Studies* 65 (3): 595–621.
Banerjee, A. (1992), "A simple model of herd behavior," *The Quarterly Journal of Economics* 107: 797–817.
Banerjee, A., Chandrasekhar, A. G., Duflo, E., and Jackson, M. O. (2013), "The diffusion of microfinance," *Science* 341 (6144).
Benaim, M. (1998), "Recursive algorithms, urn processes and chaining number of chain recurrent sets," *Ergodic Theory and Dynamical Systems* 18 (1): 53–87.
Benaim, M. and Weibull, J. W. (2003), "Deterministic approximation of stochastic evolution in games," *Econometrica*, 71 (3), 873–903.
Bikhchandani, J. S., Hirschleifer, D., and Welch, I. (1992), "A theory of fads, fashion, custom, and cultural change as information cascades," *Journal of Political Economy* 100: 992–1026.
Billingsley, P. (2008), *Probability and Measure* (John Wiley & Sons).
Blume, L. E. (1993), "The statistical-mechanics of strategic interactions," *Games and Economic Behavior* 5: 387–424.
Bongaarts, J. and Watkins, S. C. (1996), "Social interactions and contemporary fertility transitions," *Population and Development Review* 22 (4): 639–682.
Burke, M. A., Fournier, G. M., and Prasad, K. (2007), "The diffusion of a medical innovation: is success in the stars?," *Southern Economic Journal* 73: 588–603.
Burke, M. A., Fournier, G. M., and Prasad, K. (2010), "Geographic variations in a model of physician reatment choice with social interactions," *Journal of Economic Behavior and Organization* 73: 418–432.
Centola, D. and Baronchelli, A. (2015), "The spontaneous emergence of conventions: an experimental study," *PNAS* 112 (7): 1989–1994.
Chandra, A. and Staiger, D. (2007), "Productivity spillovers in health care: evidence from the treatment of heart attacks," *Journal of Political Economy* 115: 103–140.
Coleman, J., Katz, E., and Menzel, H. (1957), "The diffusion of an innovation among physicians," *Sociometry* 20: 253–270.
David, P. A. (1985), "Clio and the economics of QWERTY," *American Economic Review Papers and Proceedings* 75: 332–337.
Ellison, G. (1993), "Learning, social interaction, and coordination," *Econometrica* 61: 1047–1071.
Ellison, G. and Fudenberg, D. (1993), "Rules of thumb for social learning," *Journal of Political Economy* 101: 612–643.
Ellison, G., Fudenberg, D., and Imhof, L. (2016), "Fast convergence in evolutionary models: a Lyapunov approach," *Journal of Economic Theory* 161: 1–36.
Feige, U. (2006), "On sums of independent random variables with unbounded variance, and estimating the average degree in a graph," *Siam Journal on Computing* 35 (4): 964–984.
Feller, W (2008), *An Introduction to Probability Theory and Its Applications*, Vol. 1 (John Wiley & Sons).
Foster, A. and Rosenzweig, M. (1995), "Learning by doing and learning from others: human capital and technical change in agriculture," *Journal of Political Economy* 103: 1176–1209.
Foster, A. and Rosenzweig, M. (2010), "Microeconomics of technology adoption," *Annual Review of Economics* 2 (1): 395–424.
Gale, D. and Kariv, S. (2003), "Bayesian learning in social networks," *Games and Economic Behavior* 45: 329–346.

Garnett, B. (2020), "Small deviations of sums of independent random variables," *Journal of Combinatorial Theory, Series A* 169: 105119.

Golub, B. and Jackson, M. O. (2010), "Naive learning in social networks and the wisdom of crowds," *American Economic Journal: Microeconomics* 2: 112–149.

Golub, B. and Jackson, M. O. (2012), "How homophily affects the speed of learning and best-response dynamics," *Quarterly Journal of Economics* 1287 (1338).

Goyal, S. and Vega-Redondo, F. (2007), "Structural holes in social networks," Journal of Economic Theory, 137 (1): 460-492

Griliches, Z. (1957), "Hybrid corn: an exploration in the economics of technological change," *Econometrica* 25: 501–522.

Jackson, M. O. (2008), *Social and Economic Networks* (Princeton, NJ: Princeton University Press).

Jackson, M. O. and Watts, A. (2002), "The evolution of social and economic networks," *Journal of Economic Theory* 106: 265–295.

Jackson, M. O. and Yariv, L. (2007), "Diffusion of behavior and equilibrium properties in network games," *American Economic Review Papers and Proceedings* 97: 92–98.

Kohler, H.-P. (1997), "Learning in social networks and contraceptive choice," *Demography* 34: 369–383.

Kohler, H.-P., Behrman, J. R., and Watkins, S. C. (2001), "The density of social networks and fertility decisions: evidence from South Nyanza District, Kenya," *Demography* 38: 43–58.

Kreindler, G. E. and Young, H. P. (2013), "Fast convergence in evolutionary equilibrium selection," *Games and Economic Behavior* 80: 39–67.

Kreindler, G. E. and Young, H. P. (2014), "Rapid innovation diffusion in social networks," *Proceedings of the National Academy of Sciences* 111: 10881–10888.

Liggett, T M. (1999), *Stochastic Interacting Systems: Contact, Voter, and Exclusion Process* (Berlin: SpringerVerlag).

Mäs, M. and Nax, H. H. (2016), "A behavioral study of noise in coordination games," *Journal of Economic Theory* 162: 195–208.

Montanari, A. and Saberi, A. (2010), "The spread of innovations in social networks," *Proceedings of the National Academy of Sciences* 107: 20196–20201.

Montgomery, M. R. and J. B. Casterline (1996), "Social influence, social learning, and new models of fertility," *Population Development Review* 22: 151–175.

Morris, S. (2000), "Contagion," *Review of Economic Studies*, 67, 57–78.

Mossel, E., Sly, A., and Tamuz, O. (2015), "Strategic learning and the topology of social networks," *Econometrica* 83: 1755–1794.

Motwani, R. and Raghavan, P. (1995), *Randomized Algorithms* (Cambridge: Cambridge University Press).

Mueller-Frank, M. (2013), "A general framework for rational learning in social networks," *Theoretical Economics* 8 (1): 1–40.

Mueller-Frank, M. and Pai, M. (2016), "Social learning with costly search," *American Economic Journal: Microeconomics* 83 (109).

Munshi, K. and Myaux, J. (2006), "Social norms and the fertility transition," *Journal of Development Economics* 80: 1–38.

Phelps, C. E. and Mooney, C. (1993), "Variations in Medical practice use: causes and consequences," in R. J. Arnould, R. F. Rich, and W. White. eds., *Competitive Approaches to Health Care Reform* (Washington, DC: Urban Institute, 139–178).

Rogers, E. (2003), *Diffusion of Innovations*, 5th ed. (New York: Simon and Schuster).

Roth, G. and Sandholm, W. H. (2013), "Stochastic approximations with constant step size and differential inclusions," *SIAM Journal on Control and Optimization* 51 (1): 525–555.

Ryan, B. and Gross, N. C. (1943), "The diffusion of hybrid seed corn in two iowa communities," *Rural Sociology* 8: 15–24.

Skyrms, B. and Pemantle, R. (2000), "A dynamic model of social network formation," *Proceedings of the National Academy of Sciences of USA* 97: 9340–9346.

Solan, E., Rosenberg, D., and Vieille, N. (2009), "Informational externalities and emergence of consensus," *Games and Economic Behavior* 66 (2): 979–994.

Udry, C. and Conley, T. (2001), "Social learning through networks: the adoption of new agricultural technologies in Ghana, *American Journal of Agricultural Economics* 83: 668–673.

Valente, T. (1995), *Network Models of the Diffusion of Innovations* (Cresskill, NJ: Hampton Press).

Valente, T. (2005), "Network models and methods for studying the diffusion of innovations," in P. J. Carrington, J. Scott, and S. Wasserman. eds., *Models and Methods in Social Network Analysis* (Cambridge: Cambridge University Press, 98–116).

Vega-Redondo, F. (2007), *Complex Social Networks* (Cambridge: Cambridge University Press).

Watts, D. (2002), "A simple model of global cascades on random networks," *Proceedings of the National Academy of Sciences* 99 (9): 5766–5771.

Wennberg, J. and Gittelsohn, A. (1973), "Small area variations in health care delivery," *Science* 182: 1102–1108.

Young, H. P. (1998), *Individual Strategy and Social Structure: An Evolutionary Theory of Institutions* (Princeton, NJ: Princeton University Press).

Young, H. P. (2009), "Innovation diffusion in heterogeneous populations: contagion, social influence, and social learning," *American Economic Review* 99 (5): 1899–1924.

Young, H. P. (2011), "The dynamics of social innovation," *Proceedings of the National Academy of Sciences* 108 (4): 21285–21291.

Index

Please note that page references to Figures will be followed by the letter 'f', to Tables by the letter 't'. Notes are indicated by the letter 'n' following the number.